Campus Network Architectures and Technologies

Data Communication Series

For more information on this series, please visit: https://www.routledge.com/ Data-Communication-Series/book-series/DCSHW

Campus Network Architectures and Technologies

Ningguo Shen
Bin Yu
Mingxiang Huang
Hailin Xu

CRC Press
Taylor & Francis Group
Boca Raton London New York

CRC Press is an imprint of the
Taylor & Francis Group, an **informa** business

人民邮电出版社
POSTS & TELECOM PRESS

First edition published 2021
by CRC Press
6000 Broken Sound Parkway NW, Suite 300, Boca Raton, FL 33487-2742

and by CRC Press
2 Park Square, Milton Park, Abingdon, Oxon, OX14 4RN

© 2021 Ningguo Shen, Bin Yu, Mingxiang Huang and Hailin Xu
Translated by Baishun Yu
CRC Press is an imprint of Taylor & Francis Group, LLC

English Version by permission of Posts and Telecom Press Co., Ltd.

Library of Congress Cataloging-in-Publication Data
Names: Shen, Ningguo, author.
Title: Campus network architectures and technologies / Ningguo Shen, Bin Yu,
   Mingxiang Huang, Hailin Xu.
Description: First edition. | Boca Raton : CRC Press, 2021. |
Summary: "This book begins by describing the service challenges facing campus networks, and then details the intent-driven campus network architectures and technologies of Huawei Cloud Campus Solution. After reading this book, you will have a comprehensive understanding of next-generation campus network solutions, technical implementations, planning, design, and other know-how. Leveraging Huawei's years of technical expertise and practices in the campus network field, this book systematically describes the use of technical solutions such as virtualization, big data, AI, and SDN in campus networks. With this informative description, you will be able to quickly and efficiently reconstruct campus networks. In addition, this book provides detailed suggestions for campus network design and deployment based on Huawei's extensive project implementation experience, assisting with the construction of automated and intelligent campus networks required to cope with challenges. This is a practical, informative, and easy-to-understand guide for learning about and designing campus networks. It is intended for network planning engineers, network technical support engineers, network administrators, and enthusiasts of campus network technologies"—Provided by publisher.
Identifiers: LCCN 2020048778 (print) | LCCN 2020048779 (ebook) |
ISBN 9780367695743 (hardcover) | ISBN 9781003143314 (ebook)
Subjects: LCSH: Computer network architectures. | Metropolitan area networks (Computer networks) | Local area networks (Computer networks) | Hua wei ji shu you xian gong si—Data processing.
Classification: LCC TK5105.52 .S39 2021 (print) | LCC TK5105.52 (ebook) |
   DDC 004.6/5—dc23
LC record available at https://lccn.loc.gov/2020048778
LC ebook record available at https://lccn.loc.gov/2020048779

ISBN: 978-0-367-69574-3 (hbk)
ISBN: 978-0-367-69850-8 (pbk)
ISBN: 978-1-003-14331-4 (ebk)

Typeset in Minion
by codeMantra

Contents

Summary

THIS BOOK BEGINS BY describing the service challenges facing campus networks and then details the intent-driven campus network architectures and technologies of Huawei CloudCampus Solution. After reading this book, you will have a comprehensive understanding of next-generation campus network solutions, technical implementations, planning, design, and other know-how. Leveraging Huawei's years of technical expertise and practices in the campus network field, this book systematically describes the use of technical solutions such as virtualization, big data, AI, and SDN in campus networks. With this informative description, you will be able to quickly and efficiently reconstruct campus networks. In addition, this book provides detailed suggestions for campus network design and deployment based on Huawei's extensive project implementation experience, assisting with the construction of automated and intelligent campus networks required to cope with challenges.

This is a practical, informative, and easy-to-understand guide for learning about and designing campus networks. It is intended for network planning engineers, network technical support engineers, network administrators, and enthusiasts of campus network technologies.

Introduction

CONSTRUCTING AND MAINTAINING CAMPUS networks is a constant challenge for Chief Information Officers (CIOs) and Operations and Maintenance (O&M) professionals at companies of all sizes. Enterprises and organizations are placing increasingly higher demands on campus networks; services are more dependent on such networks than ever before; and the emergence of entirely new requirements is impacting existing campus network architecture. In particular, with Wi-Fi now established as a fundamental campus network capability, issues such as border disappearance, access control, and network security resulting from Bring You Own Device (BYOD) are emerging as primary campus network considerations. In addition, the deployment and management complexity arising from wireless networks places great strain on existing O&M systems. Within this evolving campus network landscape, how to quickly improve network performance and satisfy the ever-growing demands of enterprises and organizations, while also adhering to budget and manpower constraints, is of vital importance to the industry.

Huawei operates and maintains a super-large campus network consisting of seven large campus networks (with a total of more than 200000 concurrent access terminals) and more than 12000 small- and medium-sized campus networks, empowering the company with ubiquitous Wi-Fi and BYOD services. As Huawei's business continues to grow, its campus network faces a range of new requirements including ever-changing network services, unified user access and service assurance, open but secure access services, voice, video, and instant messaging integrated on cloud terminals, and network-wide video live broadcast. Driven by these demands, Huawei's network O&M and product design teams are constantly reevaluating what a network should be, and the direction it should evolve in.

In 2012, Huawei took the lead in proposing the groundbreaking "Agile Campus" concept and solution. Specifically, Huawei innovated on the design philosophy and implementation of campus networks from the perspective of ever-changing service requirements, resulting in the most substantial evolution of campus networks since the inception of the Virtual Local Area Network (VLAN) concept. As a result, campus networks evolved from connection-oriented to high-quality networks carrying multiple services. In particular, Huawei Agile Campus provided full support for Wi-Fi-based wireless networks, greatly enriching the application scenarios and service capabilities of campus networks, and realizing the dream of processing services anytime and anywhere.

Huawei's Agile Campus Solution has since evolved into the intent-driven campus network solution, which incorporates the cloud management automation concept introduced by Software-Defined Networking (SDN) and the Artificial Intelligence (AI) applications driven by machine learning technologies. Huawei's intent-driven campus network solution continues to rapidly develop and mature, and consistently exceeds the core requirements of campus networks of all sizes, including multiservice deployment, rapid new service rollouts, unified management of wired and wireless services, and simplified O&M.

Owing to years of continuous practice and innovation, Huawei has accumulated a wealth of experience in the areas of campus network planning and design, as well as O&M. Leveraging such knowledge, Huawei has developed this book to systematically and comprehensively introduce the architectural design and overall implementation of next-generation campus network solutions. We hope that this book can help you understand the type of campus network required, how to design and build an ideal campus network, and how to enjoy the convenience and benefits offered by advanced networks.

HOW THE BOOK IS ORGANIZED

Spread over 13 chapters, this book explains the service challenges faced by campus networks during digital transformation and describes in detail the architecture and technologies of an intent-driven campus network. After reading this book, you will have a clear picture of the next-generation campus network solutions, technical implementations, planning and design, and deployment suggestions.

We begin by describing the overall architecture and service model of an intent-driven campus network, while detailing how to build physical and virtual networks under the intent-driven network architecture.

The book then moves on to discuss how to automate the deployment and service provisioning of an intent-driven campus network, and how to integrate cutting-edge technologies such as AI and big data into the O&M and security fields in order to deliver intelligent O&M and end-to-end network security.

Finally, we will take a look at the future of campus networks with regards to the distinct service characteristics of the intelligence era. Driven by technology and services, campus networks will become more automated and intelligent, eventually achieving self-healing, fully autonomous driving networks.

CHAPTER 1: GETTING TO KNOW A CAMPUS NETWORK

This chapter opens with the definition of a campus network, then analyzes various types of campus networks from different dimensions, and ultimately abstracts the basic composition of the campus network. The chapter also describes the campus network evolution process, starting from the simple Local Area Network (LAN), and progressing through to the three-layer network architecture, the unified support for multiple services, and finally to the current intent-driven campus network.

CHAPTER 2: CAMPUS NETWORK DEVELOPMENT TRENDS AND CHALLENGES

We begin by first introducing digital transformation trends from various industry perspectives and summarizing the challenges that digital transformation presents to campus networks. The chapter then details the key concepts surrounding campus network digital transformation, and finishes by outlining the road to such transformation from the perspectives of ultra-broadband, simplicity, intelligence, security, and openness.

CHAPTER 3: OVERALL ARCHITECTURE OF AN INTENT-DRIVEN CAMPUS NETWORK

This chapter begins with a rundown of the basic architecture, key components, and main interaction protocols of the intent-driven campus network. We then proceed to describe the service models of the intent-driven campus network, including the physical network model, virtual network model, and service policy model. Finally, we explain two campus network deployment scenarios (on-premises and cloud deployment), with regard to the scale and service characteristics of campus networks.

CHAPTER 4: BUILDING PHYSICAL NETWORKS FOR AN INTENT-DRIVEN CAMPUS NETWORK

The physical network is the basic architecture for network communications services, and provides physical transmission carriers for data communications. The intent-driven campus network must build an ultra-broadband physical network in order to achieve network resource pooling and automatic service deployment. This chapter covers the key technical method for ultra-broadband forwarding, whereby the network must offer a large throughput and forward a large amount of data per unit of time. We then explain the key technical implementation of ultra-broadband access, which enables any terminal to access the network anytime, anywhere. Finally, we discuss how best to use these key technical considerations to build a physical network capable of supporting ultra-broadband forwarding and access.

CHAPTER 5: BUILDING VIRTUAL NETWORKS FOR AN INTENT-DRIVEN CAMPUS NETWORK

Services on an intent-driven campus network are carried over virtual networks and are decoupled from physical networks. As a result, service changes require the modification of virtual networks only. This chapter begins by describing why the network needs to be virtualized, and why the overlay network can shield the complexity of physical networks in order to decouple services from physical networks. We then explain why Virtual Extensible LAN (VXLAN) is selected as a key technology of network virtualization, and how to build virtual networks using VXLAN. We end the chapter by using examples to describe typical applications of network virtualization technologies on campus networks.

CHAPTER 6: AUTOMATED SERVICE DEPLOYMENT ON AN INTENT-DRIVEN CAMPUS NETWORK

The intent-driven campus network introduces the SDN concept, which transfers network complexity from hardware to software and successfully transitions from distributed to centralized. The SDN controller replaces the administrator in dealing with complex issues, freeing the administrator from complicated manual operations, and automatically deploying networks and services. This chapter describes underlay network automation, which is the process of automating device provisioning and network orchestration. Overlay network automation enables network resources to be flexibly invoked through the controller, and virtual networks to be flexibly created, modified, or deleted based on service requirements. User access automation refers to the automation of access configuration, account management, identity identification, and user policies.

CHAPTER 7: INTELLIGENT O&M ON AN INTENT-DRIVEN CAMPUS NETWORK

Faced with the ever-growing complexity and diversification of services, network O&M has become a major concern for enterprises. On an intent-driven campus network, network administrators expect to focus more on user- and application-centric experience management, and less on complex device management. Intelligent O&M is the solution, and this chapter describes how to apply AI to the O&M field. Based on existing O&M data (such as device performance indicators and terminal logs), big data analytics, AI algorithms, and an expert-level experience database combine to digitize user experience on the network and visualize the network's operational status. These capabilities enable rapid detection of network issues, prediction of network faults, correct network operations, and improved user experience.

CHAPTER 8: E2E NETWORK SECURITY ON AN INTENT-DRIVEN CAMPUS NETWORK

The global landscape of network threats is constantly shifting. New unknown threats are more complex and difficult to detect, and attack with ever-increasing frequency and severity, resulting in traditional security defense methods becoming ineffective. This chapter describes the intelligent security collaboration solution powered by big data, which evolves discrete sample processing to holographic big data analytics, manual analysis to automatic analysis, and static signature analysis to dynamic signature, full-path, behavior, and intent analysis. This solution provides a comprehensive network defense system through a series of procedures such as network information collection, advanced threat analysis, threat presentation/collaborative handling, policy delivery, and blocking and isolation, ensuring the security of campus networks and services.

CHAPTER 9: OPEN ECOSYSTEM FOR AN INTENT-DRIVEN CAMPUS NETWORK

No single vendor can provide the differentiated services required for the digital transformation of enterprises across all industries. In addition to continuously launching new products and solutions capable of meeting customer requirements, Information and Communication Technology (ICT) vendors should commit to an ecosystem focus that benefits the entire industry. This chapter describes how an intent-driven campus network adopts new, open architectures and APIs in order to achieve cloud-pipe-device synergy, gather and empower numerous developers, integrate ICT solutions across industries, and promote "ecologically coordinated" innovations to nurture new ecosystems.

CHAPTER 10: INTENT-DRIVEN CAMPUS NETWORK DEPLOYMENT PRACTICES

This chapter begins by describing the overall design process of a campus network from the perspective of application practices and then uses the school campus network scenario as an example to describe campus network deployment practices. Universities and colleges have key requirements for advanced campus network architecture, multinetwork convergence, and on-demand expansion. By focusing on these demands, this chapter describes the requirement planning, network deployment, and service provisioning of an intent-driven campus network.

CHAPTER 11: HUAWEI IT BEST PRACTICES

Huawei is currently comprised of 14 R&D centers, 36 joint innovation centers, 200000 employees, and over 60000 partners worldwide, while operating across 178 countries and regions. To effectively support business growth, Huawei operates and maintains a super-large campus network, which is world-leading in terms of network architecture, network scale, service complexity, and technical solution advancement. As such, Huawei's own campus network is an ideal showcase for the intent-driven campus network solution. Drawing on Huawei's own digital transformation journey, this chapter details the development history of Huawei's campus network, specific solutions for typical campus network scenarios, and the future prospects of Huawei's campus network.

CHAPTER 12: INTENT-DRIVEN CAMPUS NETWORK PRODUCTS

This chapter covers Huawei's intent-driven campus network products, including CloudEngine S series campus network switches, AirEngine series WLAN products, NetEngine AR series branch routers, HiSecEngine USG series enterprise security products, iMaster NCE-Campus (a campus network management, control, and analysis system), and CIS (a security situational awareness system). The application scenarios and primary functions of each product are described in this chapter.

CHAPTER 13: FUTURE PROSPECTS OF AN INTENT-DRIVEN CAMPUS NETWORK

Today, we stand on the cusp of the fourth industrial revolution, as represented by intelligent technologies. AI will lead us from the information era to the intelligence era, and this chapter describes the future development of campus networks based on service characteristics in the intelligence era. Driven by technology and services, campus networks will become more automated and intelligent, eventually achieving self-healing, fully autonomous driving networks.

Icons Used in This Book

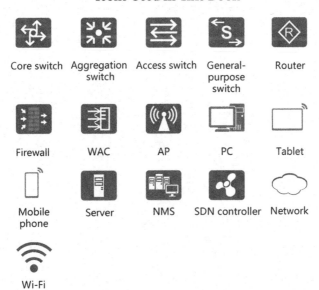

| Core switch | Aggregation switch | Access switch | General-purpose switch | Router |

| Firewall | WAC | AP | PC | Tablet |

| Mobile phone | Server | NMS | SDN controller | Network |

Wi-Fi

Acknowledgments

THIS BOOK HAS BEEN jointly written by the Data Communication Digital Information and Content Experience Department and Data Communication Architecture & Design Department of Huawei Technologies Co., Ltd. During the process of writing the book, high-level management from Huawei's Data Communication Product Line provided extensive guidance, support, and encouragement. We are sincerely grateful for all their support.

The smart campus solution proposed in this book is built on the first-of-its-kind "5G + Smart Campus" practices in Xi'an Jiaotong University (XJTU). When developing this book, we meticulously reviewed these practices together with the teachers from the network information center of XJTU. We highly appreciate relevant support and help from XJTU teachers and leaders.

The following is a list of participants involved in the preparation and technical review of this book.

Editorial board: Ningguo Shen, Bin Yu, Mingxiang Huang, Hailin Xu, Wei Liu, Hongyan Gao, Xiaosheng Bao, Chun Cui, Xiaopan Li, Naiyin Xiang, Beibei Qiu, Zhe Chen, Yinxi Zhang, Mingjie Weng, Jianguo Liu, Zhihong Lin, Zhenwei Zhang, and Yongsheng Zhang

Technical reviewers: Chengxia Yao, Cuicui Dou, Kun Yan, Yuehui Chen, Xiang Han, and Ying Chen

Translators: Baishun Yu, Jinfang Tan, Zhihong Wang, Fang He, Tingting Dong, Samuel Luke Winfield-D'Arcy, and Kyle Melieste

Special thanks go to XJTU teachers who have participated in the "5G + Smart Campus" project and provided valuable suggestions on this book: Zhihai Suo, Wei Li, Guodong Li, Mo Xu, Jun Liu, Junfeng Luo,

Huqun Li, Ninggang An, Fan Yang, Zunying Qin, Zhe Zhang, Yonggang Cheng, Feilong Wu, Xiaomang Zhu, Chen Liu, Xin Zhang, and Yuetang Wei

While the writers and reviewers of this book have many years of experience in ICT and have made every effort to ensure accuracy, it may be possible that minor errors have been included due to time limitations. We would like to express our heartfelt gratitude to the readers for their unremitting efforts in reviewing this book.

Authors

Mr. Ningguo Shen is Chief Architect for Huawei's campus network solutions. He has approximately 20 years worth of experience in campus network product and solution design, as well as a wealth of expertise in network planning and design. He previously served as a system engineer for the campus switch, data center switch, and WLAN product lines, and led the design of Huawei's intent-driven campus network solution.

Mr. Bin Yu is an architect for Huawei's campus network solutions. He has 12 years of experience in campus network product and solution design, as well as extensive expertise in network planning and design and network engineering project implementation. He once led the design of multiple features across various campus network solutions.

Mr. Mingxiang Huang is a documentation engineer for Huawei's campus network solutions. He has three years of technical service experience and four years of expertise in developing campus network product documentation. He was previously in charge of writing manuals for Huawei router and switch products. Many of the technical series he has authored, such as *Be an OSPF Expert, Insight into Routing Policies,* and *Story Behind Default Routes,* are very popular among readers.

Mr. Hailin Xu is a documentation engineer for Huawei's campus network solutions. He has two years of marketing experience in smart campus solutions, and six years of expertise in developing network products and solution documentation. Extremely familiar with Huawei's campus network products and solutions, he was previously in charge of writing manuals for Huawei routers, switches, and campus network solutions. In addition, he has participated in smart campus marketing projects within such sectors as education, government, and real estate.

CHAPTER 1

Getting to Know a Campus Network

C OMMUNICATION NETWORKS ARE UBIQUITOUS in today's information society, and campus networks are their strategic core. Before discussing campus networks, let us define what a campus is. A campus is typically a fixed area in which infrastructure is deployed. Some examples of campuses include factories, government agencies, shopping malls, office buildings, school campuses, and parks. It can be said that everything in our cities, except roads and residential areas, is composed of campuses.

According to statistics, campuses are where 90% of urban residents work and live; everyone spends an average of 18 hours every day in a campus of some sort; 80% of a country's Gross Domestic Product (GDP) is created inside campuses. The key to campuses lies in their underlying infrastructure, of which campus networks are an indispensable part. Campus networks are a key enabler to bridge campuses of all types to the digital world. They are playing an ever-increasingly significant role in our daily office work, R&D, production, and operational management. In this chapter, we will talk about the basic concepts and evolution of campus networks.

1.1 CAMPUS NETWORK OVERVIEW

As the name implies, a campus network is a network within a campus that we use for work and life. Campuses vary in size and industry attributes. Likewise, campus networks vary in type and form. However, despite the variety among them, campus networks share unified component models at different layers.

What is a campus network? How do they differ? What do they consist of? Let us find these answers together.

1.1.1 What Is a Campus Network?

All of us access various networks throughout our daily life, of which we are not actively aware. For example, when we return home at the end of the day, most of our phones automatically connect to the Wi-Fi network.

These home networks can be simple or complex. A simple home network may have just one wireless router that provides Internet access. A complex home network, however, may serve more devices and be designed for the smart life that we are experiencing now. Specifically, a complex home network can provide high-speed network services for many intelligent terminals at home, including televisions, sound systems, mobile phones, and personal computers. It can also connect to a Network Attached Storage (NAS) system to offer services such as secure data storage, automatic content acquisition, and information sharing. Similarly, the home network can interwork with an intelligent security protection system to remotely monitor the home environment, intelligently detect threats, and generate alarms accordingly. By interconnecting with an Internet of Things (IoT) system, the home network can provide automatic or remote control of various home appliances and intelligent devices. For example, an air conditioner can be turned on in advance while on the way home, so we can get comfortable the moment we open the door.

In most cases, the home network is connected externally to a carrier's Metropolitan Area Network (MAN). Through the MAN, the carrier provides enterprise and individual users with a wide range of telecommunication Internet services, typically including Internet connections, private lines, and Virtual Private Networks (VPNs), as well as various value-added services based on Internet services, such as Internet TV services. A MAN covers cities and towns and generally has three layers: core, aggregation, and access layers. The core layer is composed of routers that use Wide Area Network (WAN) technologies; the aggregation layer is formed of Ethernet switches adopting Local Area Network (LAN) technologies; and the access layer consists of Ethernet switches or alternatively Optical Line Terminals (OLTs) and Optical Network Units (ONUs) that use Passive Optical Network (PON) technologies. MANs around the globe are interconnected through WANs to form a global Internet.

As such, no matter where or when we take out mobile phones to access the Internet, we are using a mobile communication network. Generally speaking, a mobile communication network is constructed and operated by a carrier. It is made up of a series of base stations, Base Station Controllers (BSCs), a backhaul network, and a core network. With a mobile communication network, users in a wide geographical area can enjoy high-speed wireless Internet access and voice call services at ease.

Apart from the preceding networks, there is another type of network that we often encounter.

When we walk onto a campus for study, step into offices for work, go shopping, go sightseeing, or check into a hotel, we may notice that these places are also covered by networks. On the campus, we have a closed office network for teachers and also a semi-open network for students to access learning resources and browse the Internet. Inside an enterprise, we have a closed internal network for employees, facilitating their office work while ensuring security. In a shopping mall or hotel, we have not only a closed office network for employees, but also an open network for customers that provides high-quality services to enhance enterprise competitiveness. All of these networks belong to campus networks.

A campus network is a fully connected LAN in a continuous and limited geographical area. If a campus has many discontinuous areas, the networks at these discontinuous areas are considered to be multiple campus networks. Many enterprises and schools have multiple campus networks that are connected through WAN technologies.

Campus networks can be large or small. Small Office Home Office (SOHO) is a typical example of a small campus network, while school campuses, enterprise campuses, parks, and shopping malls are examples of large campus networks. Regardless, the scale of a campus network is limited. Typically, a large campus network, such as a university/college campus or an industrial campus, is constrained to a few square kilometers. Within this scope, we can use LAN technologies to construct the network. A campus beyond this scope is usually considered a metropolitan area, and the network is regarded as a MAN, involving related MAN technologies.

Typical LAN technologies used on campus networks include Institute of Electrical and Electronics Engineers (IEEE) 802.3 Ethernet technologies (for wired access) and IEEE 802.11 Wi-Fi technologies (for wireless access).

Typically, a campus network is managed by one entity only. If multiple networks within an area are managed by multiple entities, we generally consider these networks as multiple campus networks. If these networks are instead managed by the same entity, these networks are regarded as multiple subnets of the same campus network.

1.1.2 How Do Campus Networks Differ?

A campus network serves a campus and organizations inside the campus. Due to the diversity of campuses and their internal organizations, campus networks differ in size and form. These differences are detailed in the five aspects below:

1. Network scale

   Campus networks can be classified into three types: small, midsize, and large campus networks, each differing in the number of terminal users or Network Elements (NEs). This is described in Table 1.1. Sometimes, the small campus network and midsize campus network are collectively called small- and medium-sized campus networks.

   A large campus network generally has complex requirements and structures, resulting in a heavy operations and maintenance (O&M) workload. To handle this, a full-time professional O&M team takes charge of end-to-end IT management, ranging from campus network planning, construction, and O&M to troubleshooting. This team also builds comprehensive O&M platforms to facilitate efficient O&M. In contrast to large campus networks, small/midsize campus networks are budget-constrained and usually have no full-time O&M professionals or dedicated O&M platforms. Typically, only one part-time employee is responsible for network O&M.

2. Service targets

   If we look at campus networks from the perspective of service targets, we will notice that some campus networks are closed and restrictive, only allowing internal users, while others are open to

TABLE 1.1 Campus Network Scale Measured by the Number of Terminal Users or NEs

Campus Network Category	Terminal Users	NEs
Small campus network	<200	<25
Midsize campus network	200–2000	25–100
Large campus network	>2000	>100

both internal and external users. The source of network security threats differs between closed campus networks and open campus networks. Therefore, they both have distinct network security requirements and solutions.

Users on a closed campus network are typically internal employees. Their online behaviors are relatively fixed and can be effectively controlled through internal rules and regulations as well as reward and punishment. Therefore, the threats to a closed campus network mainly come from external intrusion. For this reason, a closed campus network usually uses a stronghold model to prevent unauthorized access from external and internal networks. Specifically, network admission control (NAC) is introduced to authenticate user names, accounts, tokens, certificates, and other credentials in order to prevent non-internal users from accessing the network. Additionally, firewalls are deployed at the borders of different security zones, for example, at the network ingresses and egresses.

Open campus networks paint a different picture. An open campus network aims to serve the public as much as possible. To this end, network access authentication needs to accommodate both convenient public access and effective user identification. A viable solution is to use a mobile number plus a short message service (SMS) verification code or adopt social account authentication. These approaches can simplify account management. However, public network access is unpredictable, and there may be many network security threats. As such, a user behavior control system is often deployed inside the network to prevent intentional and unintentional illegal behaviors. For example, if a user terminal is infected with a network virus, the virus may spread to attack the entire network system. To contain attacks, the user behavior control system must be able to identify user behaviors as well as isolate and clean up traffic from these users. This ensures that users can access the Internet as normal, without affecting other users on the network.

In real-world situations, a campus network usually has both closed and open subnets. A typical campus network that serves the public always has a closed subnet for internal office and administration purposes. Likewise, a campus network designed for internal personnel is typically partially open to outsiders. For example, an enterprise campus network opens up some portions of the network to guests for improved communication and collaboration. Some parts of an e-Government campus network are open for citizens who will enjoy convenient government services. In these cases, the closed subnet and open subnet belong to different security zones and must be isolated from each other. Typical isolation methods include physical isolation, logical network isolation, and firewall isolation. For networks that require strong security, physical isolation is generally used. That is, the closed and open subnets cannot communicate with each other at all.

3. Service support

Campus networks can be classified into single-service and multi-service campus networks, depending on the services carried. The complexity of services carried on the campus network determines the network complexity.

In the beginning, campus networks carried only data services, and other services were supported by disparate dedicated networks. Currently, most small- and medium-sized enterprises have a limited number of network services. For example, a small enterprise that rents offices in an office building uses the network infrastructure provided by the office building owner. Therefore, small enterprise campus networks typically require only internal data communication services. Generally speaking, a single-service campus network has a simple architecture.

An advanced large campus network is a completely different story. It usually serves an independent large campus, where various basic services, such as firefighting management, video surveillance, vehicle management, and energy consumption control, are provided. If a dedicated network was deployed for each basic service, the cost would be prohibitively high and O&M would be terribly complex. To change this, digital and Ethernet technologies are gradually introduced for these basic services. Doing so facilitates the use of mature Ethernet, and a campus network is gradually made capable of supporting multiple services. The network carries multiple services with different requirements, which are isolated from each other and effectively ensured. Due to this, the campus network architecture is becoming more and more complex and virtualized.

4. Access modes

Campus networks have two access modes: wired and wireless. Most of today's campus networks are a hybrid of wired and wireless. Wireless is free from restrictions on port locations and cables, and therefore can be flexibly deployed and utilized.

A traditional campus network is a wired campus network. From the perspective of users, each device on the wired campus network is connected to an Ethernet port in the wall or on the desktop through an Ethernet cable. These connections typically do not affect each other. As such, wired campus network architecture is usually structured and layered, with clear logic, simple management, and easy troubleshooting.

The wireless campus network, however, differs greatly from the wired campus network. It is usually built on Wi-Fi standards (WLAN standards). WLAN terminals connect to WLAN access points (APs) using IEEE 802.11 series air interface protocols. Network deployment and installation quality determine the effect of network coverage. Network optimization must be performed periodically based on network service situations to ensure network quality. The wireless network is also vulnerable to interference from external signal sources, causing a series of abnormalities that are difficult to locate.

Given that wireless connections are invisible and discontinuous, abnormalities occur abruptly and are difficult to reproduce. For these reasons, O&M personnel for wireless networks must have sufficient knowledge and expertise related to air interfaces.

5. Industry attributes

Campus networks adapt to the industries they serve. To meet the distinct requirements of different industries, we should design the campus network architecture in line with the requirements for each industry. The ultimate goal is to develop campus network solutions with evident industry attributes. Typical industry campus networks include enterprise campus networks, educational campus networks, e-Government campus networks, and commercial campus networks, just to name a few.

a. Enterprise campus network: Strictly speaking, an enterprise campus network covers the largest scope of any network type and can be further segmented by industry. The enterprise campus network described herein refers to the enterprise office network constructed based on Ethernet switching devices. In an enterprise, the office network architecture generally aligns with the internal organizational structure, as shown in Figure 1.1. The network architecture should be highly reliable and advanced enough to continuously improve employee experience and deliver uncompromised production efficiency and quality.

b. Educational campus network: There are two types of educational campus networks: primary/secondary education and higher education campus networks. Primary/secondary education campus networks are intended for primary and secondary schools, and their internal network structures and functions are similar to those of enterprise campus networks. Higher education campus networks have to cater to university/college students, and are thus more complex than primary/secondary education campus networks. Higher education campus networks often consist of multiple different networks, including a teaching and research network, a student learning network, and an operational accommodation network. The higher education campus network also has extremely high requirements for network deployment and manageability.

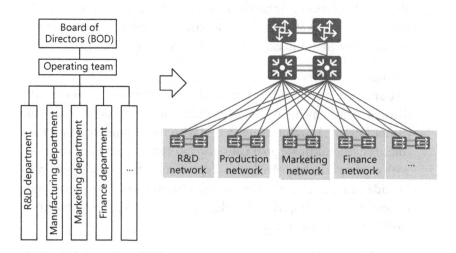

FIGURE 1.1   Enterprise campus network.

The network not only transmits data but also manages online behaviors of students to avoid extreme and aggressive actions. Supporting research and teaching is another top priority for the network. For these reasons, the network must be highly advanced and support the intense demands of cutting-edge technologies.

c. e-Government campus network: The internal networks of government agencies are a good example of this type of network. Due to stringent requirements on security, the internal and external networks on an e-Government campus network are generally isolated to ensure the absolute security of confidential information.

d. Commercial campus network: This type of network is deployed in commercial organizations and venues, such as shopping malls, supermarkets, hotels, museums, and parks. In most cases, a commercial campus network facilitates internal office work via a closed subnet but, most importantly, serves a vast group of consumers, such as guests in shopping malls, supermarkets, and hotels. In addition to providing network services, the commercial campus network also serves business intelligence systems. The resulting benefits include improving customer experience, reducing operational costs, increasing business efficiency, and mining more value from the network.

1.1.3 What Do Campus Networks Consist Of?

Campus networks are diverse but can be abstracted into a unified model that consists of components at different layers according to the service architecture, as shown in Figure 1.2.

This unified model offers network architects a chance to simplify campus networks when they become more complex due to technological innovation. With this unified model, campus networks become pliable and can flexibly adapt to their campuses. Next, we will describe the different functional components of the unified model and predict their changes in the future campus network architecture.

The unified model typically consists of a campus data network, access terminals, a network management platform, a security platform, and a service application platform.

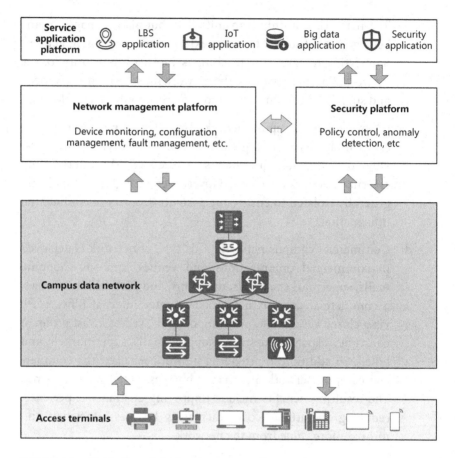

FIGURE 1.2   Unified model of campus networks.

1. Campus data network

A campus data network is constructed using Ethernet or WLAN technologies. It consists of all data communications devices on the campus, including various Ethernet switches, wireless APs, wireless access controllers (WACs), and firewalls. All internal data traffic is forwarded through the campus data network.

Generally speaking, the campus data network is composed of multiple subnets that carry different services. For example, all campuses have office subnets for daily office work, and many reserve independent video conference subnets, where dedicated links are used to ensure the quality of video conferences. Network subnets are typically managed by one administrator. IoT subnets will also be available in the future as IoT devices are connected to the campus data network. Due to the variety of IoT technologies, multiple IoT subnets will coexist.

In addition, a campus usually has an internal data center. The subnet responsible for internal forwarding of the data center is called a data center network. In the future, campus data networks may be capable of multiple services at the same time. That is, a converged campus data network will carry various services simultaneously.

2. Access terminals

On most networks, terminals are not considered as a part of the entire network, but as network consumers. This is because the owner of the terminal is not typically the network administrator. However, this is not true for campus networks. Access terminals are often considered as an indispensable part of a campus network. This is because the owner of many access terminals is also the owner of the campus network, or because the campus network administrator can obtain terminal permissions through management to manage terminals on the campus network. In this way, the campus data network and access terminals can fully interact with each other, forming an end-to-end network.

When access terminals are considered as a part of the campus network, the campus network administrator can manage and restrict terminals more actively, simplifying the entire solution. For example, the specified antivirus software is forcibly installed on terminals. In this way, terminals are forcibly checked when they attempt to access the campus network. This approach greatly reduces virus threats and simplifies the antivirus solution used on the entire campus network.

Device-pipe synergy enables the network to provide better services for terminals. For example, in the future, when a Wi-Fi network is used in smart classrooms, the network will be optimized by specifying APs for access terminals. This, in turn, will improve the concurrent access rate, bandwidth, and roaming performance of APs, meeting users' requirements for video quality.

3. Network management platform

The network management platform is a traditional component. In the latest campus network architecture, however, the positioning and enablers of the network management platform have changed greatly. This change is key to simplifying the network.

The traditional network management platform is typically home to a variety of network management systems (NMSs), which provide a limited number of remote management and maintenance functions for networks or devices. The newest generation of network management platform, however, takes on a new role. It still has all NMS functions but, most importantly, boasts radically changed management and maintenance functions; for example, automated management of common scenarios or processes. The newest network management platform is the foundation for service applications. It provides open northbound and southbound Application Programming Interfaces (APIs), through which service systems can invoke network resources.

The latest network management platform can also be used to build a next-generation security platform. Specifically, the security platform invokes the southbound and northbound APIs provided by the network management platform and then collaborates with the network management platform to collect network-wide big data through telemetry. In this way, the security platform provides more intelligent security management than ever before.

4. Security platform

An advanced security platform utilizes network-wide big data provided by the network management platform to defend against Advanced Persistent Threats (APTs). It can also invoke the northbound APIs provided by the network management platform to isolate and automatically clean up threat traffic for APT defense.

5. Service application platform

In the future, many service applications will be developed by using a base provided by the network management platform. In this way, a service application platform will take shape on top of the campus network.

For example, on a commercial campus network, it is easy to invoke the APIs of the network management platform to obtain the required Wi-Fi network positioning data and then develop customer heatmap applications. Such applications provide reference for adjustment within commercial venues.

1.2 PAST AND PRESENT OF CAMPUS NETWORKS

Since campus networks were first envisioned, they have used many technologies such as token ring and Asynchronous Transfer Mode (ATM). Although the origin of campus networks is not marked by a widely recognized epoch event, they have developed rapidly with the birth of Ethernet and the rise of Ethernet switches.

Over the past few decades of development, campus networks have mainly been constructed using Ethernet, with Ethernet switches being used as their core components. The following sections start by exploring the origins of Ethernet, then summarize the developments and changes of campus networks over the past 40 years, and finish by outlining the evolution milestones of campus networks.

1.2.1 First-Generation: From "Sharing" to "Switching"

In 1980, IEEE released the IEEE 802.3 standard to define the physical-layer connection, electrical signal, and Media Access Control (MAC) protocols. The release of this standard signaled the birth of Ethernet technology. By using twisted pair connections, Ethernet was more cost-effective and easier to implement than previous networking technologies. Consequently, Ethernet quickly became the mainstream technology for campus networks.

During the early days, campus networks used hubs as access devices. A hub was a shared-medium device that worked at the physical layer. It was limited in the number of users it could support for concurrent access. If many users connected to a hub simultaneously, network performance degraded severely due to the expanded collision domain.

As a rule of thumb, the size of a collision domain directly limited the scale of a LAN. Originally, the number of concurrent users on a single LAN never exceeded 16. A campus network had to be divided into multiple LANs, which were interconnected through expensive and low-speed routers. Under this architecture, a LAN allowed only a few users to go online concurrently. Typically, only one computer in a department could access the Internet, and users within the department had to take turns in using the computer to send or receive emails or browse the Bulletin Board System (BBS).

In the late 1980s, Ethernet switches emerged. Early Ethernet switches worked at the data link layer and therefore were called Layer 2 switches. A Layer 2 switch provided multiple ports, and each terminal was connected to one of these ports. The Layer 2 switch used a storage and forwarding mechanism that enabled terminals to send and receive packets simultaneously without affecting each other.

Layer 2 switches eliminated link conflicts and expanded the LAN scale. However, because all Layer 2 switch ports were in the same bridge domain (BD), the size of the BD directly limited the scale of the LAN. Broadcast storms therefore occurred frequently if the LAN was excessively large.

A LAN consisting of Layer 2 switches supported up to 64 concurrent users, significantly more than when hubs were used. Layer 2 switches greatly expanded the scale of a LAN while reducing interconnection costs between networks and therefore quickly replaced hubs to become the standard components of campus networks.

At this time, campus networks had a simple structure and carried simple services, but they lacked a dedicated management and maintenance system. Consequently, the role of highly skilled network engineers emerged, as professionals were required to perform all of the management and maintenance tasks.

Back in those days, campus networks provided only basic network services for campuses, and only certain computer terminals were allowed to access the network. The campus networks of this time period were expensive and inefficient as they typically used routers as backbone nodes and were therefore able to offer only non-real-time services such as email.

1.2.2 Second-Generation: Layer 3 Routed Switching

In the 1990s, two ground-breaking inventions emerged in the network field: World Wide Web (WWW) and instant messaging software.

WWW was invented in 1989 and its use became widespread in the 1990s. Numerous websites and homepages attracted much attention from the public, and people talked enthusiastically about multimedia. Since its debut in 1996, instant messaging software quickly grew and found widespread use because it incorporated the advantages of both telephone and email, and people increasingly depended on various types of instant messaging software for their work and personal life.

WWW required sufficient bandwidth, which was unable to be provided on a router-based campus backbone network. Instant messaging services had innate requirements for privacy, and people wanted to access the network and communicate with others by using their own computers. However, not all computers could be connected to the network due to the limited network scale, resulting in the inability to meet these instant messaging requirements. All of these pain points posed higher requirements on campus networks.

Against this backdrop, Layer 3 switches were developed in 1996. A Layer 3 switch, also called a routed switch due to it integrating both Layer 2 switching and Layer 3 routing functions, came with a simple and efficient Layer 3 forwarding engine designed and optimized for campus scenarios. Its Layer 3 routing capabilities were approximately equivalent to Layer 2 switching performance.

Layer 3 switches introduced the Layer 3 routing function to LANs for the first time, enabling each LAN to be divided into multiple subnets that would be connected to different Layer 3 interfaces. As shown in Figure 1.3, the Layer 3 switch-based networking structure eliminated the limitation that BDs posed on the scale of the network, meaning that LANs were no longer constrained by collision domains or BDs. Instead, the network scale could be expanded on demand simply by adding subnets, and each user or terminal could easily access the network as required.

In addition, the campus network backbone consisting of Layer 3 switches greatly increased bandwidth on the entire network, allowing users to smoothly access multimedia brought by WWW. Various office systems were also migrated to the campus network, and offices gradually transitioned toward being paperless as they became fully networked.

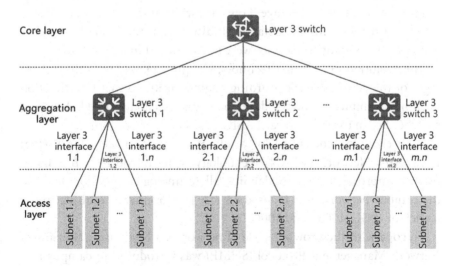

FIGURE 1.3 Three-layer structured campus network.

The development of chip technologies at this time was rapid and resulted in Application Specific Integrated Circuit (ASIC) chips, which were capable of hardware-based route lookup, quickly appearing in the industry. Layer 3 switches that integrated these ASIC chips were able to offer a Layer 3 forwarding engine and implement hardware-based forwarding at a higher performance but with lower costs. These benefits enabled ASIC chip-based Layer 3 switches to be quickly adopted on a wide scale.

By that time, Layer 3 and Layer 2 switches coexisted and began to dominate campus networks. Benefiting from low-cost technologies, campus networks had experienced an explosive growth. At the same time, Ethernet technologies were also booming. In 1995, the IEEE released the IEEE 802.3u Fast Ethernet (FE) standard, signifying the advent of the 100 Mbit/s Ethernet era. A few years later in 1999, the IEEE unveiled the IEEE 802.3ab Gigabit Ethernet (GE) standard, which defines an Ethernet rate of up to 1000 Mbit/s over a twisted pair.

Within just a few years, the Ethernet rate had increased tenfold, and this dramatic increase had far-reaching impacts on campus networks. On the one hand, Ethernet ruled out the equally competitive ATM technology and became the de facto choice for campus networks. On the other hand, the growth of Ethernet bandwidth far outpaced the development of campus network services.

In most cases on a campus network, the usage of link bandwidth did not exceed 50%, and expansion of the link capacity was required if the usage exceeded this percentage. Low bandwidth utilization paid off, as it simplified network implementation. It also meant that complex mechanisms such as Quality of Service were not as important as expected.

The evolution of Ethernet technology aligned closely with the advantages of the ASIC-based forwarding engine, meaning that the direction in which campus networks developed was essentially fixed: Ethernet switches began to evolve toward ASIC-based high performance.

Such evolution continued for a number of years. Except for Virtual Local Area Network (VLAN) technology being introduced in 2003 to further improve network scalability, all technological developments in the campus network field focused on rate improvement, resulting in the launch of 10GE, 40GE, and 100GE.

To cope with the growing scale of networks, a NMS using the Simple Network Management Protocol (SNMP) was introduced to campus networks. However, as its name implied, SNMP did not offer as much help to the network maintenance team as was expected. Maintenance and troubleshooting on the campus network still relied heavily on the skills of network engineers.

The three-layer campus network enabled easy access of computer terminals and high-performance network connections, fully meeting the needs of abundant WWW-based multimedia services and office systems. It also facilitated the expansion of the network scale. However, despite these benefits, it still did not improve the flexibility or manageability of the campus network.

1.2.3 Third-Generation: Multiservice Converged Support

Smart mobile terminals first emerged in 2007. Since then, they quickly became popular and reached a wide audience. Driven by this, Wi-Fi technology also developed rapidly.

Back in 1997, the IEEE released the 802.11 standard, marking the first attempt to standardize Wi-Fi. Although Wi-Fi technology appeared very early, its usage was limited to home and other small-scale networks for more than 10 years. The industry always expected to introduce Wi-Fi into campus networks as a supplement to office networks. This expectation, however, was not achieved mainly due to the following reasons:

- Insufficient requirements: Initially, only a few laptops had integrated Wi-Fi network interface cards, while many already supported the near-ubiquitous RJ45 wired network ports. Consequently, the demand for Wi-Fi networks was not strong.

- Concerns about security threats: The security of traditional wired networks was protected partially through isolated physical spaces. For example, security systems such as access control systems were used to restrict entry of external personnel into office spaces. As a result, there was little need to prevent unauthorized access on network ports, nor was it needed to deploy the NAC function. With Wi-Fi, however, the air interfaces broke the space constraints, meaning that the NAC function was required to improve security. Due to the limited software and hardware support, NAC deployment and O&M were extremely complex and difficult for network administrators. Because of this, network administrators usually rejected large-scale deployment of Wi-Fi networks.

- Unsuitable architecture: The traditional network architecture was not suitable for large-scale deployment of Wi-Fi networks. A Piggy-Back networking mode was therefore used to speed up the adoption of Wi-Fi networks. In this mode, VPNs, independent of the physical network, were constructed between the WAC and APs by using Control and Provisioning of Wireless Access Points (CAPWAP) tunnels. Although VPNs could theoretically be carried on top of any fixed network, there were problems in large-scale deployment. In the case of large-scale Wi-Fi user access, for example, all data traffic would be north-south traffic if the Wi-Fi network uses centralized deployment of high-specification WACs. This was inconsistent with the traffic model (mainly east-west traffic) and network model of the campus network, and resulted in the core devices being overloaded. Conversely, if distributed deployment of low-specification WACs was adopted instead, the WAC would be overloaded in large-scale roaming scenarios due to heavy detoured traffic.

Before the emergence of smart mobile terminals, Wi-Fi technology was primarily used for hotspot coverage and seldom adopted on campus networks. For example, many campuses used Wi-Fi networks only in conference rooms.

As smart mobile terminals quickly became more popular and entered the productivity field, the demand for full Wi-Fi coverage became stronger. However, two pain points hindered the large-scale deployment of Wi-Fi. One was the issue with large-scale Wi-Fi networking, and the other was the issue with NAC deployment.

To address these two pain points, leading network equipment vendors in the industry successively launched their multiservice converged network solutions. In 2012, Huawei unveiled its multiservice converged network solution: Agile Campus Network Solution.

Huawei's solution offers an innovative wired and wireless convergence feature that resolves the preceding two pain points and makes large-scale Wi-Fi deployments a reality. In the solution, an optimized WAC is deployed on an agile switch at the aggregation layer to unify the management plane across wired and wireless networks, implementing unified NAC functions. This solution also combines the innovative "local forwarding mode" of the Wi-Fi network with the unified wired and wireless forwarding feature of agile switches in order to efficiently resolve the networking problem of large-scale Wi-Fi user access.

Technological advances have made it possible for Ethernet switches to solve many problems that previously seemed like impossible challenges. As mentioned earlier, Ethernet switches have become hugely popular largely due to the emergence of ASIC forwarding engines.

Huawei takes this to a new level by combining the ASIC forwarding engines with Network Processors originally used for high-performance routers, launching high-performance chips. These chips, used as the forwarding engine, enable agile switches to efficiently support not only multiservice implementation, but also Wi-Fi features, laying a solid foundation for wired and wireless convergence.

Wi-Fi has subsequently become deeply integrated into and a typical feature of third-generation campus networks. Software-Defined Networking (SDN) has also been introduced to campus networks in order to simplify services. This generation of campus networks generally meets the requirements of enterprises that are in the early stages of all-wireless transformation. However, a number of problems still exist. For example, Wi-Fi networks cannot deliver a high enough service quality, but can instead only be used as a supplement to wired networks. Other examples of this are that maintenance difficulties related to the large-scale deployment of Wi-Fi are not completely eliminated, the network architecture is not optimized, and multiservice support still depends heavily on VPN technology. All these lead to insufficient network agility. Despite these problems, the launch of agile switches provides a good hardware basis for further evolution of campus networks.

In summary, campus networks have constantly evolved and made dramatic improvements in terms of bandwidth, scale, and service convergence. However, they face new challenges in terms of connectivity, experience, O&M, security, and ecosystem as digital transformation sweeps across all industries. For example, IoT services require ubiquitous connections; high-definition (HD) video, Augmented Reality (AR), and Virtual Reality (VR) services call for high-quality networks; and a huge number of devices require simplified service deployment and network O&M. To address these unprecedented challenges, industry vendors gradually introduce new technologies such as Artificial Intelligence (AI) and big data to campus networks. They have launched a series of new solutions, such as SDN-based management, fully virtualized architecture, all-wireless access, and comprehensive service automation.

Campus networks are embarking on a new round of technological innovation and evolution, and they are expected to gradually incorporate intelligence and provide customers with unprecedented levels of simplified service deployment and network O&M. Such a future-proof campus network is called an "intent-driven campus network", which we will explore in more detail in subsequent sections.

CHAPTER 2

Campus Network Development Trends and Challenges

A S THE DIGITAL WORLD realizes a number of remarkable break-throughs, digital transformation is spreading across all walks of life. The use of new technologies such as the Internet of Things (IoT), big data, cloud computing, and Artificial Intelligence (AI) is not only transforming the operational and production models across industries but also inspiring the creation of new business models.

Against this backdrop, enterprises and organizations need to start digital transformation in order to keep pace with the competition. Campus networks are not only the cornerstone of this digital transformation but also a bridge between the physical and digital worlds. Enterprises and organizations, faced with the constant emergence of new technologies and applications, urgently need simple methods to deploy intelligent and reliable campus networks quickly.

2.1 INEVITABLE INDUSTRY DIGITAL TRANSFORMATION

Digital transformation has had a profound impact on people's work and life.

Mobile social networking has changed the way we communicate. For example, Facebook, famous the world over, has more than two billion users, far more than the population of any single country.

Shared transportation has redefined the way we travel. Today, there are more than 100 million bike-sharing users in about 150 cities, and over 400 million online car-hailing users across more than 400 cities.

Video intelligence has revolutionized the man-machine interaction model. Technologies used for recognition, such as facial and license plate recognition, are becoming more widespread in fields such as attendance management, permission control, and commercial payment.

AI has innovated the operation model. In today's world, intelligent robots can be found in high-value fields such as customer service, language translation, production, and manufacturing, replacing from 3 to 5.6 positions that were once performed manually.

As digital technologies become more mature and used on a larger scale, they offer us unprecedented convenience both at home and at work. They also inspire us to expect unlimited possibilities for the future. The digital era — or more accurately, the intelligence era — is coming. Are you ready for this new era?

2.1.1 Large Enterprises

According to interviews held by Gartner with Chief Information Officers (CIOs) of enterprises in 2018, digital transformation has become a strategic priority of large enterprises. It is driven by both technological advances and user demands as well as external competition and internal talent pressure. Enterprise digital transformation has shifted from a conceptual phase to the key implementation phase. Large enterprises are carrying out comprehensive digital transformation centered on experience, cost, and efficiency.

1. Mobile and cloud greatly improve collaborative office efficiency.

     Enterprises are building a one-stop office platform that integrates abundant services such as instant messaging, email, video conferencing, and to-do approval. Once this platform is in place, work can be done using mobile devices anytime and anywhere, leading to greater productivity. Mobile and cloud greatly simplify cross-region collaboration and communication.

     As shown in Figure 2.1, technical experts can perform Operations and maintenance (O&M) remotely via video conferencing, without the need for a site visit. Wearing Augmented Reality (AR) glasses, engineers can remotely transmit onsite fault images to technical experts in real time.

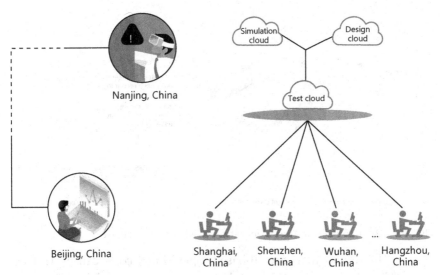

Nanjing, China

Simulation cloud

Design cloud

Test cloud

Beijing, China

Shanghai, China

Shenzhen, China

Wuhan, China

Hangzhou, China

Remote O&M via video conferencing

R&D resource sharing on multiple clouds

FIGURE 2.1 Example of a collaborative office scenario.

In the past, technical experts were required to visit the site and generally needed one week to solve all problems. Now, however, the required time is greatly reduced. R&D resources are shared on the cloud, and cloud services such as the simulation cloud, design cloud, and test cloud are available. In this way, IT resources and hardware resources can be centrally allocated in order to support cross-region joint commissioning by different teams. The time needed to prepare the environment has been more than halved from the one month it used to take.

2. Digital operations are first implemented inside enterprise campuses.

Many enterprises have chosen their internal campuses as a starting point for their digital transformation. Employees are the first to experience the digital services available at their fingertips.

As shown in Figure 2.2, all resources are clearly displayed on the Geographic Information System (GIS) map for employees to easily obtain. This involves offices, conference rooms, printers, and service desks. In a conference room, for example, people can experience intelligent services at any time. Such services include online reservation, automatic adjustment of light, temperature, and humidity before the conference, wireless projection, facial recognition sign-in, and automatic output of meeting minutes via speech recognition.

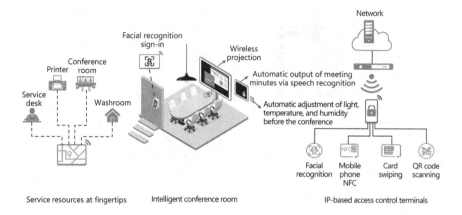

FIGURE 2.2   Example of digital operations inside an enterprise.

If IoT terminals are IP-based, operational costs can be further reduced to some extent. Take the access control system as an example. The traditional access control system is a closed system that contains the terminals, main control panel, and auxiliary control devices, and access to rooms is usually controlled via card swiping. If this access control system was to become IP-based, we would no longer need to configure professional access control management devices onsite. Instead, we could deploy the management platform on the cloud or web server. This would slash the average cost for installing an access control system by nearly half. The IP-based access control system can also support multiple IP-based access control services, such as facial recognition and Quick Response (QR) code scanning.

3. Production and manufacturing processes gradually become automated and intelligent.

Enterprises start to reduce manual intervention in each production phase, ranging from procurement, logistics, and warehousing to manufacturing. This aims to make the production process more controllable, as shown in Figure 2.3. Enterprises can automatically calculate when to purchase raw materials and the required quantity based on customer orders and supplier information. The entire logistics process is visualized by performing scheduled goods receiving and delivering, implementing goods loading simulation, and using sensors.

In the warehousing sector, automation devices such as automated guided vehicles (AGVs) and automatic barcode scanners are

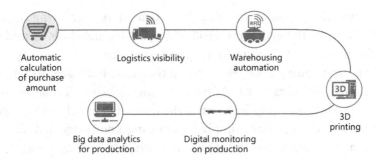

FIGURE 2.3   End-to-end automated, intelligent production, and manufacturing processes.

connected to forklifts through the IoT. These facilitate automatic movement of materials inbound and outbound, automatic stocktaking, and positioning and tracking.

Quick modeling can be achieved through 3D printing. In workshops, the key data required for quality control are collected online through sensors. The production management system performs machine learning and big data analytics based on the data collected in real time to streamline the production process and improve product quality.

Campus networks are becoming more and more important in the digital transformation of enterprises, covering nearly all aspects from office automation (OA) and internal campus services, to production and manufacturing. They no longer focus on wired IP networks in office areas, but instead extend the boundaries until they carry digital services of the entire campus. This poses higher requirements on campus networks, especially for the main production systems. These networks are expected to provide ultra-large bandwidth and ultra-low latency, as well as quickly recovering from anomalies within minutes or even seconds, in order to minimize enterprise losses to a tolerable range.

2.1.2 Education

In March 2016, the World Economic Forum in Davos released a research report entitled *New Vision for Education: Fostering Social and Emotional Learning through Technology*. The report points out that a serious gap exists between the demand for talents in the future intelligent society and the industrialized education system.

The traditional education model is focused on achieving student employment. By imparting professional knowledge and skills to students of different majors, this educational paradigm helps train talents specialized in corresponding fields. However, in the future intelligent society, the fast technological advances and new technologies such as AI may result in everyone's work changing frequently. The future talent cultivation model will therefore change from simply transferring knowledge and skills to improving students' social and emotional skills, with a much greater focus on personalized education.

The education industry has started digital transformation, with the aim of promoting converged education development, achieving co-construction and sharing of education resources, and developing creative talents. This ongoing digital transformation requires campus networks to provide more extensive connections, deliver better service experience, and offer more basic data for upper-layer applications.

1. A ubiquitous learning environment promotes high-quality education for all.

    Mobile and multimedia teaching has started to change traditional teaching models by providing students with richer and more flexible learning methods.

    Figure 2.4 shows an example of an all-new cloud classroom. The cloud is home to all teaching service resources, including courseware,

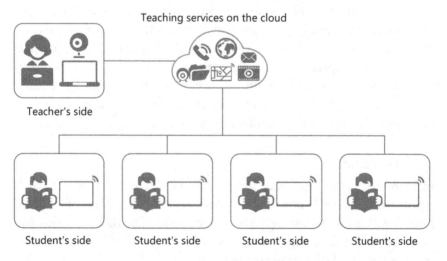

FIGURE 2.4   Example of a cloud classroom scenario.

recording, video, reservation, and scheduling systems. Students can access these resources from their mobile terminals or in remote classrooms. In this way, teaching is no longer constrained by space, and students are free to learn anywhere. Augmented Reality/Virtual Reality (AR/VR)-assisted and holographic teaching offers students more engaging, intuitive, and immersive experience.

A ubiquitous learning environment, however, is dependent on a high-quality campus network. In the case of AR/VR-assisted teaching, for instance, each user terminal requires 250 Mbit/s bandwidth, and the latency must not exceed 15 ms. If these requirements are not met, users may go through dizziness and other discomforts.

2. IoT applications help create a fully connected smart campus.

A growing number of IoT applications are used on the campus, making a fully connected smart campus a reality. For example, all-in-one cards have become "electronic ID cards" for students, enabling them to access a wide range of services such as electronic access control, attendance check-in, book borrowing, and medical treatment. These all-in-one cards can be physical cards or QR codes.

Another example is an asset management solution based on Radio Frequency Identification (RFID). For instruments in school labs, this solution enables one-click automatic stocktaking, online sharing, and generating of exception alarms, as shown in Figure 2.5. IoT applications, as their use becomes more prevalent, can deliver greater convenience in terms of student services and manage teaching resources in a more refined manner.

Full Wi-Fi coverage is a vital aspect for most campuses. To achieve a fully connected smart campus, IoT and Wi-Fi convergence must be considered during network construction. This is necessary to achieve lower cabling costs, easier deployment, and more efficient O&M management.

3. Teaching, learning, administration, and research data convergence paves the way for designing intelligent growth paths for students.

New teaching and talent cultivation paradigms fully leverage big data analytics to achieve truly personalized learning, unleash the full expertise and potential of each student, and deliver education that is oriented to essential qualities.

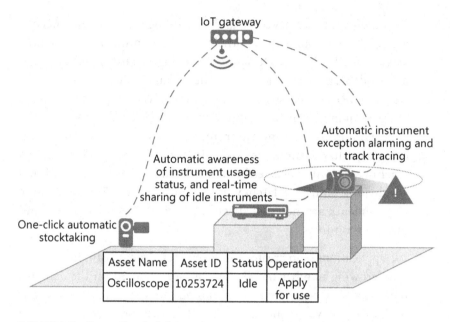

FIGURE 2.5   Example of a lab instrument management scenario.

As shown in Figure 2.6, a development path can be customized for each student based on a wide range of data, such as college entrance examination data, expertise data, exercise data, personality assessment data, career interest assessment data, face-to-face communication data,

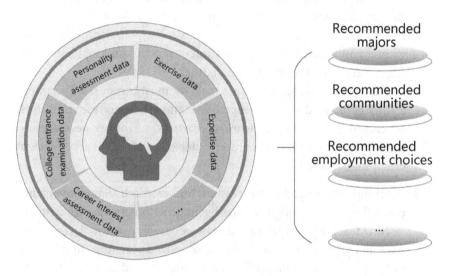

FIGURE 2.6   Design of students' intelligent development path.

and external talent demand data. This development path provides recommended majors, communities, and employment choices, giving students a strong foundation from which to plan their development.

In addition, an accurate teaching evaluation can be performed for each student based on their self-learning data, attendance data, credit data, and book borrowing data. Suitable coaching suggestions are then offered to students, helping them efficiently complete their studies.

2.1.3 e-Government Services

Building a transparent, open, and efficient service-oriented government is the goal of all governments around the world. To attain this goal, they are expected to comprehensively transform their administrative procedures, functions, and management mechanisms. So far, governments worldwide have reached a consensus that building a digital government should be one of their major priorities.

Governments' digital transformation involves reforming the traditional e-Government IT models. Specifically, digital-driven new government mechanisms, platforms, and channels will be set up to comprehensively enhance how governments fulfill their responsibilities in fields such as economic regulation, market supervision, social governance, public services, and environment protection. Such transformation will create a modern governance model featuring "data-informed dialogs, decision-making, services, and innovation". This will ultimately promote citizen-centric public services, improve management efficiency, and enhance service experience.

To keep pace as e-Government services undergo digital transformation, campus networks must be more secure and more reliable in order to achieve both service isolation and efficient use of network resources.

1. e-Government services achieve one-window approval and multi-channel service handling.

   The first effective measure governments take in their digital transformation journey is to constantly improve their ability in offering digital e-Government services to citizens. Traditional e-Government services face typical issues such as scattered services and repeated IT construction. As a result, data resource sharing or cross-department service collaboration is difficult.

In the past, citizens needed to visit government agencies multiple times due to the limited channels these agencies provided for handling service requests. Nowadays, governments deliver digital e-Government services over campus networks and build e-Government big data centers based on data exchange and sharing platforms. These new facilities achieve more efficient data exchange and sharing of public resources among departments. In particular, most public services are available to citizens through a number of channels, greatly accelerating the handling of service requests. A "one number, one window, one network" service model is taking shape, as shown in Figure 2.7. This new service model enables data to run errands for people across government agencies.

In the traditional e-Government network solution, private networks are built to achieve complete physical isolation and ensure

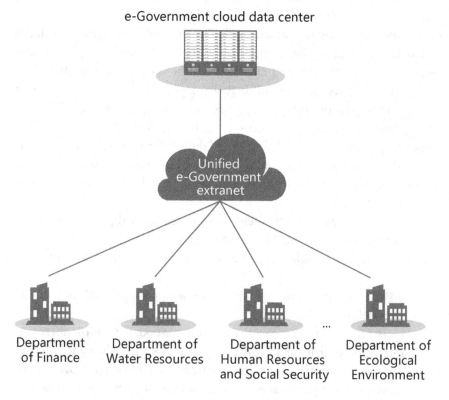

FIGURE 2.7   Example of e-Government big data networking.

absolute security of private data, meaning that services sensitive to information security continue to be carried on the original private networks. For a unified e-Government service network — one that opens up multiple service handling channels to citizens — network security will be a major challenge and something that must be addressed.

2. Digital city governance is proven productive for global coordination and rapid response.

Digital city governance primarily covers three aspects.

a. The first is to horizontally streamline service systems of government agencies and integrate audio and video communication capabilities. Doing so enables service linkage and efficient collaboration between multiple departments, ensuring quick emergency response. For example, in terms of firefighting response, a unified command platform can be used to centrally coordinate and dispatch firefighting and medical resources as well as provide notifications about fire sources and real-time road conditions.

b. The second is visualization. By integrating technologies such as IoT and GIS, the physical world can be virtualized into a digital world — one in which the overall urban running status can be constantly monitored. For example, this would enable personnel to learn about the distribution of sanitation resources and rate of garbage collection in real time for each area in a city.

c. The third is using AI to analyze the status of a city and provide alerts in real time, transform passive response into proactive defense, and comprehensively optimize the governance process. Intelligent video analysis has already played a significant role in the handling of traffic violations and case analysis. An example of this was the successful rescue of an abducted child within 10 hours thanks to video synopsis, facial recognition, and big data collision, as shown in Figure 2.8.

Digital city governance is impossible without the supportive campus networks throughout the entire city.

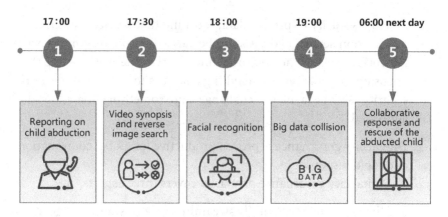

FIGURE 2.8   Example of an intelligent video analysis scenario.

2.1.4 Retail

Over the past 20–30 years, e-commerce has developed explosively due to unique advantages such as low costs and no geographical restrictions. E-commerce has shaken up the brick-and-mortar retail industry.

However, as time goes by, the online traffic dividend of e-commerce is reaching the ceiling. It has become impossible to penetrate further into the consumer goods retail market. Facing this, the retail industry has started to enter a new retail era where online and offline retail are converged.

E-commerce companies want to achieve business breakthroughs by streamlining online and offline practices and expanding traffic portals into omni-channels through campus networks. Traditional retail players are also actively embracing this change, in the hopes of taking a place in the future new retail era based on new Information and Communication Technology (ICT).

Digital transformation of the retail industry requires campus networks to be open at all layers in order to build an enriched digital application ecosystem. Further to this, traditional network O&M needs to be radically overhauled and innovations made in order to provide users with network O&M methods that are far simpler.

1. In-depth analysis and mining from massive amounts of consumer access data facilitates precision marketing.

    Traditional marketing models lack targeted insights into user needs. As a result, retailers recommend similar products to all users,

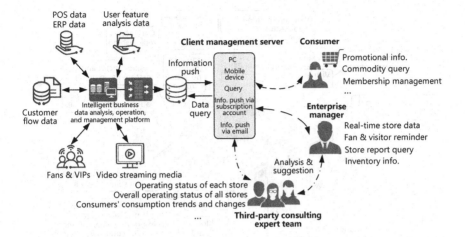

FIGURE 2.9   Example of a precision-marketing scenario.

failing to meet users' personalized needs and leading to poor customer conversion rates in marketing events.

Every time users access retailers' websites or apps, they are generally presented with the same products, leading to poor user experience. In the new retail era, campus networks are used to streamline online and offline channels as well as deeply analyzing and mining value from massive amounts of consumer access and transaction data. These practices help retailers to build precise user models, perform digital profiling for each user, and provide personalized services, as shown in Figure 2.9.

2. Abundant digital value-added applications are provided in a way similar to those in the app stores on smartphones.

Figure 2.10 illustrates a smart store scenario, in which a wide variety of value-added applications are provided to greatly improve consumer experience, reduce operating costs, enhance brand competitiveness, and increase operational efficiency.

To enable this scenario, the network architecture should be open enough to interconnect with applications from third-party application customization vendors. In this way, value-added applications such as customer flow analysis and asset positioning can be quickly developed by using Application Programming Interfaces (APIs) provided by the network. Additionally, the store network must integrate multiple IoT access capabilities, including Bluetooth, RFID, and ZigBee.

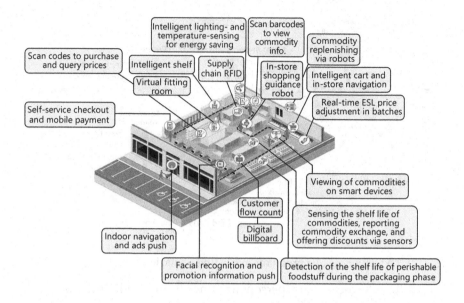

Intelligent lighting- and temperature-sensing for energy saving

Scan barcodes to view commodity info.

Commodity replenishing via robots

Scan codes to purchase and query prices

Intelligent shelf

Supply chain RFID

In-store shopping guidance robot

Intelligent cart and in-store navigation

Virtual fitting room

Self-service checkout and mobile payment

Real-time ESL price adjustment in batches

Viewing of commodities on smart devices

Customer flow count

Sensing the shelf life of commodities, reporting commodity exchange, and offering discounts via sensors

Indoor navigation and ads push

Digital billboard

Facial recognition and promotion information push

Detection of the shelf life of perishable foodstuff during the packaging phase

Source: Accenture's editorial article — *The Internet of Things: Revolutionizing the Retail Industry*

FIGURE 2.10   Example of a smart store scenario.

2.2 CHALLENGES TO CAMPUS NETWORKS PRESENTED BY DIGITAL TRANSFORMATION

Digital transformation has become increasingly popular across industries. Now, we hear of a growing number of digital transformation success stories from large enterprises, smart campuses, digital governments, and more. With this development, digital transformation poses higher requirements and greater challenges to campus networks that carry campus-wide data and services. Reflecting on the future of the digital world, we can predict that campus networks will face unprecedented challenges in the following areas:

2.2.1 Ubiquitous Connections

The pervasive use of Wi-Fi technologies in campus networks frees users from cable restrictions and enables them to enjoy convenient mobile workstyles. It has become evident that going wireless greatly improves both enterprise productivity and efficiency, and that all-wireless campus networks are becoming widespread.

In addition, the rise of the IoT drives profound campus network transformation. Huawei's latest global connectivity index shows that by 2025, there will be 100 billion connections worldwide and IoT will develop at an accelerated pace.

Campus networks of the future are expected to not only converge wired, wireless, and IoT but also provide ubiquitous connections while delivering always-on services. However, these expectations will bring the following challenges to campus networks:

1. Wi-Fi will become the mainstream access mode and will need to support large-scale, high-performance, and scenario-tailored networking.

    Since its debut, the Wi-Fi network has been positioned only as a supplement to the wired network, and so it does not innately support large-scale networking and high-density access. Due to the self-interference of air interfaces on the Wi-Fi network, large-scale networking causes adjacent-channel interference, and high-density access leads to co-channel interference. As a result, the performance and quality of the entire Wi-Fi system deteriorate quickly.

    Wi-Fi networks use Piggy-Back architecture; the wireless access controller (WAC) and access points (APs) communicate with each other through Control and Provisioning of Wireless Access Points (CAPWAP) tunnels; and data have to be aggregated and centrally processed by the WAC. As such, the Wi-Fi network is actually overlaid on top of the wired network. Piggy-Back architecture makes it difficult to improve the performance of the Wi-Fi backbone network. For example, if large-capacity WACs are deployed in centralized mode, there will be high requirements on WAC performance and costs. Further, if multiple WACs are deployed in distributed mode, complex roaming between WACs is one of the top concerns, and subsequently traffic on the network is detoured.

    Wi-Fi networks are originally deployed in a relatively simple environment (such as office spaces). However, in the future, they will be deployed in various complex environments, for example, large stadiums requiring high-density access and hotels with many rooms where inter-room interference is severe. As such, various deployment solutions will be needed in different scenarios, which will pose high requirements on vendors' products and solutions.

2. Large-scale IoT construction should address problems such as high costs and severe interference.

IoT development is application-driven rather than technology-driven. No single vendor can solve the problems in all scenarios. In each single domain, there are few high-performing vendors who can provide their disparate end-to-end (E2E) solutions. This disparity leads to incompatible IoT deployment and repeated investments for different services. As a result, multiple siloed networks, such as office, production, and security protection networks, coexist, doubling the construction and maintenance costs.

IoT networks generally use short-range wireless communications technologies, such as Bluetooth and ZigBee. These technologies use similar frequency bands (similar to those on Wi-Fi networks) and also adopt similar networking practices. Consequently, severe self-interference may occur, and if there is no advanced anti-interference technology, the quality of each network may deteriorate.

2.2.2 On-Demand Services

Over the course of digital transformation, applications are rapidly changing and new services constantly emerging. Enterprises hope that the upper-layer service platform can easily obtain all necessary information from the underlying network in order to provide more value-added services and create greater business outcomes.

To achieve business success, enterprises strive to shorten the time to market of new services and implement on-demand provisioning of campus network applications. This requires campus networks to solve the following pain points:

1. Traditional deployment and management practices cannot support fast network deployment and expansion.

Traditional deployment approaches have low-level automation and must go through multiple phases, including onsite survey, offline network planning, installation and configuration, and onsite acceptance. Professionals need to visit the site multiple times and manually configure each device separately.

Take a network project for chain stores as an example. In this project, 292 branch networks need to be added, more than 10000 devices should be installed, and networks for seven stores are to be deployed each week during peak times. In this case, if we use traditional deployment methods, chain stores may have missed business opportunities even before network deployment is complete.

2. Services and networks are tightly coupled, failing to support frequent service adjustment.

On traditional campus networks, service adjustment inevitably leads to network adjustment. For example, if a new video surveillance service is to go live, configurations (including route, virtual private network (VPN), and Quality of Service (QoS) policies) should be adjusted at different nodes such as the access layer, aggregation layer, core layer, campus egress, and data center. Such new service rollout takes at least several months or even one year.

In addition, due to the strong coupling between services and networks, the rollout of new services easily affects existing services. To prevent this, a large amount of verification is required before new services are brought online.

3. Policies and IP addresses are tightly coupled, which cannot ensure policy consistency in mobile office scenarios.

On traditional campus networks, service access policies are implemented by configuring a large number of Access Control Lists (ACLs). Administrators usually plan service policies based on information such as IP addresses and VLANs.

However, in mobile access scenarios, information like user access locations, IP addresses, and VLANs can change at any time, bringing great challenges to service access policy configuration and adjustment, and affecting user service experience during mobility.

2.2.3 Perceptible and Visible Experience

A unified physical network should carry all services throughout the campus, achieving one network for multiple purposes. This will hopefully shorten the campus network deployment period and reduce costs.

Nowadays, two types of services requiring high-quality experience are emerging on a single physical network, posing great challenges for network vendors.

The first type of service is the interaction between people. The sensing boundaries between the physical and digital worlds have become increasingly blurred. We are pursuing digital communication to be as close to "face-to-face communication" as possible in order to improve production efficiency and enhance employee and customer experience. For example, abundant media applications represented by 8K videos, as well as VR and AR require higher network bandwidth (at least 200 Mbit/s per machine) and lower latency (E2E latency within 30 ms, and less than 10 ms on the network side). We believe that these requirements will become more stringent as users pursue a better experience.

The second type of service is the interaction between things and applications, namely, campus IoT services. These services vary with quality requirements in diverse application scenarios; many IoT services may have less stringent requirements on bandwidth, but are highly sensitive to packet loss or latency. Take an unmanned warehouse where AGVs are used as an example. During wireless roaming, if more than two packets are lost for an AGV, the AGV will automatically stop, causing a series of chain reactions. As a result, all the other AGVs will run abnormally, seriously affecting services.

When the two distinct services run on the same network, they will compete for resources and so conflicts may occur. To prevent this, a large-capacity, high-reliability, and ultra-low-latency network infrastructure must be used. In addition, intelligent differentiated services must be offered to provide optimal network resources for each service. In this way, each service on the same network is not contradicted, and can deliver the optimal experience, despite having limited resources.

As well as delivering high-quality service experience to users, enabling the network to perceive users' service experience in real-time remains a challenge. The O&M personnel on traditional networks are often not aware of poor service experience until users submit complaints. Based on this, future network systems will avoid this post-event O&M approach. Instead, they should continuously and digitally measure and evaluate user experience, and automatically adjust and re-allocate network resources through AI technologies, so that each service obtains an optimal SLA (covering the packet loss rate, delay, jitter, and more). This "autonomous and self-optimizing" network will truly be a qualitative leap forward compared with the traditional network that statically deploys QoS policies.

2.2.4 Efficient and Intelligent O&M

With the constant development of campus networks, the number of nodes and bandwidth has increased exponentially. In 2010, there were only

10 billion terminals accessing the network, whereas, it is predicted that by 2020, this figure will reach up to 50 billion. Further, with the convergence of multiple services, campus networks are carrying more content and services, accommodating higher requirements on service quality, and becoming more complex than ever.

We have noticed that difficulties in network management are closely related to network complexity; the two are in direct variation. Traditional network management methods become either inadequate or unsustainable to cope with changes in campus networks due to these key reasons:

- The device-centric management mode is outdated; O&M personnel struggle with a heavy workload and managing hundreds of devices. Configurations may be inconsistent between devices, which are difficult to locate.

- O&M based on command lines is inefficient; command lines are not graphic, and so they cannot reflect the hierarchical relationships between services. Service configurations need to be translated into atomic configurations for delivery. Consequently, network configurations and service configurations are easily confused; conflicts can occur easily and are difficult to locate. Additionally, the command line files are oversized after a period of time and cannot be read due to lack of logic. For these reasons, it is more difficult to modify and expand the network.

- There is no tool to manage the running configurations and resources, or to manage conflicts that occur during system running. As a result, it is difficult to manually locate network or service faults.

- Fault information based on alarms and logs cannot efficiently support fault locating. According to statistics, more than 99% of alarms and logs are meaningless. In addition, network faults are individual and can occur suddenly; therefore, instant information is required to locate faults. Due to alarm and log recording limitations, it is impossible to indicate the root cause or record the fault scene. As such, fault locating is heavily reliant on the skills and expertise of O&M personnel.

- There is no key performance indicator system or evaluation tool for services. Deteriorating network performance leads to faults among a group of users, and due to a lack of preventative mechanisms, faults are only rectified after they have caused negative impacts. And so, maintenance work is passive.

Traditional configuration tools or network management tools cannot effectively solve these problems, particularly the latter three issues. Resolving them means being strongly dependent on the skills of network O&M professionals. If not resolved efficiently, the impact of these problems will be further amplified in future campus networks.

Noticeably, the introduction of Wi-Fi and IoT technologies to campus networks puts great pressure on O&M. However, most O&M teams lack in-depth understanding of invisible and unperceivable wireless networks.

The wireless network itself is fragile and its status is uncertain. It is also vulnerable to external interference that is often bursty. Therefore, the operational quality of the wireless network should be monitored continuously, and network optimization be performed accordingly. Otherwise, the network quality continues to deteriorate with the increase of service traffic. If this happens, O&M teams that lack professional tools will face tremendous challenges.

In order to address campus network O&M pain points, new concepts, technologies, network architectures, and O&M systems must be introduced, without increasing O&M workload or the number of personnel.

2.2.5 Strong and Unbreakable Security

Data are the core of digital transformation across industries, and it is also the key to future business competition. In today's digital world, increasingly severe security issues have gained global attention. In recent years, infamous ransomware attacks have served as a wake-up call for enterprises to the potential direct and serious losses caused by security vulnerabilities. Campus networks primarily face the following security challenges:

1. The pervasive use of wireless terminals blurs the campus network borders, making traditional border defense methods ineffective.

   Wireless Local Area Networks (WLANs) use radio-frequency signals as transmission media. Wireless channels are open, making it easy for attackers to eavesdrop and tamper with service data transmitted on the channels. Therefore, the wireless network faces various network threats, such as wireless user spoofing, AP spoofing, and Denial of Service (DoS) attacks from malicious terminals.

Additionally, Bring Your Own Device (BYOD) breaks the trust boundary for the controlled terminals to access the network. For example, in the past, all employees used company-issued PCs to access the network. These PCs were monitored by the company's dedicated security software, and so they also had to follow corporate rules and regulations. In this way, these PCs were trustworthy and controllable. However, after BYOD was introduced, terminals accessing the network became both uncontrollable and unreliable. For instance, if a user's terminal has the Dynamic Host Configuration Protocol (DHCP) server function enabled and is directly allowed to access the network, other users in the same subnet range may obtain incorrect IP addresses and therefore fail to access the network. To prevent this, a new technical method must be introduced to rebuild the trustworthy attributes of access terminals.

2. The growing use of IoT devices constantly extends the network and increases the attack surface.

With the development of IoT, networks are expanding and are required to connect not only people but also things. However, security problems arising from connecting things are far greater than those caused by connecting people, because IoT devices are exposed to external danger and their security defenses are relatively weak. Therefore, networks should provide additional protection for IoT devices. For example, the infamous botnet virus Mirai searches for IoT devices on the Internet. When it scans and spots an IoT device (such as a network camera or smart switchgear), it attempts to log in using either default or weak passwords. Once login is successful, the IoT device is added to the "broiler" list, where hackers can then control this IoT device and attack other networked devices.

3. New network attacks, such as APT attacks, make traditional security defense methods ineffective and unsustainable.

Traditional security defense methods can only schedule separate network or security devices, and so when a threat occurs, devices must counteract separately. Once a threat gets past security, it can be easily spread within the enterprise intranet, where control is lost. Meanwhile, traditional security tools provide inadequate threat detection, resulting in a high miss rate for detecting gray traffic. Concerning malicious threats hidden in encrypted traffic, there is no effective detection method for detection without decryption.

To address the preceding issues, a new campus network security solution should be introduced. The new solution must not only provide borderless defense for future ubiquitous campus networks but also effectively identify and handle constantly changing security attacks using the latest methods and technologies.

2.2.6 Open and Sharing Ecosystem

With the deepening of digital transformation, campus networks have become the crucial infrastructure of modern enterprises. Service data carried on campus networks are increasingly important for enterprises' operations and decision-making.

Acquiring network-side data conveniently through an open network platform in order to maximize data value has become a top priority for all enterprises and their partners. In the future, the relationship between campus networks and upper-layer services will become inseparable, bringing the following challenges to the openness of the network platform:

1. Solutions from single vendors cannot meet enterprises' diversified service requirements in various scenarios.

   In the digital era, enterprises pursue applications fully suitable for service scenarios to achieve business success. For this reason, no single company can provide differentiated application services for all industries. Future campus networks will have a complex ecosystem based on the integration of products and solutions from multiple vendors.

2. The use of nonstandard protocols and solutions makes service interconnection and commissioning difficult.

   To retain customers over the past few years, many vendors have developed products using proprietary protocols. Admittedly, new features of products may deliver tangible benefits to vendors in a short term. However, devices from different vendors are not interoperable with each other and Network Management System (NMS) interconnection is complex. Most enterprises have no choice but to separately deploy and manage multivendor devices, doubling network construction costs. To make matters worse, some vendors' devices may even have nonstandard platform interfaces, making service interconnection and commissioning difficult.

3. There is no complete network development and testing environment for the fast development and rollout of new services.

The network operation platform is different from the standalone operation platform. For the latter, we can easily set up a development and testing environment, whereas the network operation platform is oriented to the entire network. For individual developers and even small- and medium-sized development teams, the cost of setting up a development and testing environment is very high.

2.3 THE INDUSTRY'S VIEW OF NETWORKS IN THE DIGITAL WORLD

The development of ICT has two core drivers that complement and facilitate each other: service requirement changes and technical solution innovations.

Today's digital wave — driven by technologies such as AI — is having an unprecedented impact on networks and will push the entire world forward to the next era while dramatically increasing network complexity. If we continue to rely on traditional practices, network O&M costs will remain prohibitively high.

We are stepping into the intelligence era from the information era. As we do so, what kind of networks will we need? Numerous standards-defining organizations, industry leaders, and tech influencers have long been searching for the answer to this question, and have finally reached a consensus. They agreed upon autonomous driving networks as the best option.

An autonomous-driving campus network is intent-driven and stands out by delivering fully converged access, multiservice support, and high quality. It also comes with automatic and intelligent network O&M as well as unbreakable security, and an open ecosystem architecture.

2.3.1 Autonomous Driving Networks May Be the Ultimate Solution

We have experienced three industrial revolutions since the middle of 18th century and are about to start a fourth, as illustrated in Figure 2.11.

1ˢᵗ industrial revolution   2ⁿᵈ industrial revolution   3ʳᵈ industrial revolution   4ᵗʰ industrial revolution

Steam engine   Electric power   Information technology   Intelligence technology

FIGURE 2.11   Four industrial revolutions.

- The first industrial revolution (1760–1870) pushed human society from agricultural civilization to industrial civilization. The wide use of steam engines greatly improved productivity, freed human beings from heavy manual labor, accelerated the manufacturing industry, and ultimately drastically enhanced productivity.

- The second industrial revolution (1870–1950) saw us shift from the industrial era to an electricity era, where the pervasive use of electric power expanded upon the advancements made in the first industrial revolution and increased productivity in more industries. The rise of electric power accelerated transportation with railways and automobiles. This, in turn, led to the rapid development of the transportation industry and more frequent people-to-people and country-to-country communication. As such, a global international social system took shape.

- The third industrial revolution (1950 to now) has lasted nearly 70 years, advancing human society from the electricity era to the information era. This era has further expanded on the advancements of the first and second industrial revolutions. In particular, automated manufacturing has appeared, doubling productivity in various industries. In today's information era, electronic computers and data networks are widely used, greatly enriching our work and life. People can communicate with each other anytime and anywhere. The information society we are living in has led to globalization, where up-to-date information can be transmitted to any corner of the globe within seconds.

Today, we are standing on the cusp of the fourth industrial revolution represented by AI. New ICT will lead us from the information era to the intelligence era. By that time, new ICT will be widely applied to all aspects of our society. AI, the engine of the fourth industrial revolution, will promote the development of all industries around the world.

Figure 2.12 shows four different phases along the AI productivity/adoption curve. AI, as a general purpose technology (GPT), has left the first phase, where exploration of AI technology and application takes place on a small scale. It is now in the second phase where tech development and social environment are colliding. Driven by continuous collision, AI is finding ever-increasing and more comprehensive use in industry application scenarios. This, in turn, generates greater productivity. Data networks are a key driving force in the IT era, and will be further developed and optimized in the AI era.

1. Reference to Self-Driving Cars

   When we hear the term autonomous driving, we may initially cast our minds to self-driving cars. And actually, there are many similarities between maintaining a network and driving a car. For example, just like maintaining a network, driving a car is a complex skill. We often make mistakes while driving, especially if we are tired or not concentrating. These errors may cost lives, and for this reason, we have long been trying to introduce robots that never make these fatigue-driven mistakes. In fact, companies have been experimenting with self-driving cars for a long time. They are no longer sci-fi, but a reality thanks to recent breakthroughs with AI technologies.

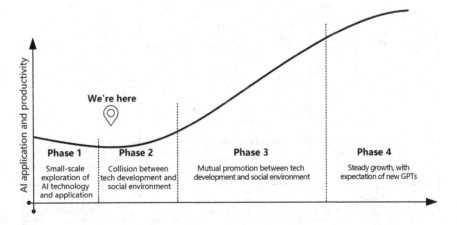

FIGURE 2.12 Productivity/application developmental curve of GPTs.

The international society has developed strict classification standards for autonomous driving technologies. Typical examples include the criteria defined by the Society of Automotive Engineers (SAE) and National Highway Traffic Safety Administration (NHTSA), as shown in Figure 2.13.

Most automobile companies are currently at Level 0 or Level 1 of autonomous driving, a few pioneers have achieved Level 2, and several frontrunners have succeeded in implementing Level 3 in strictly limited scenarios. To achieve Level 3 or higher autonomous driving, the following three conditions must be met:

a. Real-time map: A map is a digital simulation of the physical world. By superimposing real-time information, the map can be regarded as a digital twin of a road network.

b. Environment awareness: Vehicle-mounted sensors can detect the surrounding environment in real time. These sensors cover laser radars, optical radars, computer vision, and Global Positioning System (GPS). When necessary, Internet of Vehicles (IoV) based on 5G is used to share environment information between other vehicles. All of these give an overall perception of the physical world.

c. Control logic: As AI technology is maturing, a complete control logic should be developed for AI.

Levels of autonomous driving as defined by NHTSA and SAE						
Level NHTSA	Level 0	Level 1	Level 2	Level 3	Level 4	
SAE	Level 0	Level 1	Level 2	Level 3	Level 4	Level 5
Name (by SAE)	No automation	Driver assistance	Partial automation	Conditional automation	High automation	Full automation
Definition (by SAE)	The full-time performance by the human driver of all aspects of the dynamic driving task, even when enhanced by warning or intervention systems	The driving mode-specific execution by a driver assistance system of either steering or acceleration/deceleration using information about the driving environment and with the expectation that the human driver perform all remaining aspects of the dynamic driving task	The driving mode-specific execution by one or more driver assistance systems of both steering and acceleration/deceleration using information about the driving environment and with the expectation that the human driver perform all remaining aspects of the dynamic driving task	The driving mode-specific performance by an automated driving system of all aspects of the dynamic driving task with the expectation that the human driver will respond appropriately to a request to intervene	The driving mode-specific performance by an automated driving system of all aspects of the dynamic driving task, even if a human driver does not respond appropriately to a request to intervene	The full-time performance by an automated driving system of all aspects of the dynamic driving task under all roadway and environmental conditions that can be managed by a human driver
Entity Execution of steering and acceleration/ deceleration	Human driver	Human driver and system	System			
Monitoring of driving environment	Human driver			System		
Fallback performance of dynamic driving task	Human driver				System	
System capability (driving modes)	N/A	Some driving modes				All driving modes

FIGURE 2.13   Levels of autonomous driving.

2. Autonomous Driving of Networks

Similar to the requirements of self-driving cars, today's networks should evolve toward autonomous driving to radically change manual O&M. Only by doing so can we effectively operate and maintain our ever-increasingly complex networks without extra labor costs.

With self-driving cars, all we need to do is enter our destination, and the car then will automatically take us there. The car itself will also avoid any road hazards by itself without manual intervention. Similarly, when it comes to autonomous driving networks, we are not required to know the network in detail. Instead, we just need to tell the network what we want to do, and the network can automatically adjust itself to meet our requirements. In other words, network planning and deployment are automated. When a fault occurs on the network or in its external environment, the network can automatically handle and recover. This means that network fault troubleshooting is automated.

The road to autonomous driving networks will be a long one. It cannot be accomplished overnight. Huawei has defined five levels of autonomous driving networks that vary in levels of customer experience, network environment complexity, and degree of automation, as shown in Figure 2.14.

a. Level 0: Only assisted monitoring capabilities are provided. All dynamic O&M tasks are completed manually.

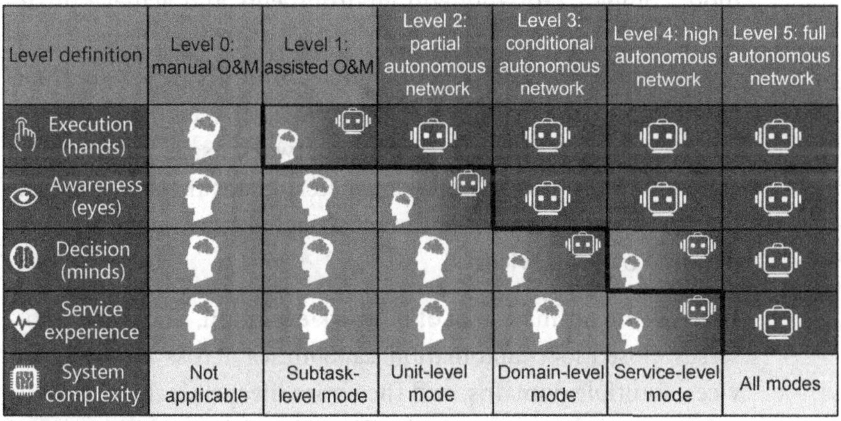

Level definition	Level 0: manual O&M	Level 1: assisted O&M	Level 2: partial autonomous network	Level 3: conditional autonomous network	Level 4: high autonomous network	Level 5: full autonomous network
Execution (hands)						
Awareness (eyes)						
Decision (minds)						
Service experience						
System complexity	Not applicable	Subtask-level mode	Unit-level mode	Domain-level mode	Service-level mode	All modes

FIGURE 2.14 Five levels of autonomous driving networks defined by Huawei.

b. Level 1: The system repeatedly executes a certain subtask based on known rules to increase execution efficiency. For example, Graphical User Interface (GUI)-based configuration wizards and batch configuration tools greatly simplify manual operations, lower skill requirements for O&M personnel, and improve the efficiency of repeated operations.

c. Level 2: The system enables closed-loop O&M for certain units under certain external environments, lowering the bar for O&M personnel regarding experience and skills. For example, a network device provides configuration APIs and automatically executes network configuration according to scheduling instructions issued by a network management platform. The entire process is automated and requires no manual intervention.

d. Level 3: Building on Level 2 capabilities, the system can sense real-time changes in the network environment, and in certain domains, optimize and adjust accordingly to enable intent-based closed-loop management. Level 3 is defined as the system's ability to continuously perform control tasks for a given goal. For example, when a fault occurs on a campus network, the system can use AI to aggregate alarms and identify fault scenarios, then locate the alarm, quickly determine rectification measures, and automatically dispatch work orders to network administrators.

e. Level 4: By enhancing Level 3, a Level 4 system can accommodate more complex cross-domain environments and achieve proactive closed-loop management. In this way, network faults can be resolved prior to customer complaints, minimizing service interruptions and greatly improving customer satisfaction. For example, a campus network can detect network usage environment changes and adjust network parameters in advance by analyzing the changes to network parameters and external environments detected by IoT devices.

f. Level 5: The ultimate goal of network evolution. The system possesses closed-loop automation capabilities across multiple services, multiple domains, and the entire lifecycle, achieving truly autonomous driving networks. With Level 5 capabilities, campus networks are completely self-maintained. They proactively predict application changes and automatically dispatch work orders to network administrators for network upgrade or expansion, ensuring that networks meet ever-changing service requirements.

Most traditional networks are still at Level 1. Specifically, the NMS collects information and implements batch or template-based device configuration using graphical tools. Network O&M relies heavily on the skills of O&M experts.

Huawei Agile Campus Solution has so far taken the automation of campus networks to Level 2. In Huawei's solution, SDN is introduced to agile campus networks to automatically deploy the most common services based on network models, greatly simplifying network service deployment.

There is still a long way to go if we want to achieve Level 3 to Level 5. Huawei's highly skilled campus network architects are currently working on building Level 3 autonomous driving networks. The intent-driven campus network solution herein is at Level 3.

3. "Digital Twins" of Campus Networks

A digital twin is a mapping of a physical entity in the digital space. A physical entity and its virtual replica form a pair of twins. A complete digital twin model consists of physical entities, virtual replicas, and mapping functions.

With regards to digital twins, programs and algorithms in the digital space can truly "understand" entities in the physical space, and can reversely affect these entities by using the mapping functions, as shown in Figure 2.15.

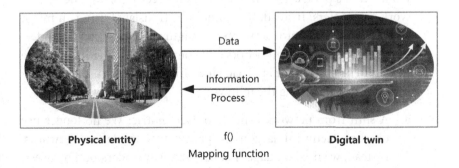

Physical entity   f()   **Digital twin**
Mapping function

FIGURE 2.15   Physical entity and digital twin.

Just as the digital twin concept is required for self-driving cars, it is also required to create more advanced autonomous driving networks.

When it comes to campus networks, if we want to achieve Level 3 autonomous driving, we should build digital twins primarily for automation scenarios, processes, and objectives. We need to expand on this when aiming for Level 4 autonomous driving, building digital twins for all scenarios, processes, and objectives. Ultimately, at Level 5, we must build digital twins for the entire campus, rather than just the campus network. This is because the campus network is only a small part of the digital twin for the entire campus.

2.3.2 Intent-Driven Networks Achieve Level 3 Autonomous Driving

By analyzing the autonomous driving network criteria, it is clear that Level 3 is a turning point.

Networks at Level 2 and lower are seen as passive networks, whereas Level 3 networks are autonomous and can proactively adjust resources. For this reason, we call them autonomous networks (ANs). Subsequent Level 4 and Level 5 networks go even further toward larger scale and more comprehensive autonomy.

Level 3 automation is a short-term goal of Huawei's campus network solution. The core idea of Level 3 automation is to automatically adjust network resources based on a given objective or intent, while also adapting to external changes. ANs with this capability are called intent-driven networks (IDNs).

1. What is an intent?

An intent is a business requirement and service objective described from a user's perspective. Intents do not directly correspond to network parameters. Instead, they should be translated using an intent translation engine into parameter configurations that can be understood and executed by networks and components. After introducing intents to networks, our approach to managing networks will change significantly, including the following:

a. A shift from network-centric to user-centric: We no longer use obscure technical jargon or parameters to define networks. Instead, we use language that can be easily understood by users. For example, we used to describe the Wi-Fi network quality according to signal strength, bandwidth, and roaming latency. These descriptions were difficult for users to understand and could not directly reflect user experience. With an IDN we can define network requirements in a simple and easy to understand format, for example, "25 employees, office use".

b. A change from fragmented management to closed-loop management: Intents can be validated. For example, the "25 employees, office use" network can perform closed-loop verification on the intent. Specifically, the network monitors the intent validity in real time, including whether the number of access users exceeds 25 or whether there is a large amount of non-office service traffic. If the intent changes, the network can proactively adjust itself, implementing closed-loop management.

c. A transformation from passive response to proactive prediction: The intent objective can also be checked proactively. For example, the "25 employees, office use" network checks in real time whether the service quality is in line with the intent. If any deviations are found, the network can proactively analyze and quickly rectify the deviations to prevent user complaints caused by the deteriorated service quality.

d. An evolution from technology dependency to AI/automation: With a given objective and digital twins established for the network, both closed-loop management and proactive prediction can be automatically executed through programs. Especially after AI technology is introduced, automatic detection and closed-loop management can be implemented in almost all scenarios.

2. IDN architecture

As shown in Figure 2.16, IDN consists of an ultra-broadband, simplified network, and a smart network brain.

The ultra-broadband, simplified network is the cornerstone of IDN. It follows an ideal in the same vein as Moore's law. That is, it is driven by continuous innovation and breakthroughs in key communications technologies. In addition, node capacity doubles every 24 months to meet the bandwidth and transmission needs of new services such as 4K/VR, 5G, and cloud computing.

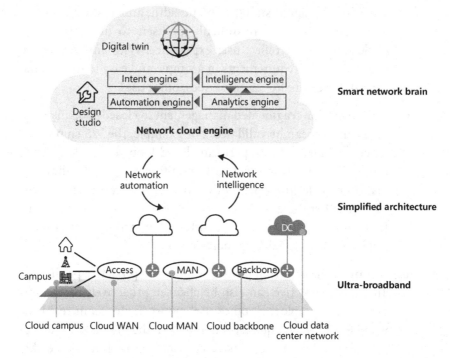

FIGURE 2.16   Intent-driven network architecture.

In line with IDN, network architecture is continuously reshaped and optimized to become more modularized, standardized, and simpler than ever. Therefore, the network boasts elastic scalability and plug-and-play capabilities, as well as the deterministic low latency required for bandwidth-hungry, latency-sensitive services, and special industrial applications of the future. Additionally, network automation protocols are used to reduce network complexity, decouple services from network connections, quickly provision services, and improve network programmability to meet different service requirements.

The smart network brain is the engine of IDN that implements intelligent management and control of the entire network. It mainly performs the following tasks:

a. Accurately understands users' business intents, translates them into network configurations, and automatically deploys the network configurations onto specific physical devices, ensuring that the network meets service intents.

b. Detects the health status of the physical network in real time, identifies faults, and gives warnings in a timely manner.

c. Offers handling suggestions for faults and quickly troubleshoots or optimizes the network based on a library of past experiences.

d. Implements real-time visibility of SLAs and performs predictive maintenance using AI technologies.

e. Accurately predicts and generates alerts about component failures through continuous modeling, behavior learning, and AI training.

f. Identifies nodes that are prone to temporary congestion, migrates important services in advance, and optimizes scheduling policies.

IDN consists of the following four engines:

a. Intent engine: translates business intents into network language and simulates network design and planning.

b. Automation engine: converts the network design and plan data into network commands, and enables network devices to automatically execute these commands through standard interfaces.

c. Analytics engine: collects and analyzes user network data (including uplink and downlink rates, latency, and packet loss, but excluding personal user data) using technologies such as Telemetry.

d. Intelligence engine: provides risk prediction and handling suggestions using AI algorithms and a continuously upgraded library of past experiences. These functions are provided in conjunction with the Analytics Engine.

2.4 ROAD TO CAMPUS NETWORK RECONSTRUCTION

Digital transformation is a qualitative leap forward. It requires multiple new enablers, such as big data, cloud computing, AI, AR, and VR. Despite appearing unrelated to each other, all of these enablers have a common ground: they are all based on networks. This means that networks are crucial to the success of digital transformation.

To facilitate digital transformation, campus networks need to be transformed from five perspectives: ultra-broadband, simplicity, intelligence, security, and openness. The following sections explore each of these perspectives in detail.

2.4.1 Ultra-Broadband

Ultra-broadband refers to two concepts: wide coverage and large bandwidth. The former means that the network must be sufficiently extensive to allow any terminal to access the network at any time and from any location. The latter means that the network must provide sufficient bandwidth and throughput so that as much data as possible can be forwarded within a given period of time.

If we compare the network to a highway, wide coverage refers to the highway covering an extensive area and having sufficient entrances and exits that enable vehicles to enter and leave at any time. High bandwidth refers to the highway being sufficiently wide so that enough lanes exist for vehicles to quickly pass.

1. Wide coverage achieves connectivity of everything.

    Campus networks initially contained only wired terminals. But as WLAN technology began to emerge, WLAN terminals started to connect to these networks, freeing users from constraints imposed by network cables and implementing mobile network access.

    The expansion from wired to wireless terminals has extended the scope of campus networks. These networks will be further extended in the near future as IoT terminals gain popularity, leading to a time when everything is connected — campus networks will then be pervasive, delivering ubiquitous connectivity.

    The realization of wide coverage on a campus network requires us to adopt a number of innovative ideas:

    a. All-scenario Wi-Fi: Wireless network coverage is generally complex due to the diversity of scenarios such as large outdoor public spaces, high-density stadiums and wireless offices, and electronic classrooms. Differentiated wireless network coverage solutions are required in order to cover these varied scenarios. For example, Huawei's all-scenario Wi-Fi solution flexibly adapts to diversified service models and coverage modes — it delivers optimal

Wi-Fi coverage in high-density environments by using triple-radio and built-in small-angle antenna technologies. Powered by an industry-leading smart antenna technology, the solution reduces signal interference between APs, increases Wi-Fi network capacity, and improves user experience.

b. Wi-Fi and IoT convergence: Short-range wireless access technologies offer a coverage range similar to that of Wi-Fi technologies, meaning that it is possible to technically converge Wi-Fi and IoT networks in order to slash deployment and O&M costs. This is why industry players continue to explore the most efficient Wi-Fi and IoT convergence solutions. One example is to integrate Bluetooth positioning technology into IoT APs in order to achieve high-precision indoor navigation and positioning. Another example is to combine short-range IoT technologies, such as RFID and ZigBee, with IoT APs in order to support a wide array of IoT applications including electronic shelf labels (ESLs) and smart waistbands.

2. High bandwidth is suitable for undersubscribed networks.

Traffic on a traditional campus network is mostly east-west traffic (between servers), while north-south traffic volumes (between clients and servers) are relatively small. Services are scattered and servers are decentralized across the campus.

In an enterprise, for example, each department has its own file server used to share documents within the department. Various services are presented in the form of standalone software, and various types of data are stored locally on terminals. Communication traffic between users or between users and scattered servers must traverse the network while having certain local assurances. Accordingly, the network has a large oversubscription ratio, which is likely to exceed 1:10 between layers.

As the reliance on IT and cloud-based architecture has become more pervasive in recent years, the main services and data in a campus have been centralized onto the cloud and presented in the form of network as a service (NaaS). Enterprises use web disks running on public or private clouds to provide file services, and users can access these services directly through a web browser without installing clients. Departments and groups no longer need to deploy their own physical servers, as they can now access virtual file services at the department level simply by applying for resources via a web portal. This means that far less data need to be stored locally.

Consequently, north-south traffic (from terminals to the cloud) becomes more prevalent, with the proportion of east-west traffic becoming increasingly smaller. In this case, a network with a large oversubscription ratio is unable to meet service requirements in the cloud era. The oversubscription ratio between layers of the campus network must therefore be decreased or, if possible, even lowered to approximate an undersubscribed network. This means that each layer of the campus network is required to provide sufficient bandwidth resources.

In order to achieve high bandwidth on the campus network, the industry primarily adopts the following practices:

a. Wireless access network: The mainstream wireless standard currently used is IEEE 802.11ac, also known as Wi-Fi 5. To meet bandwidth-hungry services such as 8K video and cloud VR, IEEE 802.11ax — also known as Wi-Fi 6 — has emerged and will be widely adopted in the future. Compared with Wi-Fi 5, Wi-Fi 6 has significantly improved the transmission rate, up to 9.6 Gbit/s, which is the highest so far. Designed with new technical solutions and incorporating multiple 4G and 5G network practices, particularly Orthogonal Frequency Division Multiple Access (OFDMA) and Multiple-Input Multiple-Output (MIMO), Wi-Fi 6 efficiently resolves many pain points facing a conventional wireless network.

b. Wired access network: Data transmission and terminal power supply are two issues that require attention on a wired access network. In order to address these issues, wired access primarily uses copper cables as the transmission medium, with the transmission rate evolving from 2.5 to 5 Gbit/s, and on to 10 Gbit/s. Typically, APs, high-performance servers, and other devices and terminals used for edge services connect to the campus network. Because IEEE 802.11ac has been widely adopted, the maximum transmission rate of the Wi-Fi network has exceeded 1 Gbit/s, meaning that 1 Gbit/s uplinks are no longer adequate or sustainable for IEEE 802.11ac-compliant APs. To address this, the industry mainly uses IEEE 802.3bz — multi-gigabit Ethernet technology — to achieve a data transmission rate of 2.5 Gbit/s over CAT5e cables. This technology helps to maximally reuse existing CAT5e cables widely deployed in buildings. In the future, the widespread use of Wi-Fi 6 will require a further increase of wired access bandwidth to 10 Gbit/s or even higher.

c. Wired backbone network: As wired and wireless access bandwidths greatly increase, the oversubscription ratio of the campus network is gradually reducing. It is therefore essential to dramatically increase the bandwidth on the wired backbone network, which mainly uses optical transmission media and evolves from 25 to 100 Gbit/s, and on to 400 Gbit/s. A typical application scenario is to interconnect campus switches to provide an ultra-broadband campus backbone network.

2.4.2 Simplicity

An ultra-broadband campus network — one that offers wide coverage and high bandwidth — functions as the "information highway", efficiently connecting everything and implementing a diverse array of services. How to quickly operate this information highway is the next key perspective involved in digital transformation.

In the future, administrators of large or midsize campus networks will handle thousands, potentially even hundreds of thousands of terminals, and will perform network planning, design, configuration, and deployment for each data forwarding service. The use of traditional methods in the future would make it extremely difficult to manage the campus network. Compounding this is mobility, which leads to frequent network policy changes, and digit transformation, which requires faster provisioning of new services in order to adapt to changing market requirements.

With all of these factors in mind, we need to transform campus networks in an E2E manner, from planning and management to deployment and O&M. This will ensure that the new campus network can overcome the problems caused by network complexity, provide ultra-broadband campus network services on demand, and greatly simplify E2E deployment.

1. Complex tasks are processed by machines, enabling administrators to stay informed of the network more easily.

   The traditional approach network administrators take when configuring network devices, for example, setting common routing protocols or VPN parameters, is to use the command line interface (CLI) or web NMS in a language understood by network devices. This direct human-machine interaction mode is inefficient and requires administrators to possess high technical skills.

   As the network complexity increases exponentially, this human-machine interaction mode is transformed into a machine-to-machine interaction mode by shifting the processing of complex tasks from "human" to "machine".

   Today's software capabilities already allow us to design a network management platform that can accurately identify network administrators' service intents. An administrator needs only to fill in the service requirements into the network management platform by using a natural language, and then the platform automatically translates these service requirements into languages that can be understood by network devices before delivering the translated results to these devices.

2. Centralized resource control greatly simplifies management.

   Traditional network O&M and management use a distributed management method that requires personnel to perform configuration, deployment, and O&M on each device separately. Such a method is not only inefficient but also unsuitable for a smart campus network that contains tens of thousands of network elements.

   To address these shortcomings, a smart campus network introduces the SDN concept and adopts a centralized O&M management and control solution. This solution completely separates the control rights of network devices and centrally manages all network devices on a unified network management platform.

   Functioning as the "brain" of network management, the unified network management platform differs from traditional network management platforms that are only able to implement simple batch configuration and upgrade of devices. The SDN-based unified network management platform, in contrast, can pool network devices so that users can schedule resources as required and customize network route configurations and service policies, eliminating the need for users to repeatedly configure devices one by one.

3. Services are decoupled from the network, eliminating the impact of network complexity.

Traditional campus networks are closely coupled with services, but this design is no longer suitable for today's fast-changing upper-layer service applications, let alone those of tomorrow. In order to realize fast and flexible service rollouts, the most effective option is to decouple upper layer service planning and deployment from the network, eliminating the impact of network complexity.

The decoupling of upper-layer services from the network is achieved using one of two methods currently available. The first method involves shielding the underlying physical networks, typically done by using network virtualization. Specifically, this method builds a logical network on top of the existing physical networks. Administrators can consider these underlying physical networks as "black boxes" and disregard them, focusing only on the service configurations of the logical network.

The other method involves reducing interactions between service applications and network devices. For example, Huawei's free mobility solution introduces security groups in order to decouple policies from IP addresses. In this way, administrators define service policies based only on user identities, greatly simplifying service configurations.

2.4.3 Intelligence

Technical advances and service upgrades complicate network planning, design, deployment, and O&M, increasing the workload for network engineers. Enterprises therefore have to invest more labor resources to deploy networks, analyze network problems, and rectify network faults, leading to expensive and inefficient O&M. The ideal way to overcome these burdens is to add intelligence and build an intelligent network.

1. From difficult perception of user experience to visibility at any time

SNMP-based network management tools are unable to reflect the network status in real time, because they require an interval of minutes to collect data. This approach is no longer effective as the network scale increases and services become increasingly diversified.

New technologies that achieve real-time collection of network performance data are therefore required. These new technologies, combined with multidimensional big data analytics, can visually reflect the service running status.

Specifically, 360° network profiling can be created for the service running status of a single user. This would allow us to clearly see the user's network experience on the timeline according to key indicators, such as the wireless access authentication, latency, average packet loss rate, and signal strength.

We can also view the running status of the entire regional network from a global perspective, based on metrics including the access success rate, access time consumption, roaming fulfillment rate, coverage, and capacity fulfillment rate. Further to this, we can also verify whether user experience in the local regional network is better or worse than that of other regional networks, providing guidance for subsequent O&M.

2. From passive response to AI-based predictive maintenance

Using traditional network O&M approaches that rely on passive response and post-event troubleshooting, enterprise O&M personnel remain unaware of service faults until they receive customer complaints. In other words, the network is unable to detect service exceptions in advance.

An IDN revolutionizes network O&M by integrating AI into the device, network, and cloud layers, and performing continuous AI training through an intelligent learning engine deployed on the cloud to enhance the fault knowledge database. In this way, IDN can automatically predict, rectify, and close the majority of network faults, improving network experience while reducing customer complaints.

3. From manual network optimization to AI-based intelligent network optimization

The management of Wi-Fi networks has always been a complex task to perform because they are self-interference systems, where different APs interfere with each other due to factors such as channel, power, and coverage. In most cases, a huge amount of labor and time is required for network optimization at the deployment and O&M phases.

Take a large office campus, one with tens of thousands of employees, as an example. It would usually take 3–5 weeks to plan, deploy, commission, and accept thousands of Wi-Fi APs if traditional approaches were used. Furthermore, after the Wi-Fi network is deployed, the network quality and user experience may deteriorate along with service and environment changes (for example, the number of access terminals increases, or the office environment is reconstructed).

At the opposite end of the spectrum is the intent-driven campus network. Using AI, it intelligently optimizes the Wi-Fi network in minutes by detecting changes across a plethora of access terminals and network services in real time, identifying potential faults and risks, and simulating and predicting the network.

2.4.4 Security

Traditional campus networks have clear security boundaries and trustworthy access terminals, and can leverage security devices such as firewalls to defend against attacks and ensure security. However, as these security boundaries become increasingly blurred or even disappear entirely, new security concepts and technical approaches must be developed in order to implement borderless security defense. To this end, the following ideas are prevalent in the industry.

1. Security is deeply integrated into the network.

   Information and data are transmitted on communications networks. From a traditional security defense perspective, it is often overlooked that numerous network devices seem to act as "onlookers" during security attack and defense.

   An analogy of this is urban public security. To reduce public security incidents, with the aim of eliminating them entirely, authorities usually consider increasing the police force, but often ignore the power of citizens. Regardless of how many police officers are added, there will always be a criminal element giving rise to illegal incidents. A better approach would be to turn to citizens for help and rely on more auxiliary means. In so doing, it would be possible to almost completely eliminate illegal incidents.

   If we compare security devices to the police force, network devices can be considered as citizens. Network devices are "trump cards" that would enable unimaginable results if they were mobilized to

collaborate with security devices. Without them, unbreakable security will remain merely an illusion.

2. Technologies such as big data and AI are introduced into the security
   analysis field to build an intelligent security threat detection system.

   Security probes are installed onto all devices on the network so
   that a security analysis engine can collect network-wide data, monitor the security status of the entire network, and perform in-depth
   analysis on network behavior data.

   Deep neural network algorithms and machine learning technologies are also used to promptly detect, analyze, and pinpoint security
   threats, while network devices collaborate with security devices to
   quickly handle security threats.

   Although remarkable achievements have been made regarding the performance of various security devices and the use of new
   threat defense methods, no enterprise can establish a complete threat
   defense system by focusing exclusively on limited security.

   In the future, campus networks will become borderless. To
   embrace them, enterprises should build an integrated and multidimensional security defense system from the perspective of the entire
   network. Only in this way can they benefit from unbreakable campus network security.

2.4.5 Openness

To upgrade the entire network from the current to the next generation
generally takes at least five years, far longer than that for applications. As
such, the campus network and applications running on it develop at different speeds.

Take WLAN as an example. Each generation of WLAN standard is usually released five years after the preceding one, but WLAN applications are
updated more quickly, usually every 2–3 years and even shorter nowadays.

For this reason, the current campus network is unable to support the constant emergence of innovative applications if it is not adequately open. The
openness of the campus network can be achieved from the following aspects.

1. Campus network architecture that is open in all scenarios and at all
   layers facilitates easy integration.

   The architecture employed on a campus network is decoupled into
   different layers and is fully open both within and between layers.

In the horizontal direction (within each layer), network components use products that comply with general-purpose standards and adopt standard protocols instead of proprietary ones. This makes it easier to achieve multivendor networking and interoperability while avoiding single vendor lock-in that leads to insufficient supply and poor service.

In the vertical direction (between layers), standards-compliant and open northbound and southbound interfaces are used, allowing more developers to create innovative applications on the campus network.

2. A network platform backed by a sound ecosystem ensures that more third-party applications are preintegrated.

Solution vendors prefer to capitalize on their own open capabilities in order to build a favorable ecosystem. For example, Huawei uses its open intent-driven campus network solution to develop ESL, ultra-wideband positioning, and other joint solutions together with its partners. A sound solution ecosystem, one that encompasses more partners, means that solutions can preintegrate more third-party functions and applications.

3. Vendors provide a comprehensive developer platform to fully support third-party developers.

In addition to continuously launching new products that meet ever-changing user requirements, campus network vendors are also expected to become ecosystem companies. They can provide complete developer platforms and related lab environments to help attract, identify, and cultivate high-value core developers across the product ecosystem. They can also offer developers E2E services — covering experience, learning, development, testing, verification, and marketing — to assist in incubating innovative solutions. All of these efforts pay off and accelerate the digital transformation of enterprises.

Digital transformation cannot rely purely on a single company or organization, as it requires the concerted efforts of all stakeholders. With its intent-driven campus network solution, Huawei is an innovator and contributor to building reliable digital infrastructure, aiming to build a global open campus ecosystem platform through which all stakeholders across all industries can collaborate more closely. The open ecosystem platform, once in place, will stimulate cross-industry innovations, create a digital culture, and bring huge value for global partners and customers.

CHAPTER 3

Overall Architecture of an Intent-Driven Campus Network

To cope with the challenges brought by digitalization to campus networks, Huawei has launched its intent-driven campus network solution that is ideal for building next-generation campus networks of all sizes. This solution builds ultra-broadband, simplified, intelligent, secure, and open campus networks for customers to better tackle the challenges in connectivity, experience, O&M, security, and ecosystems during their digital transformation. This chapter describes the basic architecture, key interaction interfaces, service models, and cloud-based deployment modes of the intent-driven campus network.

3.1 BASIC ARCHITECTURE OF AN INTENT-DRIVEN CAMPUS NETWORK

As shown in Figure 3.1, the basic architecture of an intent-driven campus network consists of the network layer, management layer, and application layer.

1. Network layer

    An intent-driven campus network uses virtualization technology to divide the network layer into the physical network and virtual

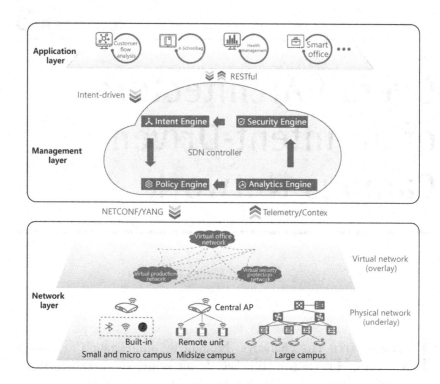

FIGURE 3.1   Basic architecture of an intent-driven campus network.

network, which are completely decoupled from each other. The physical network keeps evolving in compliance with Moore's Law to deliver ultra-broadband forwarding and access capabilities. The virtual network, on the other hand, uses virtualization technology to rise above the complex networking of physical devices and provide a fabric that ensures reachability between any two network elements (NEs), resulting in a simplified virtual network.

a. A physical network is also known as an underlay network, which provides basic connection services for campus networks. The underlay network of an intent-driven campus network has a hierarchical structure. That is, just like a traditional campus network, it can be divided into four layers: access layer, aggregation layer, core layer, and egress area. The intent-driven campus network provides converged access for three networks, allowing simultaneous access of wired, wireless, and Internet of Things (IoT) terminals.

b. A virtual network is also known as an overlay network. With virtualization technology, one or more overlay networks can be built on top of the underlay network, and service policies can be deployed on these overlay networks. As service policies are decoupled from the underlay network, service complexity is also decoupled from network complexity.

Multiple overlay networks can be constructed to serve different services or different customer segments.

Common network layer components include switches, routers, firewalls, wireless access controllers (WACs), and access points (APs). These network components can be connected to form an interoperable physical network.

2. Management layer

The management layer acts as the campus network's brain by providing network-level management capabilities such as configuration management, service management, network maintenance, fault detection, and security threat analysis. Traditional campus networks are managed using a network management system (NMS), and although the NMS can display the network's status, it cannot manage the network from the entire network perspective and lacks automatic management capabilities. Every time service requirements change, the network administrator has to re-plan services and manually modify configurations on network devices. Such an approach is inefficient and error-prone.

Therefore, maintaining network flexibility is critical in rapidly changing service environments, which requires the use of automation tools to assist in network and service management. Unlike traditional campus networks that manage services on a per-device basis, an intent-driven campus network implements automatic management using a Software-Defined Networking (SDN) controller. The controller abstracts the campus network from users' perspective into the four engines — Intent Engine, Policy Engine, Analytics Engine, and Security Engine — which are used together with big data and Artificial Intelligence (AI) technologies to realize automatic campus network management.

a. Intent Engine: works with the Policy Engine to implement the simplified management capability of an intent-driven campus network. In contrast to traditional device management and O&M that use professional interfaces, the Intent Engine abstracts common operations on campus services and allows network administrators to perform management and O&M using natural languages. This is especially needed for complex networks, such as a Wi-Fi network capable of delivering real-time video conferencing for a conference room that can accommodate 50 participants. Configuring such a network can be extremely complicated and time-consuming in a traditional campus network solution. First, a network administrator needs to plan the Wi-Fi network and set the radio frequency (RF) parameters to ensure the coverage and quality of wireless signals. Second, to deliver real-time video conferencing with sufficient bandwidth as well as minimal delay and packet loss, the network administrator also needs to set complex QoS parameters. Finally, the network administrator has to manually configure commands on each device. If a configuration error is detected, the network administrator needs to check all the commands one by one, which is very laborious. However, the Intent Engine in the intent-driven campus network solution can substantially reduce the workload as a network administrator only needs to describe the requirements on the SDN controller in a natural language. The Intent Engine then automatically identifies the description, translates it into the configuration using a language that can be understood by network devices, and delivers the configuration to network devices. For example, the Intent Engine can automatically adjust the AP transmit power and subsequent RF parameters based on the size of the conference room. The engine can also automatically calculate QoS settings and the bandwidth to be reserved for video conferencing based on the number of participants. Moreover, it can automatically simulate the client to verify its configuration. In short, the network administrator does not need to worry about network implementation; rather, they only need to describe the requirements on the graphical user interface (GUI) of the SDN controller, which automatically identifies the intent and implements it.

b. Policy Engine: works with the Intent Engine to implement the simplified management capability of an intent-driven campus network. The Policy Engine abstracts the frequently changing parts of campus services and focuses on the access rules and policies between people, people and applications, and applications themselves. The operation objects of the Policy Engine are people and applications, which on an intent-driven campus network, are fully decoupled from the campus network to maximize flexibility and ease of use. Assume that there are two departments (R&D and marketing) on a campus, and the network administrator needs to add an application that is only accessible to the R&D department. Implementing this application will be time-consuming and labor-intensive in a traditional network solution. More specifically, the network administrator needs to record the IP network segments of the two departments and the IP address of the new application. They then configure a policy on each device so that the IP address of the new application is accessible only to the IP address segment of the R&D department. Additionally, configuration errors can be identified only after the R&D department attempts to access the application. However, implementation in an intent-driven campus network solution is simplified significantly. After the Policy Engine is abstracted, the network administrator simply needs to add the new application information on the SDN controller and specify only the R&D department that can access the application. The Policy Engine will then dynamically deliver configurations to network devices based on the access requirements for different departments.

c. Analytics Engine: provides the intelligence capability of an intent-driven campus network. The Analytics Engine uses Telemetry to collect network-wide information (not private user data but device status data, performance data, and logs) with which it uses big data and AI technologies for correlative service analysis. In this way, network problems can be quickly located and resolved. Troubleshooting is usually time-consuming in a traditional network solution and it is difficult to fix problems that cannot be reproduced. For example, to locate application access failure reported by an end user, a network administrator needs to find the access location of the terminal, check the access path between the terminal and application, and collect traffic statistics on a per-device basis. This situation would differ significantly on an intent-driven campus network. On such a network, the Analytics Engine constantly collects statistics about network performance and the status of each user and application. With its predictive maintenance capability, the Analytics Engine can notify the network administrator of a fault even before an end user can detect, helping improve user satisfaction and experience.

d. Security Engine: implements the security capability of an intent-driven campus network. By using big data analytics and machine learning technologies, the Security Engine extracts key information from massive amounts of network-wide data collected by Telemetry. The Security Engine then analyzes the information from multiple dimensions to detect potential security threats and risks on the network and determine the situation of a network's security. When a security risk is detected, the Security Engine automatically works with the Intent Engine to instantly isolate or block the threat.

In addition to the core component SDN controller, the management layer also includes network management components such as the authentication server, Dynamic Host Configuration Protocol (DHCP) server, and Domain Name System (DNS) server. These components may be integrated into the SDN controller or independently deployed on general-purpose servers.

3. Application layer

The application layer of a traditional campus network consists of various independent servers that provide value-added services (VASs) for the network, such as office automation (OA) application servers, email servers, instant messaging servers, and video conferencing servers. Most service servers do not have special network requirements and will therefore be provided with best-effort services. Some applications that have high requirements on network quality, such as video conferencing applications, usually work on private networks. Moreover, some service servers use exclusive network control interfaces to connect to network devices, creating a network that provides assured quality for the services.

The intent-driven campus network uses an SDN controller to provide standards-compliant northbound interfaces, through which service servers can program network services. In this way, applications are explicitly abstracted and presented on the controller as application components. Through the standards-compliant northbound interfaces, the network can be fully used to develop applications. To be more specific, network functions can be orchestrated to implement complex services and network resources can be invoked to ensure quality of service.

The application layer typically includes standard applications provided by network vendors, such as network service control applications, network O&M applications, and network security applications. These applications provide the standard functions of a controller. However, campus networks may have custom requirements, which need to be addressed using specific applications. To develop such applications, the SDN controller must be open. It can be considered as such if either of the following is achieved: Network vendors can partner with third parties to develop applications and solutions (for example, IoT solutions and business intelligence solutions), allowing customers to select applications and solutions that meet their needs and quickly deploy services. Alternatively, network vendors can provide standards-compliant interfaces, through which customers can program network services and customize applications on demand. Openness does not depend solely on standardized interfaces and documents, but also requires network vendors to provide efficient service development and verification environments. For example, a network vendor may provide a public cloud-based application development and simulation environment for customers, or release a developed application to the application store of the SDN controller, thereby boosting efficiency in application development and testing as well as lowering the difficulties and costs in application development.

3.2 KEY INTERACTION INTERFACES ON AN INTENT-DRIVEN CAMPUS NETWORK

On an intent-driven campus network, the SDN controller provisions services to network devices and connects to application-layer software. The SDN controller connects to application-layer software in the northbound direction through the RESTful application programming interface (API), and interacts with network devices in the southbound direction through the Network Configuration Protocol (NETCONF). This section explains the NETCONF protocol, Yet Another Next Generation (YANG) model (a data modeling language used in NETCONF), and RESTful API.

3.2.1 NETCONF Protocol

1. Overview

NETCONF is an XML-based (XML refers to Extensible Markup Language) network management protocol that provides a programmable network device management mechanism. With NETCONF, users can add, modify, and delete network device configurations and obtain the configurations and status of network devices.

As the network scale keeps expanding and new technologies such as cloud computing and IoT rapidly develop, the traditional way to manage network devices through the command line interface (CLI) and Simple Network Management Protocol (SNMP) are phasing out due to their inability to support fast service provisioning and innovation.

CLI is a man-machine interface, through which administrators enter commands provided by the network device system. The CLI then parses the commands to configure and manage the network device. However, the CLI model of different equipment vendors may vary, in addition to the CLI not providing structured error messages and output. Therefore, administrators need to develop adaptable CLI scripts and network management tools for different vendors, which complicates network management and maintenance.

SNMP is a machine-to-machine interface that consists of a set of network management standards (application layer protocols, database models, and a set of data objects) for monitoring and managing devices connected to the network. SNMP, based on the User Datagram Protocol (UDP), is the most commonly used network management protocol on TCP/IP networks. It is not a configuration-oriented protocol and lacks a secure, effective mechanism for committing configuration transactions. Therefore, SNMP is typically used for performance monitoring, but not for network device configuration.

In 2002, the Internet Architecture Board (IAB) summarized the problems of network management at a network management workshop. Participants subsequently proposed 14 requirements for the next-generation network management protocol, of which eight key requirements are easy-to-use, differentiation of configuration data from status data, service- and network-oriented management, configuration data import and export, configuration consistency verification, standard data models, multiple configuration sets, and role-based access control. The meeting minutes eventually became the Requirement For Comments (RFC) 3535, which formed the basis of the subsequent NETCONF protocol. The Internet Engineering Task Force (IETF) set up the NETCONF working group in May 2003, with the purpose of developing a new XML-based NETCONF protocol. The first version of the base NETCONF protocol (NETCONF 1.0) was published in 2006, followed by the publication of several extensions (notification mechanism, the YANG model, and access control standards) in the following years. These finally became the revised version available today.

NETCONF uses the client/server network architecture. The client and server communicate with each other using the remote procedure call (RPC) mechanism, with XML-encoded messages. NETCONF supports the industry's mature secure transport protocols and allows equipment vendors to extend it with exclusive functions, therefore achieving flexibility, reliability, scalability, and security. NETCONF can work with YANG to implement model-driven network management and automated network configuration with programmability, simplifying O&M and accelerating service provisioning. In addition to this, NETCONF allows users to commit configuration transactions, import and export configurations, and flexibly switch between predeployment testing, configuration, and configuration rollback. These functions make NETCONF an ideal protocol in SDN, Network Functions Virtualization (NFV), and other cloud-based scenarios.

2. NETCONF protocol specifications

a. NETCONF network architecture

Figure 3.2 shows the basic NETCONF network architecture. At least one NMS must act as the entire network's management center.

The NMS runs on the NMS server and manages devices on the network. Listed below are the main elements in the NETCONF network architecture:

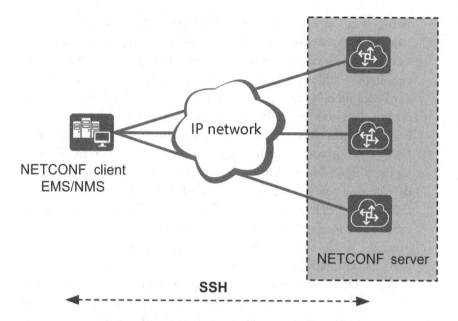

FIGURE 3.2 NETCONF network architecture.

i. NETCONF client: manages network devices using NETCONF. In most cases, the NMS assumes the role of NETCONF client. A client sends an RPC request to the servers to query or modify one or more specific parameter values, and also receives alarms and events from the servers to obtain the status of managed devices.

ii. NETCONF server: maintains information about managed devices, responds to requests from clients, and reports management data to the clients. NETCONF servers are typically network devices (such as switches and routers), which, after receiving a client request, parses data and processes the request with the assistance of the Configuration Manager Frame, and then returns a response to the client. If a fault alarm is generated or an event occurs on a managed device, the server sends an alarm or event notification to the client, so the client can learn the status change of the managed device.

To exchange requests, a client and a server first set up a connection using Secure Shell (SSH) or Transport Layer Security (TLS), exchange their respective functions using <hello> messages, and then establish a NETCONF session. A NETCONF client obtains configuration data and status data from a NETCONF server but has different permissions on these two types of data.

i. The client can manipulate configuration data to change the state of the NETCONF server to an expected one.

ii. The client cannot manipulate status data, including the running status and statistics of the NETCONF server.

b. NETCONF protocol structure

Similar to the Open System Interconnection (OSI) model, the NETCONF protocol also adopts a hierarchical structure. Each layer encapsulates certain functions of NETCONF and provides services for its upper layer. The hierarchical structure enables each layer to focus only on a single aspect of NETCONF and reduces the dependencies between different layers, minimizing the impact of each layer's internal implementation on other layers. NETCONF can be logically divided into four layers, as shown in Figure 3.3.

Table 3.1 describes each layer of NETCONF.

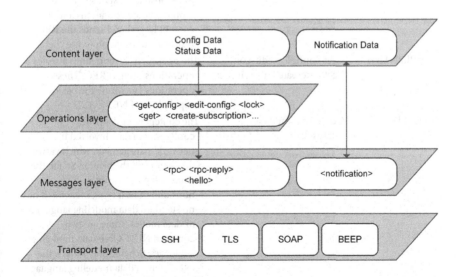

FIGURE 3.3 NETCONF protocol layers.

TABLE 3.1 NETCONF Protocol Layers

Layer	Example	Description
Transport layer	SSH, TLS, Simple Object Access Protocol, Blocks Extensible Exchange Protocol	The transport layer provides an interaction path between a client and a server. NETCONF can run over any transport layer protocol that meets these basic requirements: Connection-oriented: a permanent connection must be established between a client and a server, and this connection must provide reliable, sequenced data delivery The transport layer provides user authentication, data integrity, and confidentiality for NETCONF The transport protocol provides a mechanism to distinguish the session type (client or server) for NETCONF
Messages layer	<rpc>, <rpc-reply>, <hello>	The messages layer provides a simple, transport-independent RPC request and response mechanism. The client uses the <rpc> element to encapsulate operation request information and sends the RPC request information to the server. The server uses the <rpc-reply> element to encapsulate response information (content at the operations and content layers) and sends the RPC response information to the client
Operations layer	<get-config>, <edit-config>, <get>, <create-subscription>	The operations layer defines a series of operations used in RPC. These operations constitute the basic capabilities of NETCONF
Content layer	Configuration data, status data	The content layer describes configuration data involved in network management, which varies depending on the vendor. So far, the content layer is the only nonstandardized layer, meaning it has no standard data modeling language or data model. There are two common NETCONF data modeling languages: Schema and YANG, the latter being a data modeling language specially designed for NETCONF

c. NETCONF packet format

NETCONF requires that messages exchanged between a client and server be XML-encoded (a data encoding format of the NETCONF protocol). XML allows complex hierarchical data to be expressed in a text format that can be read, saved, and edited with both traditional text tools and XML-specific tools. XML-based network management leverages the strong data presentation capability of XML to describe managed data and management operations, transforming management information into a database that can be understood by computers. XML-based network management helps computers efficiently process network management data for enhanced network management. Below is the format of a sample NETCONF packet:

```
<?xml version="1.0" encoding="utf-8"?>
<rpc message-id="101" xmlns="urn:ietf:params:xml:ns:ne
tconf:base:1.0">
//Messages layer; RPC operation; capability set
<edit-config>//Operations layer; edit-config operation
<target>
<running/> //Data set
</target>
<config xmlns:xc="urn:ietf:params:xml:ns:netcon
f:base:1.0">
<top xmlns="http://example.com/schema/1.2/config/xxx">
//Content layer
<interface xc:operation="merge">
<name>Ethernet0/0</name>
<mtu>1500</name>
            </interface>
    </top>
</config>
</edit-config>
    </rpc>
```

d. NETCONF capability set

A capability set is a collection of basic and extended functions of NETCONF. A network device can add protocol operations through the capability set to expand existing operations on configuration objects. A capability is identified by a unique uniform resource identifier (URI), presented in the following format as defined in NETCONF:

```
urn:ietf:params:xml:ns:netconf:capability:
{name}:{version}
```

In the format, *name* and *version* indicate the name and version of a capability, respectively.

The definition of a capability may depend on other capabilities in the same capability set. Only when a server supports all the capabilities that a particular capability relies on, will that capability be supported too. In addition, NETCONF provides specifications for defining the syntax and semantics of capabilities, based on which equipment vendors can define nonstandard capabilities when necessary.

The client and server exchange capabilities to notify each other of their supported capabilities. The client can send only the operation requests that the server supports.

e. NETCONF configuration database

A configuration database is a collection of complete configuration parameters for a device. Table 3.2 lists the configuration databases defined in NETCONF.

3. NETCONF application on an intent-driven campus network

a. The controller uses NETCONF to manage network devices.

Huawei's intent-driven campus network solution supports cloud management and uses the SDN controller to manage network devices through NETCONF, realizing device plug-and-play while implementing quick and automated deployment of network services. The following example describes how NETCONF is used when a device registers with the controller for management. After the device accesses the network and obtains an IP address through DHCP, it establishes a connection with the controller following the NETCONF-based interaction process shown in Figure 3.4.

TABLE 3.2   NETCONF-Defined Configuration Databases

Configuration Database	Description
<running/>	It stores a device's running configuration, as well as its status information and statistics
	Unless the server supports the candidate capability, <running/> is the only mandatory standard database for a device
	To support modification of the <running/> configuration database, a device must have the writable-running capability
<candidate/>	It stores the configuration data that will run on a device
	An administrator can perform operations on the <candidate/> configuration database. Any change to the <candidate/> database does not directly affect the device
	To support the <candidate/> configuration database, a device must have the candidate capability
<startup/>	It stores the configuration data loaded during a device's startup, which is similar to the configuration file saved on the device
	To support the <startup/> configuration database, a device must have the distinct startup capability

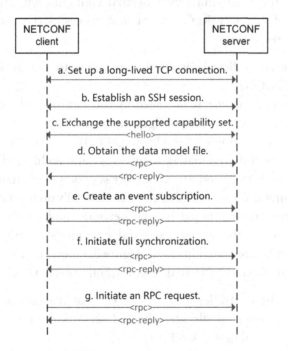

FIGURE 3.4   NETCONF-based interaction process.

i. The device functions as the NETCONF server and proactively establishes a long-lived TCP connection with the controller that functions as the NETCONF client.

ii. After the long-lived TCP connection is set up, the server and client establish an SSH session, verify each other's certificates, and establish an encrypted channel.

iii. The client and server exchange <hello> messages to notify each other of their supported capability set.

iv. The client obtains the data model file supported by the server.

v. (Optional) The client creates an event subscription. This step is required only when alarm and event reporting is supported.

vi. (Optional) The client initiates full synchronization to ensure data consistency between the client and server.

vii. The client initiates a normal configuration data or status data query RPC request and processes the corresponding response.

b. The controller delivers configurations to the device through NETCONF.

After a device is managed by the controller, the controller orchestrates service configurations and delivers them to the device. Leveraging the advantages of the NETCONF protocol in model-driven management, programmability and configuration transaction, the controller supports network-level configuration. The controller can automatically orchestrate service configurations based on the network model built by users as well as the available site and site template configurations, thereby implementing SDN. Figure 3.5 shows how the controller uses NETCONF to deliver configurations to devices.

i. The controller abstracts services at the network layer. For example, it allocates physical network resources and creates virtual network subnets.

ii. The controller orchestrates network configurations and abstracts the configurations into YANG models of devices.

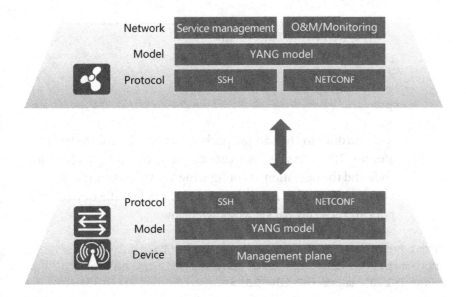

FIGURE 3.5   Controller delivering configurations to devices through NETCONF.

iii. The controller encapsulates control packets using NETCONF and sends the packets to each device.

iv. The device decapsulates the NETCONF packets and instructs the management plane to deliver configurations based on the YANG model.

For example, to create a VLAN on a device, the controller sends the device a NETCONF packet similar to the following.

```
<?xml version='1.0' encoding='UTF-8'?>
  <rpc message-id="25" xmlns="urn:ietf:params:xml:ns
:netconf:base:1.0">
    <edit-config>
    <target>
        <running/>
    </target>
    <config>
        <huawei-vlan:vlans
xmlns:huawei-vlan="urn:huawei:params:xml:ns:
yang:huawei-vlan">
            <huawei-vlan:vlan>
```

```
        <huawei-vlan:id>100</huawei-vlan:id>
      </huawei-vlan:vlan>
     </huawei-vlan:vlans>
  </config>
 </edit-config>
</rpc>
```

According to the sample packet content, the controller initiates an RPC message that carries the <edit-config> operation code and the operation of configuring VLAN 100 for the huawei-vlan YANG model. After the VLAN is created, the device compiles an rpc-reply packet as shown below:

```
<?xml version='1.0' encoding='UTF-8'?>
<rpc-reply xmlns="urn:ietf:params:xml:ns:netcon
f:base:1.0" message-id="25">
<ok/>
</rpc-reply>
```

c. The controller obtains device status through NETCONF.

The controller needs to display the running status of its managed devices, such as CPU and memory usage, the equipment serial number (ESN), registration status, and alarm information. The controller detects the device status through either its proactive query (using the <get> operation) query or device notification. The following is a sample NETCONF packet through which the controller obtains device status.

```
<?xml version='1.0' encoding='UTF-8'?>
<rpc message-id="0" xmlns="urn:ietf:params:xml:ns:netc
onf:base:1.0">
<get>
<filter type="subtree">
<dev:device-state xmlns:dev="urn:huawei:params:xml:ns:
yang:huawei-device"/>
</filter>
</get>
</rpc>
```

Assume that the Huawei-device YANG model is used for obtaining device status data. After receiving the <get> request, the device replies with an <rpc-reply> message similar to the following:

```
<?xml version='1.0' encoding='UTF-8'?>
    <rpc-reply xmlns="urn:ietf:params:xml:ns:netconf
:base:1.0" message-id="0">
        <data>
        <device-state xmlns="urn:huawei:params:xml:n
s:yang:huawei-device">
<clock>
<boot-datetime>2019-11-03T02:32:58+00:00</
boot-datetime>
<current-datetime>2019-11-03T02:45:07+00:00</
current-datetime>
<up-times>698</up-times>
</clock>
<vendor>huawei</vendor>
<esn>2102350DLR04xxxxxxxx</esn>
<mac-address>00:10:00:20:00:04</mac-address>
<model>S5720S-52X-SI-AC</model>
<name>huawei</name>
<patch-version/>
<performance>
<cpu-using-rate>9</cpu-using-rate>
<memory-using-rate>10</memory-using-rate>
<upstream-interfaces>
<interface>GigabitEthernet0/0/16</interface>
<management-vlan-id>1</management-vlan-id>
<management-vlan-ip>192.168.50.112</
management-vlan-ip>
</upstream-interfaces>
<user-define-info>
<local-manage-ip>192.168.10.8/24</local-manage-ip>
<stack-status>single</stack-status>
<system-mac-address>00:0b:09:ef:5f:03</
system-mac-address>
</user-define-info>
<version>V200R019C00SPC200</version>
```

```
</device-state>
</data>
</rpc-reply>
```

When an alarm is generated on a device or the device status changes, the device reports the information in a NETCONF packet to the controller through the notification mechanism. After receiving the NETCONF packet, the controller displays the information on its GUI in real time. For instance, to report an alarm, the device compiles the following packet and sends it to the controller:

```
<?xml version='1.0' encoding='UTF-8'?>
<alarm-notification xmlns="urn:huawei:params:xml:ns:yan
g:huawei-system-alarm">
<resource>OID=1.3.6.1.4.1.2011.5.25.219.2.5.1
index=67108873</resource>
<alarm-type-id>equipmentAlarm</alarm-type-id>
<alarm-type-qualifier>hwPowerRemove</
alarm-type-qualifier>
<alt-resource>1</alt-resource> <event-time>
2016-09-13T07:31:20Z</event-time>
<perceived-severity>4</perceived-severity>
<alarm-text>Power is absent.(Index=67108873,
EntityPhysicalIndex=67108873,
PhysicalName="MPU Board 0",
EntityTrapFaultID=136448)</alarm-text>
</alarm-notification>
```

3.2.2 YANG Model

1. YANG model overview

When the NETCONF protocol was developed, no data modeling language was defined for operations. Traditional data modeling languages such as Structure of Management Information (SMI), Unified Modeling Language (UML), XML, and Schema, however, all failed to satisfy the requirements of NETCONF. Therefore, a new data modeling language was urgently needed, more specifically, one that is decoupled from the protocol mechanism, easily parsed by computers, easy to learn and understand, and compatible with the existing

SMI language. It must also be able to describe the information and operation models.

As a solution to this, YANG was proposed by the NETMOD working group in 2010 and published in IETF RFC 6020 as a data modeling language used to model configuration and state data manipulated by NETCONF, NETCONF RPCs, and NETCONF notifications. In short, YANG is used to model the operations and content layers of NETCONF. YANG is a modular language that presents data structures in a tree format, which is similar to Abstract Syntax Notation One (ASN.1), a language that describes the management information base (MIB) of SNMP. However, SNMP defines the hierarchy of the entire tree in a rigid manner and therefore has its limitations. YANG, on the other hand, is more flexible and is expected to be compatible with SNMP.

2. YANG model development

The YANG 1.0 specification was published as RFC 6020 in 2010, which defines the method of using YANG with NETCONF. In 2014, the YANG model-based draft was proposed on a large scale in standards organizations. Listed below are some standard YANG models published by the IETF for IP management, interface management, system management, and SNMP configuration.

a. RFC 6991: Common YANG Data Types

b. RFC 7223: A YANG Data Model for Interface Management

c. RFC 7224: IANA Interface Type YANG Module

d. RFC 7277: A YANG Data Model for IP Management

e. RFC 7317: A YANG Data Model for System Management

f. RFC 7407: A YANG Data Model for SNMP Configuration

In October 2016, YANG 1.1 (RFC 7950) was published. In fact, the YANG model has become an indisputable mainstream data model in the industry and is drawing growing concerns from vendors. More and more vendors require adaptations to the YANG model, and many well-known network equipment vendors have rolled out NETCONF- and YANG-capable devices. For example, Huawei SDN controllers support the YANG model and YANG model-driven development, and an increasing number of YANG model-based tools have been launched in the industry, such as YANGTools, PYANG, and YANG Designer.

3. Functions of the YANG model

A YANG model defines a data hierarchy and NETCONF-based operations, including configuration data, state data, RPCs, and notifications. This is a complete description of all data transmitted between a NETCONF client and a NETCONF server. Figure 3.6 shows the basic structure of a YANG file.

a. YANG file header and import of external model files

YANG structures data models into modules and submodules. A module can import data from other external modules. The hierarchy of a module can be augmented, allowing one module to add data nodes to the hierarchy defined in another module. This augmentation can be conditional; for example, new nodes appear only if certain conditions are met.

The YANG file header defines the namespace and model description of the local module and imports other referenced YANG modules. The following example is an interfaces YANG model defined in RFC 7223. This model names the local module and imports other modules. It also details the organization, contact information, description, and version of the local module.

```
module ietf-interfaces {
    namespace "urn:ietf:params:xml:ns:yang:
ietf-interfaces";
```

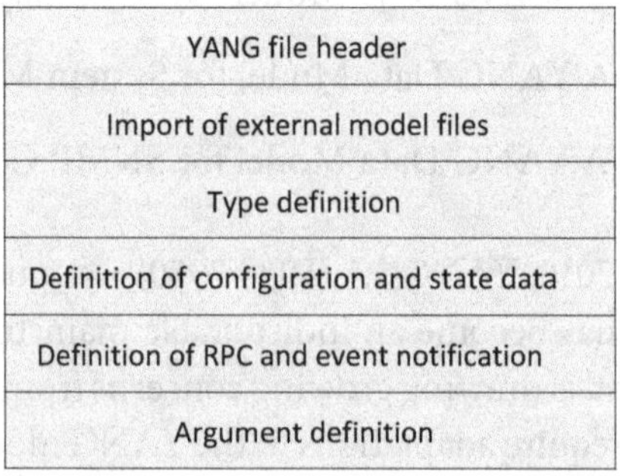

YANG file header
Import of external model files
Type definition
Definition of configuration and state data
Definition of RPC and event notification
Argument definition

FIGURE 3.6   Basic structure of a YANG file.

```
    prefix if;


prefix yang;
}
organization
"IETF NETMOD (NETCONF Data Modeling Language) Working
Group";
contact
"WG Web:<http://tools.ietf.org/wg/netmod/>
WG List:<mailto:netmod@ietf.org>
WG Chair: Thomas Nadeau
<mailto:tnadeau@lucidvision.com>
WG Chair: Juergen Schoenwaelder
<mailto:j.schoenwaelder@jacobs-university.de>
Editor:     Martin Bjorklund
<mailto:mbj@tail-f.com>";

Description
"This module contains a collection of YANG definitions
for managing network interfaces.
Copyright (c) 2014 IETF Trust and the persons
identified as authors of the code.
All rights reserved.
     Redistribution and use in source and binary
forms, with or without modification, is permitted
pursuant to, and subject to the license terms
contained in, the Simplified BSD License set forth in
Section 4.c of the IETF Trust's Legal Provisions
Relating to IETF Documents (http://trustee.ietf.org/
license-info).
     This version of this YANG module is part of RFC
7223; see the RFC itself for full legal notices.";

revision 2014-05-08 {
description
"Initial revision.";
reference
"RFC 7223: A YANG Data Model for Interface
Management";
}
```

b. Type definition of the YANG model

YANG defines base types and allows more complex types to be defined based on service requirements.

The definition of a base type includes the corresponding restriction definition, through which type definitions can be personalized. Restriction example:

```
typedef my-base-int32-type {
type int32 {
    range "1..4 | 10..20";
  }
}
```

YANG allows users to define derived types from base types using the "typedef" statement based on their own requirements. Derived type example:

```
typedef percent {
type uint16 {
range "0 .. 100";
}
description "Percentage";
}
leaf completed {
type percent;
}
```

The following example combines multiple type definitions using the union statement.

```
typedefthreshold {
description "Threshold value in percent";
type union {
type uint16 {
range "0 .. 100";
}
type enumeration
enum disabled {
description "No threshold";
}
}
}
}
```

The following example defines a reusable block of nodes using the grouping statement.

```
grouping target {
leaf address {
type inet:ip-address;
description "Target IP";
}
leaf port {
type inet:port-number;
description "Target port number";
      }
   }
   container peer {
         container destination {
      uses target;
         }
   }
```

RFC 6021 defines a collection of common data types derived from the built-in YANG data types (for example, the derived type ietf-yang-types. yang). These derived types can be directly referenced by developers.

The identity statement is used to define a new, globally unique, abstract, and untyped identity.

```
module phys-if {
identity ethernet {
description "Ethernet family of PHY interfaces";
}
identity eth-1G {
base ethernet;
description "1 GigEth";
}
identity eth-10G {
base ethernet;
description "10 GigEth";
}
```

The identityref type is used to reference an existing identity.

```
module newer {
identity eth-40G {
base phys-if:ethernet;
```

```
description "40 GigEth";
}
identity eth-100G {
base phys-if:ethernet;
            description "100 GigEth";
}
leaf eth-type {
type identityref {
            base "phys-if:ethernet";
}
}
```

Features give the modeler a mechanism for making portions of the module conditional in a manner that is controlled by the device. It facilitates software upgrade configuration.

```
feature has-local-disk {
description
"System has a local file system that can be used for
storing log files";
}

container system {
container logging {
if-feature has-local-disk;
presence "Logging enabled";
leaf buffer-size {
type filesize;
}
}
}
```

c. Definition of configuration and state data

YANG models the hierarchical organization of data as a tree, which consists of four types of nodes (container, list, leaf list, and leaf). Each node has a name, either a value or a set of child nodes.

YANG provides clear and concise descriptions of the nodes as well as the interaction between them. YANG data hierarchy constructs include defining lists where list entries are identified by keys that distinguish them from each other. Such lists may be defined as either user sorted or automatically sorted by the system. For user-sorted lists, YANG defines the operations for manipulating the list entry sequence, as shown in the following interfaces YANG model as defined in RFC 7223.

```
+--rw interfaces
|  +--rw interface* [name]
|     +--rw name                    string
|     +--rw description?                string
|     +--rw type                    identityref
|     +--rw enabled?                    boolean
|     +--rw link-up-down-trap-enable?        enumeration
+--ro interfaces-state
+--ro interface* [name]
+--ro name                    string
+--ro type                    identityref
+--ro admin-status            enumeration
+--ro oper-status             enumeration
```

Configuration and state data are defined using two separate containers, becoming configured interfaces and dynamically generated interfaces, respectively. The interfaces configuration data contain a list of interfaces, using the leaf node as the key. This example illustrates the suitability of the YANG model's hierarchical tree structure for defining the configuration and state data of network devices.

d. Definition of RPC and event notification

RPCs are introduced to cope with the syntax defects of the YANG model; for instance, one-off operations that do not need to be saved or actions that cannot be expressed using NETCONF operations (such as system reboot and software upgrade). The latest YANG 1.1 allows the definition of actions for operation objects, so RPCs should be avoided if possible. An RPC defines the input and output of an operation, whereas an event notification defines only the information to be sent. The following is a sample RPC and event notification.

```
rpc activate-software-image {
   input {
      leaf image {
         type binary;
      }
```

```
    }
    output {
       leaf status {
          type string;
       }
    }
}

notification config-change {
    description "The configuration changed";
    leaf operator-name {
       type string;
    }
    leaf-list change {
       type instance-identifier;
    }
}
```

   e. Argument definition
      An argument is used to add new schema nodes to a previously defined schema node.

```
augment /sys:system/sys:user {
    leaf expire {
       type yang:date-and-time;
    }
}
```

The foregoing description of the YANG model reveals that YANG strikes a balance between high-level data modeling and low-level bits-on-the-wire encoding, making YANG quickly become the industry's mainstream modeling language. YANG modules can be translated into an equivalent XML syntax called YANG Independent Notation (YIN), allowing applications using XML parsers and Extensible Stylesheet Language Transformations scripts to operate on the models. The conversion from YANG to YIN is semantically lossless, so content in YIN can be round-tripped back into YANG. YANG is an extensible language, allowing extensions to be defined by standards bodies, vendors, and individuals. The statement syntax allows these extensions to coexist with standard YANG statements in a natural way, while extensions in a YANG module stand out clearer to the reader. Where possible, YANG maintains compatibility with SNMP SMIv2. SMIv2-based MIB modules can be automatically translated into YANG modules for read-only access; however, YANG does not perform reverse translation. This translation mechanism allows YANG to protect or expose elements in the data model using available access control mechanisms.

3.2.3 RESTful APIs

The SDN controller interconnects with the application layer through northbound RESTful APIs, including basic network APIs, VAS APIs, third-party authentication APIs, and location-based service (LBS) APIs. Representational State Transfer (REST) is a software architectural style that defines a set of constraints to be used for creating web services. Using this architecture allows everything on a network to be abstracted as resources, with each resource identified by a unique identifier. Through standard methods, all resources can be operated without altering their identifiers, as all operations are stateless. A RESTful architecture — one that complies with the constraints and rules of REST — aims for better use of the rules and constraints in existing web standards.

In REST, anything that needs to be referenced can be considered a resource (or a "representation"). A resource can be an entity (for example, a mobile number) or an abstract concept (for example, a value). To make a resource identifiable, a unique identifier — called a URI in the World Wide Web — needs to be assigned to it.

APIs developed in accordance with REST design rules are called RESTful APIs. External applications can use the Hypertext Transfer Protocol (HTTP), as well as the secure version — HTTPS — for security purposes, to access RESTful APIs in order to implement functions such as service provisioning and status monitoring.

Standard HTTP methods for accessing managed objects include GET, PUT, POST, and DELETE, as described in Table 3.3.

A description of a RESTful API should include the typical scenario, function, constraints, invocation method, URI, request and response parameter description, and sample request and response. For example, the

TABLE 3.3 Functions of the HTTP Methods for Accessing
Managed Objects

Method	Function Description
GET	Queries specified managed objects
PUT	Modifies specified managed objects
POST	Creates specified managed objects
DELETE	Deletes specified managed objects

RESTful API provided by Huawei's SDN controller for user access authorization is described as follows:

- Typical scenario: used for user authorization

- Function: to grant authenticated users the corresponding permissions based on the user information

- Constraints: invoked only after a user session is established

- Invocation method: POST

- URI: /controller/cloud/v2/northbound/accessuser/haca/authorization

- Request parameters: see Tables 3.4 and 3.5.

- Sample HTTP request

```
POST /controller/cloud/v2/northbound/accessuser/haca/
authorization HTTP/1.1
Host: IP address:Port number
Content-Type: application/json
Accept: application/json
Accept-Language: en-US
X-AUTH-TOKEN: CA48D152F6B19D84:637C38259E6974E17788348
128A430FEE150E874752CE75
4B6BF855281219925
{
"deviceMac" : "48-digit MAC address of the device",
"deviceEsn" : "ESN of the device",
"apMac" : "48-digit MAC address of the AP",
```

TABLE 3.4 Parameters in the Body

Parameter Name	Mandatory	Type	Parameter Description
body	Yes	REFERENCE	Authorization information

TABLE 3.5 Parameters in the Body Object

Parameter Name	Mandatory	Type	Value Range	Default Value	Parameter Description
deviceMac	No	STRING	-	-	Media Access Control (MAC) address of a device. Either the device MAC address, ESN, or both must be provided
deviceEsn	No	STRING	-	-	ESN of a device. Either the device MAC address, ESN, or both must be provided
apMac	No	STRING	-	-	MAC address of an AP
ssid	Yes	STRING	-	-	Base64 code of an AP service set identifier (SSID)
policyName	No	STRING	-	-	Name of an access control policy. If this parameter is empty, policy-based access control is not performed
terminalIpV4	No	STRING	-	-	IPv4 address of a terminal. This parameter is mandatory if a terminal has an IPv4 address
terminalIpV6	No	STRING	-	-	IPv6 address of a terminal. This parameter is mandatory if a terminal has an IPv6 address
terminalMac	Yes	STRING	-	-	MAC address of a terminal
userName	Yes	STRING	-	-	User name
nodeIp	Yes	STRING	-	-	IP address of a node that performs authorization
temPermitTime	No	INTEGER	[0–600]	-	Duration a user is allowed temporary network access, measured in seconds. If this parameter is not carried in a RESTful packet or if the value is set to 0, there is no time limit

TABLE 3.6 Parameters in the Body

Parameter Name	Mandatory	Type	Value Range	Default Value	Parameter Description
errcode	No	STRING	-	-	Error code
errmsg	No	STRING	-	-	Error information
psessionid	No	STRING	-	-	Session ID

```
"ssid" : "dcd=",
"policyName" : "aa",
"terminalIpV4" : "IPv4 address of the terminal",
"terminalIpV6" : "IPv6 address of the terminal",
"terminalMac" : "MAC address of the terminal",
"userName" : "User name",
"nodeIp" : "IP address of a node that performs
authorization",
"temPermitTime" : 300
}
```

- Response parameters: see Table 3.6.

- Sample HTTP response

```
HTTP/1.1 200 OK
Date: Sun,20 Jan 2019 10:00:00 GMT
Server: example-server
Content-Type: application/json
{
"errcode" : "0",
"errmsg" : " ",
"psessionid" : "5ea660be98a84618fa3d6d03f65f47ab578ba
3b4216790186a932f9e
8c8c880d"
}
```

3.3 SERVICE MODELS OF AN INTENT-DRIVEN CAMPUS NETWORK

The basic components of an intent-driven architecture and the protocols used for the interaction between the SDN controller and devices/applications are critical for implementing an intent-driven architecture. An additional aspect that is equally critical to this architecture is an SDN controller-based abstract service model for the campus network. By using this model, network administrators can centrally plan and manage network services through GUIs to automate the deployment of these services. With a service model that is suitably configured, the SDN controller can accurately display a management GUI from the perspective of user services through northbound interfaces and deliver the configurations of user services to the corresponding NEs through southbound interfaces.

3.3.1 Service Layering of an Intent-Driven Campus Network

A campus network primarily connects end users to applications. The implementation of this connection involves three objects: a user terminal, network device, and external network.

- A user terminal, such as a PC or mobile phone, is a tool through which a user accesses the campus network.

- A network device is located between a terminal and an external network. It enables terminals to access the campus network, controls their access scope, transmits their application traffic, and ensures their quality of service.

- An external network is any network between the campus and data center, for example, the Internet, a private line network, or a transmission network.

- The intent-driven campus network adopts a hierarchical decoupling mechanism for abstracting and modeling network device management. As shown in Figure 3.7, a campus network is divided into the underlay network, overlay network, and service layer on the SDN controller. These three layers are decoupled from each other.

- Underlay network: refers to the physical network in the intent-driven campus network architecture. The underlay network only ensures Layer 3 connectivity and disregards what services are carried on the campus network.

- Overlay network: refers to the virtual network in the intent-driven campus network architecture. The overlay network uses network virtualization technology to pool network resources such as IP address segments, access ports, and accessible external resources. Campus network administrators can create multiple virtual network

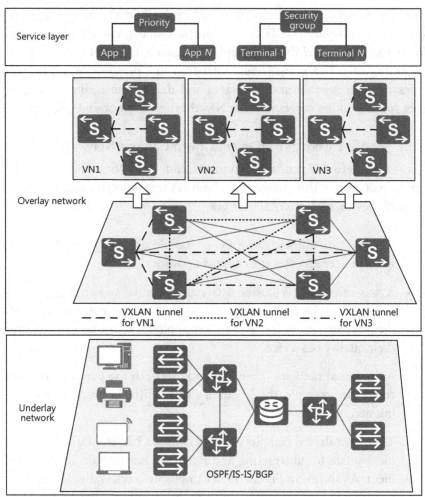

Service layer

Overlay network

VN1   VN2   VN3

VXLAN tunnel for VN1
VXLAN tunnel for VN2
VXLAN tunnel for VN3

Underlay network

OSPF/IS-IS/BGP

OSPF: Open Shortest Path First
BGP: Border Gateway Protocol
IS-IS: Intermediate System-to- Intermediate System
VXLAN: Virtual Extensible Local Area Network
VN: Virtual Network

FIGURE 3.7   Hierarchical model of an intent-driven campus network.

instances based on network resource pools, with each instance being independently managed and decoupled from other instances. Similar to a traditional campus network, a virtual network instance assigns IP addresses to user terminals and provides access to external network resources. The overlay network is completely decoupled from the underlay network, and any configuration adjustments of the overlay network have no impact on the underlay network.

- Service layer: controls access between campus users, between users and applications, and between applications, and delivers policies to network devices. The service layer closely matches real campus services and is the layer most frequently operated by campus network administrators. Focused exclusively on users and applications, the service layer is completely decoupled from the overlay network, and any configuration adjustments of the service layer have no impact on the overlay network.

3.3.2 Abstract Model of the Underlay Network

The objective of underlay network service modeling is to build a simplified underlay network. The word "simplified" conveys two meanings: simplified management and flexible networking.

- Simplified management: In an ideal scenario, a network administrator would need to focus on only IP address planning, without needing to consider the port and route configurations. Focusing on as few things as possible is desirable when managing an underlay network. We can compare the construction and management of a campus network to those of an expressway system. Building an underlay network is similar to building expressways. When managing a traditional underlay network, an administrator needs to design the start and end of each road, any crossroads, and the driving route if roads intersect. Conversely, when managing an intent-driven campus network, a network administrator needs only to determine the start and end of a road — all other management tasks are automatically performed by the network, thereby simplifying management.

- Flexible networking: The underlay network needs to not only support any topology but also allow for flexible scale in/out and replacement of any faulty components.

The underlay network of a legacy campus is usually built and adjusted over a long period of time and therefore may contain devices from multiple vendors. For example, buildings A and B may each have devices of a different vendor, or the core, aggregation, and access layers in the same building

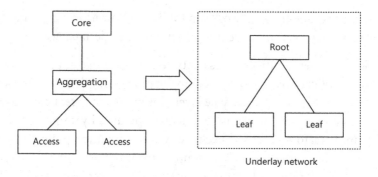

FIGURE 3.8 Service model abstracted from the underlay network.

may use devices from three different vendors. A further example is a three-layer (core/aggregation/access) tree topology that is subsequently changed to a four-layer (core/aggregation/level-1 access/level-2 access) topology due to complex and expensive cabling, leading to a network topology that may be disorganized and difficult to manage. Such examples highlight the difficulties in building a unified service model for the underlay network of a traditional campus. By using an SDN controller for modeling, the intent-driven campus network eliminates the impact of a disorganized topology, as shown in Figure 3.8.

Underlay network service modeling aims for Layer 3 interoperability between all devices, with one device assuming the root role and other devices assuming the leaf role. An administrator needs only to specify the root device (typically the core switch) and configure IP address resources used by the underlay network on the root device. The root device then automatically calculates and configures routes, discovers aggregation and access switches, and specifies them as leaf devices through negotiation. It also delivers all the underlay network configurations to the leaf devices, which can be connected in any networking mode, and maintains the routes between itself and the leaf devices as well as those between leaf devices. In effect, the root device manages the entire underlay network.

3.3.3 Abstract Model of the Overlay Network

The virtualization technology used on the campus network is derived from cloud computing, in which virtualization technologies regarding computing, storage, and network have been widely used. Due to the differences in product specifications, virtual networks constructed using different vendors' products may have varying service models. However, all of these models are based on cloud platform architectures, among which the open-source OpenStack is the most popular.

OpenStack is the general front end of Infrastructure as a Service (IaaS) resources and manages computing, storage, and network resources. Its virtual network model is typically used in cloud data center scenarios. Compared with campus scenarios, cloud data center scenarios do not have wireless networks and users, but have virtual machines (VMs). Consequently, the virtual network service model needs to be expanded and reconstructed when defined in the intent-driven campus network architecture.

This section starts by describing OpenStack service models in the cloud data center scenario and then provides details about service models for a campus virtual network.

1. Typical OpenStack service model

OpenStack uses a component-based architecture, in which computing, storage, and network resources are managed by several component modules. The core modules include Nova, Cinder, Neutron, Swift, Keystone, Glance, Horizon, and Ceilometer. Among these modules, Neutron is responsible for network management. Figure 3.9 shows the Neutron service model.

a. Components of the Neutron service model

i. Tenant: A tenant is an applicant of network resources. After a tenant applies for resources from OpenStack, all activities of the tenant can be performed using only these resources. For example, a tenant can create a VM using only the resources assigned to the tenant.

ii. Project: Neutron maintains 1:1 mappings between projects and tenants. Since Keystone V3, OpenStack recommends the use of a project to uniquely identify a tenant in Neutron.

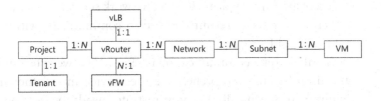

FIGURE 3.9 Neutron service model.

    iii. vRouter: A vRouter enables Layer 3 communication between network segments on a logical internal network, connects an internal network to an external network, and provides the network address translation function for intranet users who attempt to access the Internet. It needs to be created only if a tenant's logical network has multiple network segments that need to communicate with each other at Layer 3, or if the internal network needs to communicate with an external network.

    iv. Network: A network represents an isolated Layer 2 bridge domain, which can contain one or more subnets.

    v. Subnet: A subnet can be an IPv4 or IPv6 address segment that contains multiple VMs, whose IP addresses are assigned from the subnet. An administrator, when creating a subnet, needs to define the IP address range and mask, and configure a gateway IP address for the network segment to enable communication between the VMs and external networks.

    vi. Port: A port uniquely identifies an access port on a logical network in the Neutron service model. It may be a virtual network interface card (vNIC) that connects a VM to the network, or it may be a physical NIC through which a bare metal server accesses the network.

    vii. vLB: A vLB provides L4 load balancing services, as well as related health checks, for tenant services.

    viii. vFW: A vFW is an advanced service — Firewall as a Service V2.0 — defined in Neutron. It is similar to a traditional firewall and uses firewall rules on vRouters to control tenants' network data.

b. Service models that can be managed by Neutron

There are three typical Neutron network service models.

Service model 1: Terminals need to communicate with each other only at Layer 2.

If only Layer 2 communication is required between terminals, an administrator can orchestrate the service model shown in Figure 3.10. Specifically, the administrator needs to create one or more networks and mounts the compute node NICs on the same

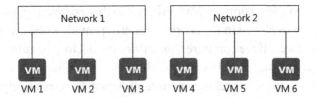

FIGURE 3.10 Service model in which only Layer 2 communication is required between terminals.

network segment to the same network. An example of where such a model may be used is in a temporary test environment created for simple service function or performance verification and the test nodes are located on the same network segment.

Service model 2: Terminals need to communicate with each other at Layer 3 but do not need to access external networks.

Consider a scenario where the deployment of users' services poses high requirements on the network. For example, user terminals need to be divided into multiple departments (such as R&D and marketing departments), and each department needs to use a different network segment to minimize the broadcast range. In this scenario, an administrator can divide the compute resources into different subnets for Layer 3 isolation and configure each subnet to correspond to only one network for Layer 2 isolation, as shown in Figure 3.11.

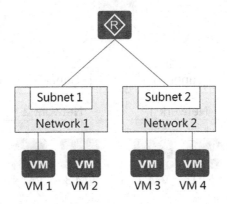

FIGURE 3.11 Service model in which Layer 3 communication is required between terminals, but access to external networks is not needed.

To allow communication between these subnets, vRouters need to be deployed in the service model to provide Layer 3 forwarding services. If we compare this service model to traditional physical networking, a network and vRouter are similar to a VLAN and the Layer 3 switching module of a switch, respectively. Further similarities exist: a subnet is similar to a network segment deployed in a VLAN, and the gateway IP address of a network segment is similar to the IP address of a Layer 3 interface (such as a VLANIF interface) on the switch.

Service model 3: Layer 3 communication is required between terminals, and the terminals need to access external networks.

In the scenario mentioned earlier for service model 2, if the terminals need to access an external network such as the Internet, a tenant needs to connect the vRouter to the external network, as shown in Figure 3.12. In Neutron, a system administrator creates external networks during system initialization. Only these networks are available for selection by a tenant when the tenant configures a service network. According to OpenStack, a vRouter can be connected to only one external network.

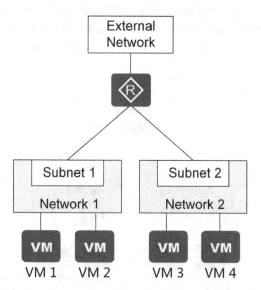

FIGURE 3.12   Service model in which Layer 3 communication is required between terminals, and the terminals need to access external networks.

2. Service model of campus virtual networks

Huawei's SDN controller reconstructs the OpenStack service models based on real-world campus network service scenarios and redesigns a service model for an intent-driven campus virtual network, as shown in Figure 3.13.

This new service model is composed of the following logical elements: tenant, site, logical router, subnet, logical port, DHCP server, external network, logical firewall, and logical WAC.

On the Huawei SDN controller, a tenant administrator can configure multiple sites and create, as well as use, multiple groups of resources at each site. The resources include logical routers, subnets, logical ports (wired and wireless), logical firewalls, logical WACs, DHCP servers, and external networks. At a site, the Fabric as a Service function is used to abstract the physical network into a variety of service resources such as logical routers, subnets, logical firewalls, and logical WACs for tenant administrators.

Listed below are the logical elements in the service model defined by the Huawei SDN controller and the corresponding components of the Neutron service model in OpenStack:

a. Tenant: corresponds to a tenant in Neutron.

b. Site: is a new concept defined in Huawei SDN controller. It is similar to an independent physical area on a tenant network and is the smallest unit for campus network management.

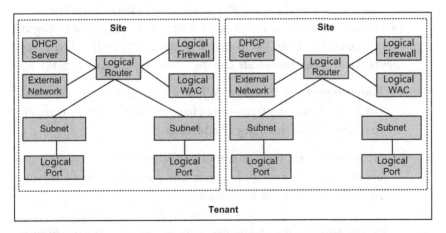

FIGURE 3.13   Service model of an intent-driven campus virtual network.

    c. Logical router: corresponds to a vRouter in Neutron.

    d. Subnet: corresponds to a subnet in Neutron.

    e. Logical port: corresponds to a port in Neutron.

Because Neutron is typically applied to cloud data centers, there is no abstraction for the WACs used on campus networks. Logical WACs (used to define access resources, such as channels and radios, for wireless users) in the service model defined by Huawei SDN controller are unavailable in Neutron.

3.3.4 Abstract Model of the Campus Service Layer

Because the service layer is the closest to real services on a campus network, construction of this layer's service model is critical. The service layer can be considered a translator between real services and network configurations on the campus network, which, in terms of service characteristics, has the following three objects: user, application, and network. Users access networks through terminals, and applications access networks through servers. The campus service model is applied to two types of policy management: access permission management between users and applications, and traffic priority management during users' access to applications.

1. Access permission management between users and applications

    Traditional network management uses IP addresses to identify terminals and servers. In order to manage access permissions between users and applications, network administrators therefore configure access control lists (ACLs) to match IP addresses or IP address segments one by one. This mode has the following disadvantages:

    a. Poor support for mobility: When a terminal moves on a network, its IP address frequently changes, but the policy is unable to migrate with changing IP addresses.

    b. Heavy ACL resources consumption: When there are a large number of terminals and applications, the consumption of hardware ACL resources can be significant.

    To address these disadvantages, Huawei proposes the use of security groups, as shown in Figure 3.14. A security group is an identity

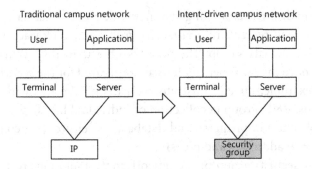

FIGURE 3.14   Service model abstracted from the policy layer.

assigned to a user or an application. This identity does not change if the user changes terminal or access location. In addition, authorization is required when a terminal attempts to access a network. Therefore, a policy needs only to focus on the user identity, irrespective of the IP address and the location of a terminal or application server. This ensures a consistent user experience in mobile scenarios and lowers the consumption of ACL resources.

As shown in Figure 3.15, the overall policy orchestration model for a campus network consists of two logical units: security group and group policy. A security group is the minimum unit to which a campus policy can be applied, and policies applied between security groups are called group policies.

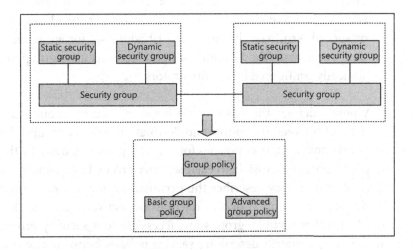

FIGURE 3.15   Policy orchestration model of an intent-driven campus network.

An administrator can group network objects that have the same access requirements into a security group, and deploy access policies for the security group. The network objects in this security group then have the same permissions as configured for the security group. For example, an administrator can define the following security groups: R&D group (a collection of individual hosts), printer group (a collection of printers), and database server group (a collection of server IP addresses and ports).

The security group-based network control solution greatly reduces the workload of network administrators by eliminating the need to deploy access policies for each individual network object. Security groups are classified into two categories:

a. Dynamic user group: contains user terminals that can access the network only after successful authentication. The IP address of a dynamic user is not fixed. Instead, it is dynamically associated with a security group after the user is authenticated, and the association is automatically canceled after the user logs out. This means that the mappings between user IP addresses and security groups are valid only when users are online. These mappings can be obtained by a network device from an authentication device or the SDN controller.

b. Static security group: contains terminals using fixed IP addresses. They can be any network object that can be identified by an IP address, such as data center servers, interfaces of network devices, or special terminals that use fixed IP addresses for authentication-free access. The IP addresses of a static security group are statically configured by administrators.

A policy defines the relationship between security groups and takes effect between the source and destination security groups. An administrator configures policies for security groups to describe the network services accessible to each security group. For example, an access control policy describes the permissions of a security group by defining the mutual access relationship between groups, that is, which other security groups are accessible to a security group. In this way, a matrix describing service policies between any two security groups can be developed, as shown in Figure 3.16. With the

Source security group	Policies for Destination Security Groups			
	R&D Group	Marketing Group	R&D Server Group	Marketing Server Group
R&D group	Enable Default permission : ● Permit Application control : ⊘ Deny (0)	Enable Default permission : ⊘ Deny Application control : ● Permit (0)	Enable Default permission : ● Permit Application control : ⊘ Deny (0)	Enable Default permission : ⊘ Deny Application control : ● Permit (0)
Marketing group	Enable Default permission : ⊘ Deny Application control : ● Permit (0)	Enable Default permission : ● Permit Application control : ⊘ Deny (0)	Enable Default permission : ⊘ Deny Application control : ● Permit (0)	Enable Default permission : ● Permit Application control : ⊘ Deny (0)
R&D server group	Enable Default permission : ● Permit Application control : ⊘ Deny (0)	Enable Default permission : ⊘ Deny Application control : ● Permit (0)	Enable Default permission : ● Permit Application control : ⊘ Deny (0)	Enable Default permission : ⊘ Deny Application control : ● Permit (0)
Marketing server group	Enable Default permission : ⊘ Deny Application control : ● Permit (0)	Enable Default permission : ● Permit Application control : ⊘ Deny (0)	Enable Default permission : ⊘ Deny Application control : ● Permit (0)	Enable Default permission : ● Permit Application control : ⊘ Deny (0)

FIGURE 3.16 Intergroup policy matrix.

intergroup policy matrix, the permissions of specific users or hosts are clearly defined.

2. Traffic priority management during users' access to applications

To ensure priority of specified user traffic on a traditional campus network, an administrator grants different priorities to users, as shown in Figure 3.17.

During user access to the network, the access switch interacts with the Authentication, Authorization, and Accounting (AAA) server through the Remote Authentication Dial-In User Service (RADIUS) protocol to assign a higher priority to User 1 and a lower priority to User 2. The access switch will then grant a higher priority to User 1's

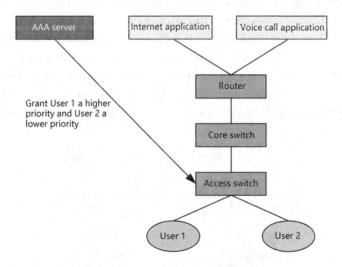

FIGURE 3.17 Method of ensuring user priorities on a traditional campus network.

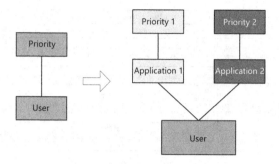

FIGURE 3.18   Traffic priority model defined for an intent-driven campus network.

packets and a lower priority to User 2's packets. If congestion occurs, the switch will preferentially forward User 1's packets.

This process ensures the network priority of an individual user or a type of users, but due to its coarse management granularity, a user's service may affect the quality of service for another user. For example, while User 1 downloads a large file such as a video, User 2 may experience undesirable issues when placing a voice call. To address this problem, an intent-driven campus network defines a traffic priority model shown in Figure 3.18.

This model defines the priorities of user application traffic, implementing refined control over network bandwidth. To implement differentiated traffic management, administrators simply need to define user templates, as shown in Figure 3.19.

When defining a service policy, an administrator can first group users, for example, divide them into R&D and marketing user groups, and then define the application priority for each user group. For example, the application priorities of the two user groups can be sorted in the following descending order: *Application 1 of R&D* > *Application*

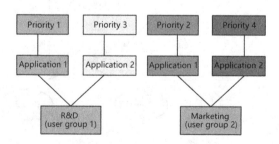

FIGURE 3.19   Differentiated traffic management model.

1 of Marketing>Application 2 of R&D>Application 2 of Marketing. This method effectively resolves the problem of a coarse priority granularity in network management, thereby improving user experience.

3.4 CLOUD-BASED DEPLOYMENT MODES OF AN INTENT-DRIVEN CAMPUS NETWORK

Cloud computing has completely changed the production mode of enterprises over the past decade. A large number of services are already deployed and operating on the cloud, enabling enterprises to quickly roll out new services. With the rapid development and wide application of cloud computing technologies, two operating modes emerge during the construction of campus networks.

- Asset-heavy mode: This mode applies to large campus networks, such as those used by large enterprises and higher education institutions. These enterprises and institutions typically have a large network scale, complex service types, sufficient capital support, and professional O&M teams. They therefore have the ability to purchase network devices, servers, and storage devices, build ICT infrastructure, and operate and maintain ICT systems by themselves.

- Asset-light mode: This mode applies to medium-, small-, and micro-sized campus networks, such as those used by retail and chain stores. These stores typically have a small network scale, multiple branches, simple service types, and no professional O&M team. They therefore can use the management service provided by the public cloud to build their own campus networks in order to reduce their initial investment in network construction and future O&M investment.

1. Three cloud-based deployment modes of intent-driven campus networks

    Based on the construction roadmap mentioned earlier, three cloud-based deployment modes are available for intent-driven campus networks. Enterprises can choose the most suitable mode for their campus networks.

    a. On-premises deployment: Enterprises purchase and maintain physical server resources and software entities such as the SDN controller. The software can be deployed in their data centers or on the public cloud IaaS platform. In this mode, enterprises have full ownership of all software and hardware resources.

b. Huawei public cloud deployment: Enterprises do not need to purchase physical server resources and software entities such as the SDN controller. Instead, they can purchase cloud management services on the Huawei public cloud to manage their networks. They can also flexibly customize cloud management services based on the network scale and service characteristics.

c. MSP-owned cloud deployment: Managed service providers (MSPs) purchase physical server resources and software entities such as the SDN controller. They can deploy the software in their data centers or on the public cloud IaaS platform and provide cloud management services for enterprises.

Figure 3.20 shows the cloud management network architecture in the on-premises and MSP-owned cloud deployment modes. Network devices distributed in different areas of a campus network can traverse the carrier network and be managed by the cloud management platform. The cloud management platform provides services for tenant networks using different components.

Figure 3.21 shows the cloud management network architecture in the Huawei public cloud deployment mode. In this mode, different from the on-premises and MSP-owned cloud deployment modes, Huawei provides the following peripheral systems to work with the cloud management platform:

a. Electronic software delivery platform: provides the license management service.

b. Public key infrastructure platform: provides the certificate management service.

c. SecCenter: provides the Deep Packet Inspection and antivirus signature database query services.

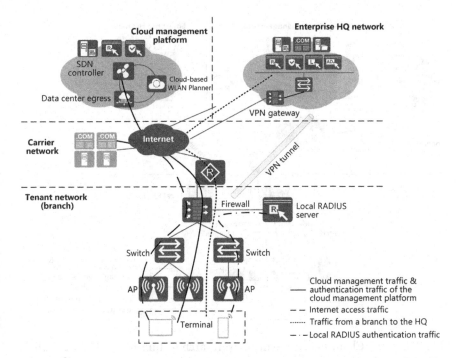

FIGURE 3.20   Cloud management network architecture in the on-premises and MSP-owned cloud deployment modes.

d. Registration center: queries and verifies devices that automatically go online.

e. Technical support website: provides system upgrade software and patches for devices.

In the three cloud-based deployment modes of the intent-driven campus network, the management channels are encrypted to ensure secure data transmission. Service data are directly forwarded locally without passing through the cloud management platform, thereby protecting the privacy of service data. To further secure data transmission, the configurations that the cloud management platform delivers to network devices using NETCONF are transmitted over SSH. Network devices upload performance data through HTTP2.0 and report alarms through NETCONF.

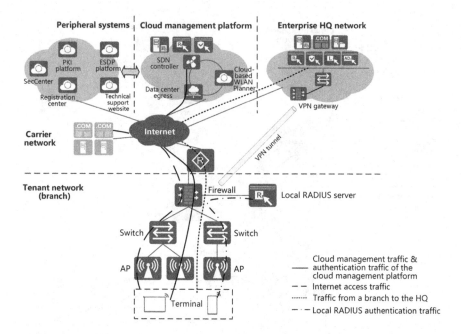

FIGURE 3.21 Cloud management network architecture in the Huawei public cloud deployment mode.

In addition, bidirectional certificate authentication is required for device registration. The device system software, signature database, and antivirus database are transmitted over HTTPS. Both HTTPS and HTTP2.0 protocol packets are encrypted using TLS1.2 to ensure the secure transmission of data. And, tenants or MSPs who log in to the cloud management platform over the Internet do so via HTTPS connections.

2. Business model and role positioning of an intent-driven campus network

In the three cloud-based deployment modes of intent-driven campus networks, the network architectures are similar, but the cloud management platform and network infrastructure are operated and owned by different entities. This means that different business models are used. Cloud-based deployment of intent-driven campus networks involves four important roles, each with its specific responsibilities.

These roles are described as follows:

a. Platform operator: also called the platform administrator or system administrator, is responsible for installing and operating the cloud management platform, managing MSPs and tenants, collecting statistics on the device quantities and services on the entire network, and providing basic network VASs.

b. MSP: has professional network construction and maintenance capabilities and is responsible for promoting and selling the intent-driven campus network solution. MSPs can construct and maintain campus networks for tenants that do not have the corresponding capabilities once authorized by the tenants to do so.

c. Tenant: manages, operates, and maintains a campus network. A tenant purchases cloud-based devices and services to build a campus network that supports the tenant's service development. The construction and maintenance of this network can be performed by the tenant or by an MSP once authorized by the tenant.

d. User: is the end user of a campus network. A user accesses and uses the network through a wired or wireless terminal.

In the on-premises deployment mode, the enterprise O&M team plays the role of a platform operator or MSP. In the Huawei public cloud deployment mode, Huawei plays the role of a platform operator, and the MSP must be certified by Huawei. In the MSP-owned cloud deployment mode, the MSP plays the role of a platform operator in order to provide network management services to tenants.

CHAPTER 4

Building Physical Networks for an Intent-Driven Campus Network

A N ULTRA-BROADBAND PHYSICAL NETWORK (including ultra-broadband forwarding and coverage) is crucial to an intent-driven campus network to eliminate limitations such as low bandwidth and high latency when planning and deploying virtual networks on top of the physical network. Ultra-broadband is also the prerequisite for decoupling the physical network from services.

With the trend toward deployment of all-wireless and IoT-integrated campus services, various sectors inevitably choose multinetwork convergence and ultra-broadband coverage when building campus networks to reduce the capital expenditure. Multinetwork convergence enables multiple previously separated services to coexist, which subsequently poses different bandwidth requirements on the physical network. However, this convergence requires a highly reliable large-capacity, and ultra-low-latency physical network, where all services can run concurrently.

Currently, ultra-broadband forwarding and coverage can be easily achieved for physical networks as the related industry and technologies are highly developed. For wired networks, the 25GE, 100GE, and 400GE technology standards have been released and ready for large-scale commercial use thanks to the improved performance of chips and productivity at lower costs. These standards enable campus networks with the ultra-broadband forwarding capability at reasonable costs. As for wireless networks, the Wi-Fi 6 standard is quickly being put into commercial use and supports larger capacity and more user density than previous Wi-Fi standards, providing comprehensive coverage of campuses. Wireless device vendors are launching next-generation Wi-Fi 6 access points (APs), and in terms of IoT, an increasing number of IoT terminals connect to campus networks based on Wi-Fi standards. IoT APs integrate multiple wireless communication modes such as Bluetooth and radio frequency identification (RFID), further expanding the scope of access terminals.

4.1 ULTRA-BROADBAND FORWARDING ON A PHYSICAL NETWORK

Ultra-broadband forwarding on physical networks involves the wired backbone as well as wired and wireless access networks. The transmission rate of the wired backbone network will evolve from 25, 100, to 400 Gbit/s, and that of the wired access network will evolve from 2.5, 5, to 10 Gbit/s. In addition to this, the wireless local area network (WLAN) standard evolves from Wi-Fi 4, Wi-Fi 5, to Wi-Fi 6.

4.1.1 Ultra-Broadband Forwarding Driven by Changes in the Campus Traffic Model

As mentioned in Section 2.4.1, the informatization and cloudification of campus networks have been insufficient for a long time. A campus network has predominantly east-west local traffic, which needs to traverse the network regardless of whether data are mutually accessed or data are transmitted between users or between users and scattered servers. Correspondingly, the network architecture has a large oversubscription ratio between layers, which may exceed 1:10, as shown in Figure 4.1.

In recent years, with the widespread use of the cloud computing architecture and increasing digitalization, major services and data on a campus network are centralized into a data center (DC). The campus network begins to mainly carry north-south traffic from terminals to the DC. Therefore, the oversubscription ratio between layers on a campus network must be reduced or even 1:1 (indicating an undersubscribed network),

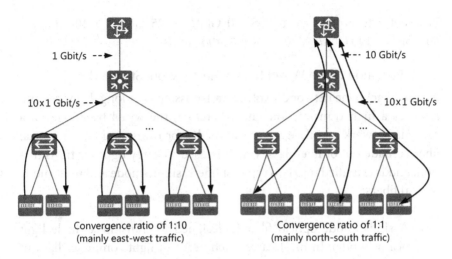

Convergence ratio of 1:10
(mainly east-west traffic)

Convergence ratio of 1:1
(mainly north-south traffic)

FIGURE 4.1 Campus network traffic model evolution.

requiring that the physical network for a campus network supports ultra-broadband forwarding.

The ultra-broadband forwarding rate of wired networks fundamentally complies with the Ethernet evolution rule. This rule stated that, before 2010, the Ethernet transmission rate increased 10-fold each time, as follows: 10 Mbit/s → 100 Mbit/s → 1 Gbit/s → 10 Gbit/s → 100 Gbit/s. However, since 2010, when a large number of wireless APs started to access campus networks, the Ethernet rate has evolved in a variety of ways, as shown in Figure 4.2.

With the copper media, the Ethernet rate evolves as follows: 1 Gbit/s → 2.5 Gbit/s → 5 Gbit/s → 10 Gbit/s, whereas with the optical media, the

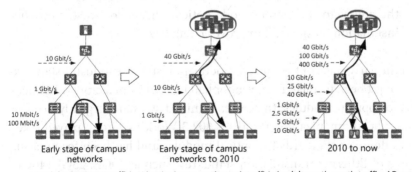

Early stage of campus networks

Early stage of campus networks to 2010

2010 to now

(mainly east-west traffic) (developing to north-south traffic) (mainly north-south traffic, AP access)

FIGURE 4.2 Evolution of the port forwarding rate on campus network devices.

Ethernet rate evolves as follows: 10 Gbit/s → 25 Gbit/s → 40 Gbit/s → 50 Gbit/s → 100 Gbit/s → 200 Gbit/s → 400 Gbit/s.

4.1.2 Evolution of the Wired Backbone Network Standard

A wired backbone network involves transmission over long distances and requires a high transmission rate. Therefore, the wired backbone on a campus network mainly uses optical media for transmission. An optical fiber is made out of fiberglass, which is used to carry light for transmission, and is classified into two types of fibers, single-mode and multimode optical fibers.

- A single-mode optical fiber is designed to transmit a single light beam (mode). In most cases, each beam of light comprises light of multiple wavelengths.

- A multimode optical fiber has multiple transmission paths, which causes a differential mode delay.

- Single-mode optical fibers are used for long-distance transmission, whereas multimode optical fibers are used when the transmission distance is less than 300 m.

- The core of a single-mode optical fiber is narrow (generally with a diameter of 8.3 μm), supporting a longer transmission distance. Therefore, more precise terminals and connectors are required. In contrast, the core of a multimode optical fiber is wider (generally with a diameter of 50–62.5 μm). In addition, a multimode optical fiber can be driven by a low-cost laser in a short distance. Moreover, the connector on a multimode optical fiber is low-cost, simplifies installation on site, and improves reliability.

In the early stage of the optical module industry, the industry standard had not yet been formed. Each vendor had its own form factor and dimensions for optical modules; therefore, a unified standard urgently needed. As an official organization, the IEEE 802.3 Working Group played a key role in defining standards for optical modules and has been formulating a series of Ethernet transmission standards, such as 10GE, 25GE, 40GE, 50GE, 100GE, 200GE, and 400GE. The Ethernet transmission rate on a campus network is developed as follows:

1. 10GE

In 2002, IEEE released the IEEE 802.3ae standard, which determined the 10 Gbit/s (10GE) wired backbone for early campus networks. 10GE achieves a leap in Ethernet standards, reaching a transmission rate of 10 Gbit/s, ten times higher than that of the earlier GE standard. Additionally, 10GE greatly extends the distance of transmission, eliminating the limitation of the traditional Ethernet on a Local Area Network (LAN). Its technology is applicable to various network structures. It can simplify a network as well as help construct a simple and cost-effective network that supports multiple transmission rates, enabling transmission at a large capacity on the backbone network.

2. 40GE/100GE

On the heels of 10GE, the 40GE and 100GE standards were released by IEEE 802.3ba in 2010. At that time, however, the 40GE rate was actually achieved by a 4-lane 10GE serializer/deserializer (SerDes) bus, and the 100GE rate by a 10-lane 10GE SerDes bus. The costs of deployment for 40GE and 100GE are high; for example, in contrast to 10GE optical transceivers that require the small form-factor pluggable (SFP) encapsulation mode, 40GE optical transceivers use the quad small form-factor pluggable (QSFP) encapsulation mode. Additionally, 40GE connections require four pairs of optical fiber; the optical transceivers and cables result in high costs.

3. 25GE

Originally applied to data center networks, 40GE is high in cost and does not efficiently leverage Peripheral Component Interconnect Express (PCIe) lanes on servers. To eliminate these drawbacks, companies including Microsoft, Qualcomm, and Google established the 25 Gigabit Ethernet Consortium and launched the 25GE standard. To ensure the unification of Ethernet standards, IEEE introduced 25GE to the IEEE 802.3 by standard in June 2016.

25GE optical transceivers use the single-lane 25GE SerDes bus and SFP encapsulation. In addition, the previous 10GE optical transceivers can be smoothly upgraded to 25GE without the need to replace existing network cables, greatly reducing the overall cost of network deployment.

With the widespread use of Wi-Fi 5 on campus networks, 25GE starts to be deployed on campus networks, in which APs transmit data at a rate of 2.5 Gbit/s in the uplink. The data rate on such networks is 25 Gbit/s between access and aggregation switches and 100 Gbit/s between aggregation and core switches, resulting in reduced use of 40GE.

4. 50GE/100GE/200GE

In December 2018, IEEE released the IEEE 802.3cd standard, which defines the 50GE, 100GE, and 200GE standards. 50GE is typically used on Fifth Generation (5G) backhaul networks to provide higher access bandwidth than that offered by 10GE/25GE.

Taking into account the cost of deployment and fully developed industry chain, 25GE is mainly used on campus networks for which it is more suitable. 100GE will become the mainstream of the wired backbone on campus networks for a long time in the future.

5. 200GE/400GE

In December 2017, IEEE released the IEEE 802.3bs standard, which defines the 200GE and 400GE standards. 400GE may not be applied on campus networks in the near future; however, with the development of Wi-Fi technologies, it is estimated that the uplink data rate of APs will reach 25 Gbit/s in the Wi-Fi 7 era, which will require a core bandwidth of 400GE. Therefore, 400GE will become the direction for evolution of the wired backbone network.

The 400GE standard uses four-level pulse amplitude modulation (PAM4) technology and 8-lane 25GE SerDes buses to achieve a transmission rate of 400 Gbit/s. The number of SerDes buses used by 400GE is only twice as many used by 100GE, but the transmission rate of 400GE is four times higher than that of 100GE, greatly reducing cost per bit.

To ensure high bandwidth on the network and low deployment costs, 25GE, 100GE, and 400GE are the targets for evolution of the wired backbone on campus networks.

4.1.3 Evolution of the Wired Access Network Standard

Based on cost-effective twisted pairs and the need for power over Ethernet (PoE), the wired access layer on a traditional campus network typically uses copper media for transmission. Legacy category 5 enhanced (CAT5e) twisted pairs can directly support 2.5GE, or even 5GE, transmission without re-cabling. Therefore, copper is still used for transmission on the wired access layer on an intent-driven campus network.

Twisted pairs are the most commonly used copper transmission cables on campus networks, formed by winding two thin copper wires wrapped with insulation materials in a certain proportion. Twisted pairs are classified into two types: unshielded twisted pairs (UTPs) and shielded twisted pairs (STPs).

STPs are wrapped with a layer of metal, which reduces leakage of radiation and prevent interception of data. In addition, STPs support a high transmission rate of data but are high cost and require complex installation. UTPs, however, are wrapped with only a layer of insulation tape without metal shielding material.

Compared with STPs, UTPs are lower in cost and support more flexible networking. In most cases, UTPs are preferentially used, except in some special scenarios (for example, scenarios where interference from electromagnetic radiation is strong or the requirement for transmission quality is high). Table 4.1 lists the frequency bandwidth, transmission rate, and typical application scenarios of common twisted pairs.

The evolution of the transmission rate at the access layer of a campus network complies with the evolution of Ethernet standards defined by the IEEE 802.3 Working Group, including physical layer connections,

TABLE 4.1 Frequency Bandwidth, Transmission Rate, and Typical Application Scenarios of Common Twisted Pairs

Category	Cable Frequency Bandwidth (MHz)	Maximum Data Transmission Rate	Typical Application Scenario
Category 5 (CAT5) cable	100	100 Mbit/s	100BASE-T and 10BASE-T Ethernets
CAT5e cable	100	5 Gbit/s	1000BASE-T, 2.5GBASE-T, and some 5GBASE-T Ethernets
Category 6 (CAT6) cable	250	10 Gbit/s (at a distance of 37–55 m)	5GBASE-T and some 10GBASE-T Ethernets
Augmented category 6 (CAT6a) cable	500	10 Gbit/s (at a distance of 100 m)	10GBASE-T Ethernet
Category 7 (CAT7) cable	600	10 Gbit/s (at a distance of up to 100 m)	10GBASE-T Ethernet

TABLE 4.2   Data Transmission Rates Defined in Different IEEE 802.3 Ethernet Standards

Standard	Data Transmission Rate
IEEE 802.3	10 Mbit/s
IEEE 802.3u	100 Mbit/s
IEEE 802.3ab	1 Gbit/s
IEEE 802.3ae	10 Gbit/s
IEEE 802.3bz	2.5 and 5 Gbit/s

electrical signals, and MAC layer protocols. Table 4.2 shows the evolution phases of Ethernet standards in chronological order.

1. Standard Ethernet (IEEE 802.3)

    Formulated in 1983, IEEE 802.3 is the first formal Ethernet standard, which defines a method for LAN access using carrier sense multiple access with collision detection (CSMA/CD) technology, with a maximum data rate of 10 Mbit/s. This early Ethernet is called standard Ethernet that enables connections through various types of transmission media, such as thick coaxial cables, thin coaxial cables, UTPs, STPs, and optical fibers.

2. Fast Ethernet (IEEE 802.3u)

    Fast Ethernet was introduced in 1995 as the IEEE 802.3u standard that raised the maximum data rate to 100 Mbit/s. Compared with the IEEE 802.3 standard, IEEE 802.3u provides a higher network bandwidth for desktop users and servers (or server clusters). IEEE 802.3u signifies that LANs are embracing the Fast Ethernet era.

3. Gigabit Ethernet (IEEE 802.3ab)

    The IEEE 802.3ab standard came into use in 1999. It supports a maximum data rate of 1 Gbit/s, 10 times the rate of Fast Ethernet. On this Ethernet, data can be transmitted through optical fibers, twisted pairs, and twinax cables. IEEE 802.3ab signifies the beginning of the Gigabit Ethernet era.

4. 10GE (IEEE 802.3ae)

    The IEEE 802.3ae standard was ratified in 2002, defining 10GE with a maximum data rate of 10 Gbit/s. Data can be transmitted through optical fibers, twisted pairs, and coaxial cables. IEEE

802.3ae signifies the beginning of the 10 Gbit/s Ethernet era and lays a foundation for end-to-end Ethernet transmission.

5. Multi-GE (IEEE 802.3bz)

   Initially released in 2016, the IEEE 802.3bz standard comes with 2.5GBASE-T and 5GBASE-T, which reach transmission rates of 2.5 and 5 Gbit/s, respectively, at a distance of 100 m. Physical layer (PHY) transmission technology of IEEE 802.3bz is based on 10GBASE-T but operates at a lower signal rate. Specifically, IEEE 802.3bz reduces the transmission rate to 25% or 50% that of 10GBASE-T, to achieve a rate of 2.5 Gbit/s (2.5GBASE-T) or 5 Gbit/s (5GBASE-T). This lowers the requirements on cabling, enabling 2.5GBASE-T and 5GBASE-T to be deployed over 100 m long unshielded CAT5e and CAT6 cables, respectively.

   The IEEE 802.3bz standard is intended to support the improvement in air interface forwarding performance of wireless APs.

   In January 2014, the IEEE 802.11ac standard was officially released. It uses signals on the 5 GHz frequency band for communication and provides a theoretical transmission rate of higher than 1 Gbit/s for multistation WLAN communication. With IEEE 802.11ac, an uplink rate of 1 Gbit/s on the backbone network can hardly meet the requirements of next-generation APs, as it is impossible to carry 10 Gbit/s traffic over CAT5e cables that have been routed alongside existing network devices. The 10GE can provide the rate required by APs, but the need for re-cabling using CAT6 or higher-specifications cables or optical fibers to implement this rate results in high costs and complex construction.

   Against this backdrop, there is an immediate solution. Specifically, an Ethernet standard that supports intermediate speeds between 1 and 10 Gbit/s while eliminating the need for re-cabling is gaining wide recognition from users and network vendors. This technology is also known in the industry as multi-GE technology.

   Two technology alliances have been established in the world to promote the development of 2.5GE and 5GE technologies on enterprise networks. In October 2014, the NBASE-T Alliance was jointly founded by Aquantia, Cisco, Freescale Semiconductor, and Xilinx, with its members now including most network hardware manufacturers. In December 2014, the MGBASE-T Alliance was founded by Broadcom, Aruba, Avaya, Brocade, and Ruijie Networks. Finally, the multi-GE technology promoted by the NBASE-T Alliance became a standard supported by IEEE.

Currently, campus networks use a wired access transmission rate of 1 or 2.5 Gbit/s; however, with the popularity of Wi-Fi 6, the uplink transmission rate of APs needs to reach 5 Gbit/s, or even 10 Gbit/s. Therefore, during network construction or capacity expansion, it is recommended that twisted pairs of CAT6 or higher specifications be used for the wired access network, as these cables deliver a transmission rate of 10 Gbit/s, protecting customer investment.

4.1.4 Evolution of Wireless Access: Wi-Fi 4, Wi-Fi 5, and Wi-Fi 6

The continuous development of wired ultra-broadband technologies facilitates construction of an ultra-broadband core network for a campus network. This provides an ultra-broadband pipe for service traffic on campus, meeting the requirements of various large-capacity and high-concurrency services on the campus network. At the same time, wireless access technologies are also developing rapidly and the continuous evolution of Wi-Fi standards results in a surge in the air interface rate from the initial 2 Mbit/s to nearly 10 Gbit/s.

The WLAN supports two technical protocol standards: IEEE 802.11 and high performance radio LAN (HIPERLAN). IEEE 802.11 series standards are promoted by the Wi-Fi Alliance. Therefore, IEEE 802.11 WLANs are also called Wi-Fi networks. IEEE 802.11 standards continuously evolve to optimize the PHY and MAC technologies of WLANs, increasing the transmission rate and improving the anti-interference performance of the WLANs. Significant standards include IEEE 802.11, IEEE 802.11a/b, IEEE 802.11g, IEEE 802.11n, IEEE 802.11ac, and IEEE 802.11ax. Figure 4.3 shows the changes in wireless transmission rates of these standards.

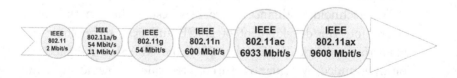

FIGURE 4.3 Wireless transmission rate changes during IEEE 802.11 evolution.

TABLE 4.3   Mappings between Wi-Fi Generations and IEEE 802.11 Standards

Released In	IEEE 802.11 Standard	Frequency Band (GHz)	Wi-Fi Generation
2009	IEEE 802.11n	2.4 or 5	Wi-Fi 4
2013	IEEE 802.11ac Wave 1	5	Wi-Fi 5
2015	IEEE 802.11ac Wave 2	5	
2019	IEEE 802.11ax	2.4 or 5	Wi-Fi 6

In 2018, the Wi-Fi Alliance introduced a new naming system for IEEE 802.11 standards to identify Wi-Fi generations by a numerical sequence, providing Wi-Fi users and device vendors with an easy-to-understand designation for the Wi-Fi technology supported by their devices and used in Wi-Fi connections. Additionally, the new approach to naming highlights the significant progress of Wi-Fi technology, with each new generation of the Wi-Fi standard introducing a large number of new functions to provide higher throughput, faster speeds, and more concurrent connections. Table 4.3 lists the mappings between Wi-Fi generations and IEEE 802.11 standards announced by the Wi-Fi Alliance.

1. IEEE 802.11a/b/g/n/ac

   a. IEEE 802.11

      In the early 1990s, the IEEE set up a dedicated IEEE 802.11 Working Group to study and formulate WLAN standards. In June 1997, the IEEE released the first generation WLAN protocol, the IEEE 802.11 standard. This standard allows the PHY of WLANs to work on the 2.4 GHz frequency band at a maximum data transmission rate of 2 Mbit/s.

   b. IEEE 802.11a/IEEE 802.11b

      In 1999, the IEEE released the IEEE 802.11a and IEEE 802.11b standards. The IEEE 802.11a standard operates on the 5 GHz frequency band and uses orthogonal frequency division multiplexing (OFDM) technology. This technology divides a specified channel into several subchannels, each of which has one subcarrier for modulation. The subcarriers are transmitted in parallel, improving the spectrum utilization of channels and bringing a PHY transmission rate of up to 54 Mbit/s. IEEE 802.11b is a direct extension of the modulation technique defined in IEEE 802.11, which still works on the 2.4 GHz frequency band but increases the maximum transmission rate to 11 Mbit/s.

Although IEEE 802.11a supports a much higher access rate than IEEE 802.11b, the IEEE 802.11b standard became a mainstream standard in the market at the time, as IEEE 802.11a is dependent on 5 GHz chips, which are difficult to develop. When the corresponding chip was launched, IEEE 802.11b had already been widely used in the market. Furthermore, the IEEE 802.11a standard is not compatible with IEEE 802.11b, and 5 GHz chips had relatively high costs, which restricted the use of the 5 GHz band in some areas. All this makes wide use of IEEE 802.11a impossible.

c. IEEE 802.11g

In early 2000, the IEEE 802.11g Working Group aimed to develop a standard that is compatible with IEEE 802.11b and can work at a transmission rate of 54 Mbit/s. In November 2001, the Working Group proposed the IEEE 802.11g draft standard, which was ratified in 2003. Similar to IEEE 802.11b, IEEE 802.11g operates on the 2.4 GHz frequency band. To achieve a rate of 54 Mbit/s, IEEE 802.11g uses the same OFDM-based transmission scheme as IEEE 802.11a on the 2.4 GHz frequency band. IEEE 802.11g met people's demand for high bandwidth at the time and promoted the development of WLAN.

d. IEEE 802.11n

The 54 Mbit/s rate, however, cannot always meet users' demand in the booming network field. In 2002, a new IEEE Working Group was established to conduct research on a faster WLAN technology, with the initial aim to achieve a rate of 100 Mbit/s. After much dispute within the group, the new protocol, known as IEEE 802.11n, was ratified in September 2009. After seven years of development, IEEE 802.11n increased the rate from the original 100–600 Mbit/s. IEEE 802.11n operates on both the 2.4 and 5 GHz bands, and is compatible with IEEE 802.11a/b/g.

Compared with previous standards, IEEE 802.11n introduces more key technologies to increase the wireless rate, such as involvement of more subcarriers, higher coding rates, shorter guard interval (GI), wider channels, more spatial streams, and MAC-layer packet aggregation.

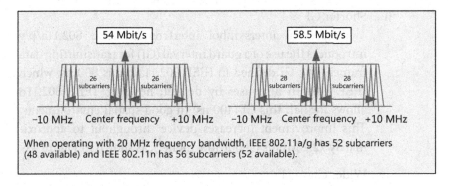

FIGURE 4.4   Comparison between subcarriers defined in IEEE 802.11a/g and IEEE 802.11n.

i. More subcarriers

Compared with IEEE 802.11a/g, IEEE 802.11n has four more valid subcarriers and increases the theoretical rate from 54 to 58.5 Mbit/s, as shown in Figure 4.4.

ii. Higher coding rate

Data transmitted on WLANs contains valid data and forward error correction (FEC) code. If errors occur in the valid data due to attenuation, interference, or other factors, the FEC code can be used to rectify the errors and restore data. IEEE 802.11n increases the effective coding rate from 3/4 to 5/6, thereby increasing the data rate by 11%, as shown in Figure 4.5.

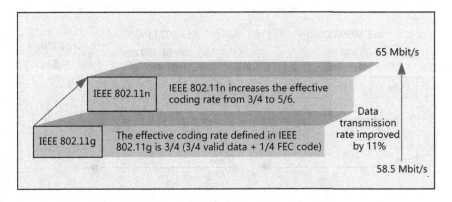

FIGURE 4.5   Coding rate defined in IEEE 802.11n.

iii. Shorter GI

To prevent intersymbol interference, IEEE 802.11a/b/g introduces the use of a guard interval (GI) for transmitting data frames. The GI defined in IEEE 802.11a/b/g is 800 ns, which IEEE 802.11n also uses by default; however, IEEE 802.11n allows for half this GI (400 ns) in good spatial environments. This improvement increases device throughput to approximately 72.2 Mbit/s (about 10% higher), as shown in Figure 4.6.

iv. Wider channel

With IEEE 802.11n, two adjacent 20 MHz channels can be bonded into a 40 MHz channel, doubling bandwidth and therefore delivering a higher transmission capability. Use of the 40 MHz frequency bandwidth enables the number of subcarriers per channel to be doubled from 52 to 104. This increases the data transmission rate to 150 Mbit/s (108% higher). Figure 4.7 shows the 40 MHz channel bandwidth as defined in IEEE 802.11n.

v. More spatial streams

In IEEE 802.11a/b/g, a single antenna is used for data transmission between a terminal and an AP in single-input single-output (SISO) mode. This means that data are transmitted through a single spatial stream. In contrast, IEEE 802.11n supports multiple-input multiple-output (MIMO) transmission, which enables it to achieve a maximum transmission

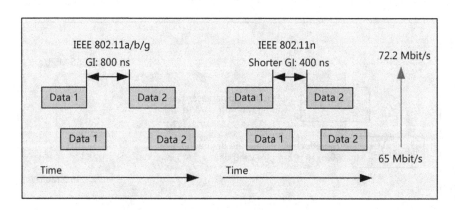

FIGURE 4.6   GI change from IEEE 802.11a/b/g to IEEE 802.11n.

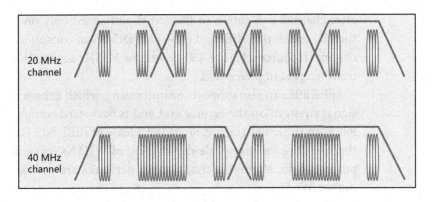

FIGURE 4.7   40 MHz channel bandwidth.

rate of 600 Mbit/s, through a maximum of four spatial streams, as shown in Figure 4.8.

vi. MAC protocol data unit (MPDU) aggregation at the MAC layer

The IEEE 802.11 MAC layer protocols define numerous fixed overheads for transmitting frames, especially acknowledgement frames. At the highest data rate, these overheads consume larger bandwidth than that required for transmitting entire data frames. For example, the theoretical transmission rate defined in IEEE 802.11g is 54 Mbit/s; however, the actual transmission rate is only 22 Mbit/s, which means that more than half is unutilized. IEEE 802.11n defines MPDU aggregation, which aggregates multiple MPDUs into one PHY packet. In this manner, N MPDUs can be sent simultaneously

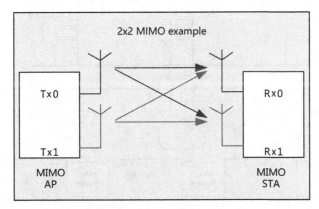

FIGURE 4.8   MIMO in IEEE 802.11n.

after channel contention or backoff is performed only once, thereby preventing different $(N-1)$ MPDUs from consuming channel resources. Figure 4.9 shows how MPDU aggregation technology is implemented.

IEEE 802.11n also supports beamforming, which enhances signal strength on the receive end and is backward compatible with IEEE 802.11a/b/g. Since the release of IEEE 802.11n, there has been a large-scale deployment of WLANs on campus networks, signifying that campus networks are embracing the Wi-Fi era.

e. IEEE 802.11ac

After IEEE 802.11n WLANs are deployed on campus networks, the rapid development of campus network services and scales poses the following new challenges:

i. Bandwidth-hungry service applications: Applications such as enterprise application synchronization, voice/video call, and enterprise video promotion pose higher bandwidth requirements on Wi-Fi networks. Furthermore, video traffic on mobile networks increases by more than 60% annually.

ii. Access from numerous STAs: As wireless access continues to grow, an increasing number of users choose to access networks through Wi-Fi. For example, the Bring Your Own

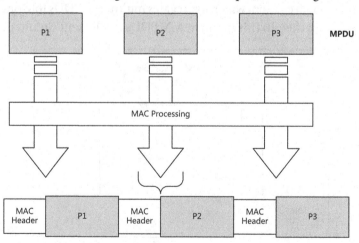

FIGURE 4.9   Implementation of MPDU aggregation technology.

Device (BYOD) trend consists of each employee using two or more Wi-Fi terminals. From a different perspective, numerous wireless users simultaneously access the network at football stadiums, new product launch venues, and campuses.

iii. Traffic distribution requirements of carriers' Third Generation (3G) or Fourth Generation (4G): With the explosive growth of cellular network data services, an increasing amount of carriers' mobile traffic is being offloaded to Wi-Fi networks to reduce cellular network load. As a result, the Wi-Fi network has been positioned as the "Nth" network, which faces high requirements.

Driven by these requirements, the fifth-generation very high throughput (VHT) Wi-Fi standard is emerging, and the IEEE 802.11ac standard was ratified in 2013. IEEE 802.11ac operates on the 5 GHz frequency band (dual-band APs and STAs still use the IEEE 802.11n standard on the 2.4 GHz frequency band) and is backward-compatible with both IEEE 802.11n and IEEE 802.11a. In addition to the multiple technologies in IEEE 802.11n, the following improvements have been made in IEEE 802.11ac to achieve a transmission rate of up to 1.3 Gbit/s:

i. It uses new technologies or extends original technologies to improve the maximum throughput and increase the number of access users. These technologies include MIMO with more spatial streams, 256-quadrature amplitude modulation (256-QAM), and multiuser MIMO (MU-MIMO).

ii. It optimizes the Wi-Fi standard and discards several optional functions provided by earlier standards to reduce complexity. For example, implicit transmit beamforming (TxBF) is not included in 802.11ac, and the channel sounding and channel feedback modes have been unified.

iii. It maintains compatibility with earlier IEEE 802.11 protocols. 802.11ac improves the PHY frame structure and channel management with different channel bandwidths.

Table 4.4 provides details of the improvements.

TABLE 4.4   Improvements Made in IEEE 802.11ac

Improvement	Description	Benefit
Channel bandwidth	Supports 80 MHz channel bandwidth Supports 160 MHz channel bandwidth Bonds two nonadjacent 80 MHz channels into a 160 MHz channel	Higher throughput
Operating frequency	Operates on frequencies bands below 6 GHz, excluding 2.4 GHz. IEEE 802.11ac mainly uses the 5 GHz frequency band	More abundant spectrum resources Less interference
MIMO	Improves single-user MIMO (SU-MIMO) and supports a maximum of eight spatial streams Uses MU-MIMO and transmits data for up to four users simultaneously	Higher throughput More access users Higher link reliability
TxBF	Supports only explicit beamforming, as opposed to implicit beamforming Improves channel sounding and feedback. IEEE 802.11ac uses null data packets for channel sounding and adopts Compressed V Matrix for channel information feedback, instead of the multiple original sounding and feedback modes	Simplified design
Modulation and coding scheme (MCS)	Uses 256-QAM (supporting the 3/4 and 5/6 coding rates) Provides 10 MCS modes	Higher throughput
Compatibility	Discards the Greenfield preamble and supports only the Mixed preamble Improves the PHY frame structure to be compatible with earlier IEEE 802.11 standards	Enhanced compatibility with earlier Wi-Fi standards
Channel management	Enhances channel management when 20, 40, 80, and 160 MHz channel bandwidths are used simultaneously	Higher channel utilization Less channel interference Higher throughput Enhanced compatibility
Frame aggregation	Improves the frame aggregation degree Supports only MPDU aggregation at the MAC layer	Improved MAC layer performance and throughput

Due to these improvements, IEEE 802.11ac is evidently more advantageous than earlier Wi-Fi standards.

i. Higher throughput

Higher throughput was pursued for Wi-Fi standards earlier than 802.11ac. Throughput has increased from a maximum of 2 Mbit/s in IEEE 802.11 to 54 Mbit/s in IEEE 802.11a, 600 Mbit/s in IEEE 802.11n, and 6.93 Gbit/s in IEEE 802.11ac. Furthermore, throughput planning and design have enabled IEEE 802.11ac to better meet high-bandwidth requirements. Table 4.5 lists the parameter changes of different IEEE 802.11 Wi-Fi standards.

ii. Less interference

IEEE 802.11ac mainly works on the 5 GHz frequency band where more resources are available. Frequency bandwidth is only 83.5 MHz on the 2.4 GHz frequency band; whereas the 5 GHz frequency band allows planning for bandwidth resources attaining hundreds of megahertz in certain countries. A higher frequency indicates a lower frequency reuse degree with the same bandwidth, reducing intrasystem interference. In addition, as numerous devices (including microwave ovens) work on the 2.4 GHz frequency band, devices in a Wi-Fi system on this frequency band face external

TABLE 4.5   Parameter Changes of Different IEEE 802.11 Wi-Fi Standards

Wi-Fi Standard	Frequency Bandwidth (MHz)	Modulation Scheme	Maximum Number of Spatial Streams	Maximum Rate
IEEE 802.11	20	Differential quadrature phase shift keying	1	2 Mbit/s
IEEE 802.11b	20	Complementary code keying	1	11 Mbit/s
IEEE 802.11a	20	64-QAM	1	54 Mbit/s
IEEE 802.11g	20	64-QAM	1	54 Mbit/s
IEEE 802.11n	20 or 40	64-QAM	4	600 Mbit/s
IEEE 802.11ac	20, 40, 80, or 160	256-QAM	8	6.93 Gbit/s

interference. Compared with the 2.4 GHz frequency band, the 5 GHz frequency band suffers less external interference.

iii. More access users

While maintaining the multiple access mode of Wi-Fi, IEEE 802.11ac provides higher throughput by introducing the MU-MIMO function, thereby increasing user access capacity. A higher rate indicates that a user occupies the air interface for a shorter period of time. Therefore, within the same period of time, an AP can allow access to more users. Furthermore, the MU-MIMO function significantly improves the concurrent access capability of APs, enabling an individual AP to transmit data for multiple users simultaneously.

2. 802.11ax (Wi-Fi 6)

As video conferencing, cloud VR, mobile teaching, and various other service applications increasingly diversify, an increasing number of Wi-Fi access terminals are being used. In addition, more mobile terminals are available due to the development of the Internet of Things (IoT). Even relatively dispersed home Wi-Fi networks, where devices used to be thinly scattered, are becoming crowded with an increasing number of smart home devices. Therefore, it is imperative that improvements are achieved in terms of Wi-Fi networks accommodating various types of terminals. This will meet users' bandwidth requirements for different types of applications running on their terminals.

Figure 4.10 shows the relationship between access capacity and per capita bandwidth in different Wi-Fi standards.

The low efficiency of Wi-Fi networks, which is caused by the access of more terminals, needs to be resolved in the next-generation Wi-Fi standard. To address this issue, the High Efficiency WLAN Study Group was established in as early as 2014, and the Wi-Fi 6 standard was ratified in 2019. By introducing technologies such as uplink MU-MIMO, orthogonal frequency division multiple access (OFDMA), and 1024-QAM high-order coding, Wi-Fi 6 is designed to resolve network capacity and transmission efficiency problems from aspects such as spectrum resource utilization and multiuser access. Compared with IEEE 802.11ac (Wi-Fi 5), Wi-Fi 6 aims to

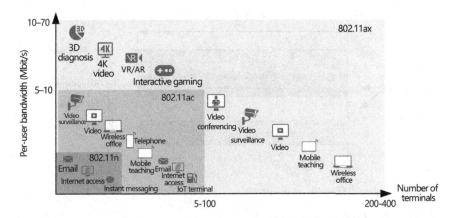

FIGURE 4.10 Relationship between the access capacity and per capita bandwidth in different Wi-Fi standards.

achieve a fourfold increase in average user throughput and increase the number of concurrent users more than threefold in dense user environments.

Wi-Fi 6 is compatible with the earlier Wi-Fi standards. This means that legacy terminals can seamlessly connect to a Wi-Fi 6 network. Furthermore, Wi-Fi 6 inherits all the advanced MIMO features of Wi-Fi 5 while also introducing several new features applicable to high-density deployment scenarios. Some of the Wi-Fi 6 highlights are described as follows:

a. OFDMA

Before Wi-Fi 6, the OFDM mode was used for data transmission and users were distinguished based on time segments. In each time segment, one user occupied all subcarriers and sent a complete data packet, as shown in Figure 4.11.

OFDMA is a more efficient data transmission mode introduced by Wi-Fi 6. It is also referred to as MU-OFDMA since Wi-Fi 6 supports uplink and downlink MU modes. This technology enables multiple users to reuse channel resources by allocating subcarriers to various users and adding multiple access in the OFDM system. To date, OFDMA has been utilized in 3rd Generation Partnership Project (3GPP) Long Term Evolution (LTE) among numerous other technologies. In addition, Wi-Fi 6 defines the smallest subcarrier as a

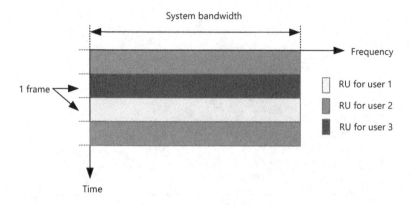

FIGURE 4.11   OFDM working mode.

resource unit (RU), which includes at least 26 subcarriers and uniquely identifies a user. The resources of the entire channel are divided into small RUs with fixed sizes. In this mode, user data are carried on each RU; therefore, on the total time-frequency resources, multiple users can send data in a time segment simultaneously (Figure 4.12).

Compared with OFDM, the following improvements have been made in OFDMA:

i. More refined channel resource allocation
    Transmit power can be allocated based on channel quality, especially when the channel status of certain nodes is below

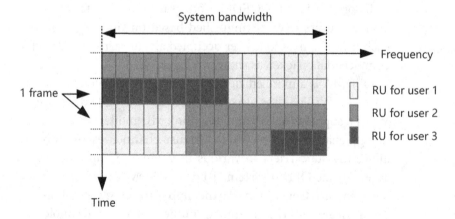

FIGURE 4.12   OFDMA working mode.

standard. This can help allocate channel time-frequency resources in a more delicate manner.

ii. Enhanced QoS

To transmit data, IEEE 802.11ac and all earlier Wi-Fi standards occupy the entire channel. Therefore, a QoS data frame can be sent only after the current transmitter releases the entire channel, which leads to long latency. In OFDMA mode, one transmitter occupies only some of the entire channel's resources. Therefore, multiple users' data can be sent simultaneously, thereby reducing the network access latency of QoS nodes.

iii. More concurrent users and higher user bandwidth

OFDMA divides the entire channel's resources into multiple subcarriers, which are then divided into several groups based on RU type. Each user may occupy one or more groups of RUs to meet various bandwidth requirements. In Wi-Fi 6, the minimum RU size and minimum subcarrier bandwidth are 2 MHz and 78.125 kHz, respectively. Therefore, the minimum RU type is 26-subcarrier RU. By analogy, RU types include 52-subcarrier, 106-subcarrier, 242-subcarrier, 484-subcarrier, and 996-subcarrier RUs. The more the number of RUs, the higher the efficiency of multiuser processing and the higher throughput.

b. Downlink/Uplink (DL/UL) MU-MIMO

MU-MIMO utilizes the spatial diversity of channels to transmit independent data streams on the same bandwidth. Unlike OFDMA, all users occupy all bandwidths, delivering multiplexing gains. Limited by the size of the antenna, a typical terminal supports only one or two spatial streams (antennas), which is less than that on an AP. Therefore, Wi-Fi 6 introduces MU-MIMO technology to enable APs to exchange data with multiple terminals simultaneously, significantly improving throughput.

Downlink MU-MIMO (DL MU-MIMO) technology: MU-MIMO has been applied since the release of IEEE 802.11ac, but only DL 4×4 MU-MIMO is supported. In Wi-Fi 6, the number of MU-MIMO spatial streams has been further increased to support DL 8×8 MU-MIMO. DL OFDMA technology can be used to simultaneously perform MU-MIMO transmission and allocate different RUs for multiuser multiple-access transmission. This increases the concurrent access capacity of the system and balances throughput, as shown in Figure 4.13.

Uplink MU-MIMO (UL MU-MIMO) is one of the key features introduced in Wi-Fi 6. Like UL SU-MIMO, UL MU-MIMO uses the same channel resources to transmit data on multiple spatial streams by utilizing the transmitter's and receiver's multiantenna technology. The only difference is that the multiple data streams of UL MU-MIMO originate from multiple users. Wi-Fi 5 (802.11ac) and earlier Wi-Fi standards use UL SU-MIMO, which is inefficient in multiuser concurrency scenarios due to the fact that one AP can receive data from only one user. After UL MU-MIMO, UL OFDMA technology is utilized to enable MU-MIMO transmission and multiuser multiple-access transmission simultaneously. This improves the transmission efficiency in multiuser concurrency scenarios while significantly reducing application latency. Figure 4.14 shows the uplink scheduling sequence in multiuser mode.

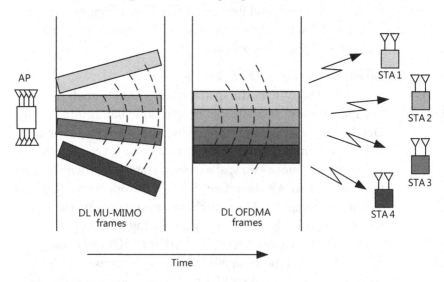

FIGURE 4.13   8×8 MU-MIMO AP scheduling sequence in the downlink multiuser mode.

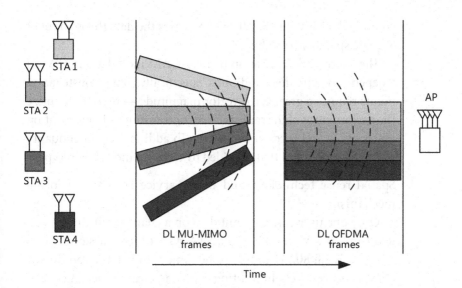

FIGURE 4.14   Uplink scheduling sequence in multiuser mode.

OFDMA and MU-MIMO used by Wi-Fi 6 both improve the concurrent users; however, these two technologies work in different mechanisms. OFDMA allows multiple users to subdivide channels to improve concurrency efficiency; whereas MU-MIMO allows multiple users to use different spatial streams to increase throughput. Table 4.6 compares OFDMA and MU-MIMO.

c. Higher-order modulation technology (1024-QAM)

The Wi-Fi 6 standard aims to increase system capacity, reduce latency, and improve efficiency in multiuser high-density scenarios. However, high efficiency does not necessarily compromise speed. Wi-Fi 5 uses 256-QAM, through which each symbol transmits 8-bit data ($2^8 = 256$). Wi-Fi 6 uses 1024-QAM, through which each symbol transmits 10-bit data ($2^{10} = 1024$). Therefore,

TABLE 4.6   Comparison between OFDMA and MU-MIMO

OFDMA	MU-MIMO
Improved efficiency	Improved capacity
Reduced latency	Increased user rate
Most suitable for low-bandwidth applications	Most suitable for high-bandwidth applications
Most suitable for small-packet transmission	Most suitable for large-packet transmission

compared with Wi-Fi 5, Wi-Fi 6 increases the data throughput of a single spatial stream by 25%.

The successful application of 1024-QAM modulation in Wi-Fi 6 depends on channel conditions. Specifically, dense constellation points require a great error vector magnitude — used to quantize the performance of the radio receiver or transmitter in modulation precision — and receiver sensitivity. In addition, channel quality must be higher than that provided by different modulation types.

d. Spatial reuse technology and Basic Service Set (BSS) coloring mechanism

Only one user can transmit data on a channel within a specified time. If a Wi-Fi AP and a STA detect transmission from a different IEEE 802.11 radio on the same channel, they automatically avoid conflicts by waiting until the channel becomes idle. Therefore, each user occupies channel resources in sequence. As channels are valuable resources on wireless networks, the capacity and stability of an entire Wi-Fi network can be significantly improved in high-density scenarios through proper channel allocation and utilization. Wi-Fi 6 devices can work on the 2.4 or 5 GHz frequency band; however, in high-density deployment scenarios, the number of available channels may be extremely small, especially on the 2.4 GHz frequency band. In this case, system throughput can be increased by improving the channel multiplexing capability.

In Wi-Fi 5 and earlier Wi-Fi standards, the mechanism for dynamically adjusting the clear channel assessment (CCA) threshold is used to alleviate co-channel interference. This mechanism identifies the co-channel interference strength, dynamically adjusts the CCA threshold, ignores co-channel weak interference signals, and implements co-channel concurrent transmission, thereby increasing system throughput. Figure 4.15 shows how this mechanism works.

STA 1 on AP 1 is transmitting data. If AP 2 wants to send data to STA 2, AP 2 needs to detect whether the channel is idle. The default CCA threshold is −82 dBm. If AP 2 discovers that the channel is occupied by STA 1, it delays data transmission due to parallel transmission being unavailable. In fact, packet

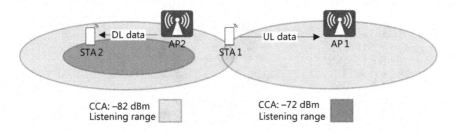

FIGURE 4.15   Dynamic adjustment of the CCA threshold.

transmission for all the STAs associated with AP 2 will be delayed. The dynamic CCA threshold adjustment mechanism is introduced to solve this issue. When AP 2 detects that the co-frequency channel is occupied, it can adjust the CCA threshold (for example, from −82 to −72 dBm) based on the interference strength to avoid interference impact. Through this process, co-frequency concurrent transmission can be implemented.

Due to the mobility of Wi-Fi STAs, co-channel interference on a Wi-Fi network is dynamic and it changes with the movement of STAs. Therefore, the dynamic CCA mechanism is effective.

Wi-Fi 6 introduces a new co-channel transmission identification mechanism referred to as BSS coloring. This mechanism adds a BSS color field to the PHY packet header to color data from different BSSs and allocate a color to each channel. The color identifies a BSS that should not be interfered with. In this manner, the receiver can identify co-channel interference signals and discontinue receiving them during an initial stage, thereby not wasting the transceiver's time. If the colors are the same, the interference signals are considered to be in the same BSS, and signal transmission is delayed. If the colors are different, no interference exists between the two Wi-Fi devices, thereby enabling them to transmit data on the same channel and frequency. In this mode, channels with the same color are maintained at significant distances from each other. The dynamic CCA mechanism is used to set such signals to be insensitive. In fact, it is unlikely that these signals interfere with each other. Figure 4.16 shows how the BSS coloring mechanism works.

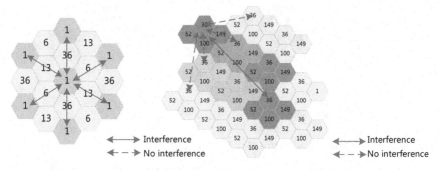

Co-channel BSS congestion   Co-channel BSS congestion only with the same color

FIGURE 4.16   How the BSS coloring mechanism reduces interference.

e. Expanding range

Wi-Fi 6 uses the long OFDM symbol transmission mechanism, which leads to an increase in the data transmission duration from 3.2 to 12.8 µs. A longer transmission time can reduce the packet loss rate of STAs. Additionally, Wi-Fi 6 can use only 2 MHz bandwidth for narrowband transmission, which reduces noise interference on the frequency band, improves the receiver sensitivity of STAs, and increases coverage distance, as shown in Figure 4.17.

The preceding core technologies can be used to verify the efficient transmission and high-density capacity brought by Wi-Fi 6. However, Wi-Fi 6 is not the ultimate Wi-Fi standard. This is just the beginning of the HEW. The new Wi-Fi 6 standard still needs to achieve compatibility with legacy devices while taking into account the development of future-oriented IoT networks and energy conservation. The other new features of Wi-Fi 6 include:

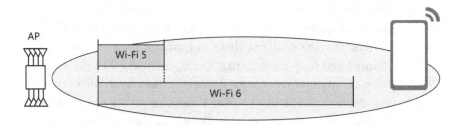

FIGURE 4.17   How long OFDM symbol and narrowband transmission increase coverage distance.

i. Support for the 2.4 GHz frequency band: Compared with the 5 GHz frequency band, the 2.4 GHz frequency band is more advantageous as it provides a longer transmission distance, wider coverage, and lower costs. In addition, a large number of IoT devices still work on the 2.4 GHz frequency band as it allows the devices to be interconnected with Wi-Fi networks.

ii. Target Wakeup Time (TWT): It is an important resource scheduling function supported by Wi-Fi 6. It allows devices to negotiate with APs for the waking schedule. APs can group STAs into different TWT periods to reduce the number of devices that simultaneously compete for the wireless medium after wakeup. In addition, the TWT increases the device sleep time. For battery-powered STAs, battery life has significantly improved.

Wi-Fi 6 is designed for high-density wireless access and high-capacity wireless services, such as in outdoor large-scale public locations, high-density stadiums, indoor high-density wireless offices, and electronic classrooms. In these scenarios, the number of STAs connected to the Wi-Fi network greatly increases within a short time. The increasing voice and video traffic also leads to adjustment in the Wi-Fi network. Some services are sensitive to bandwidth and latency, for example, 4K video streams (bandwidth of 50 Mbit/s), voice streams (latency of less than 30 ms), and virtual reality (VR) streams (bandwidth of 75 Mbit/s and latency of less than 15 ms).

An IEEE 802.11ac network can provide large bandwidth. However, as access density increases, throughput performance encounters a bottleneck. Wi-Fi 6 introduces technologies such as OFDMA, UL MU-MIMO, and 1024-QAM to ensure more reliable services. In addition to a larger access capacity, the network can balance the bandwidth of each user.

Wi-Fi 6 APs also support access of IoT devices. These additional functions and access devices inevitably increase APs' power consumption requirements, which cannot be met by the PoE function defined in IEEE 802.11af. It is estimated that the standard configurations of Wi-Fi 6 APs will incorporate

PoE+ or PoE++ in compliance with IEEE 802.11at. Therefore, the upstream PoE power supply must be taken into account during planning of future networks, and PoE switches at the access layer must be upgraded. The air interface throughput of the lowest-performing Wi-Fi 6 APs, which support 2×2 MU-MIMO, has reached almost 2 Gbit/s. This significantly increases the network capacity requirements for upstream switches. At the bare minimum, uplink switches need to provide multirate Ethernet ports of 2.5 and 5 Gbit/s. In the future, the demand for 10 Gbit/s Ethernet ports in the upstream direction will increase, and growth in wireless network capacity will drive the expansion of the overall campus network capacity. Consequently, high-quality campus networks need to provide high-bandwidth channels and refined management.

4.2 ULTRA-BROADBAND COVERAGE ON A PHYSICAL NETWORK

The development of Wi-Fi technologies enables campus networks to provide full wireless coverage over an entire campus. However, ultra-broadband coverage on the physical network is faced with another challenge, that is, connecting IoT terminals to campus networks.

4.2.1 IoT Drives a Fully Connected Campus Network

Labeled as the "Next Big Thing", IoT is emerging as the third wave in the development of the ICT industry. As network access becomes mainstream for PCs, mobile phones, cars, and electricity meters, more and more devices connect to networks in various ways, such as through cellular networks, Near Field Communication (NFC), RFID, Bluetooth, ZigBee, and Wi-Fi. According to Huawei's forecast, more than 100 billion things (excluding individual broadband users) will be connected by 2025, and devices will be able to connect to networks anytime, anywhere. For this reason, IoT is bringing about a major shift in the way we work and live.

On a campus network, there are various types of IoT terminals, such as access control devices, asset tags, and smart lights. These terminals, in addition to mobile phones, laptops, and desktop computers, constitute access devices on the campus network and allow users to interact with the network. In this case, managing a large number of access terminals on a campus network is a major concern during network construction.

On traditional campuses, different networks are constructed to support different types of terminals and run different wireless access protocols for terminal access. For example, RFID for asset management and ZigBee for smart lights both require different access devices to be deployed for terminals to access the network. This leads to complex network deployment, high networking costs, and fragmented solutions, making deployment and Operations & Maintenance (O&M) difficult.

A unified access mode for diversified IoT devices is needed to address the above-mentioned challenges and ensure the large-scale promotion of IoT applications on campus networks. To cope with these challenges, industry vendors launch IoT APs by integrating connection technologies such as Bluetooth and RFID into Wi-Fi APs that have already been widely deployed and boast mature applications. The IoT AP solution implements the same base station, shared backhaul, unified entry, and central management for various IoT connections on Wi-Fi APs. This flexible and scalable solution promotes the rapid development of campus network convergence.

4.2.2 IoT-Related Communication Protocols

In IoT access scenarios, various wireless technologies are used. These technologies can be classified into either short-range or wide-area wireless communication technologies based on their coverage areas. Commonly used short-range wireless communication technologies include Wi-Fi, RFID, Bluetooth, and ZigBee. Wide-area wireless communication technologies include Sigfox, Long Range (LoRa), and Narrowband Internet of Things (NB-IoT). On a campus network, short-range wireless communication technologies are used to connect IoT devices. The following section describes short-range wireless communication technologies.

1. Wi-Fi

    As the most important wireless access mode on a campus network, a Wi-Fi network allows access from various Wi-Fi terminals, such as laptops, mobile phones, tablets, and printers. In recent years, more and more terminal types are starting to access campus networks through Wi-Fi, including electronic whiteboards, wireless displays, smart speakers, and smart lights. Wi-Fi technologies have already been detailed in the preceding sections; therefore, they will not be mentioned here.

2. RFID

RFID is a wireless access technology that automatically identifies target objects and obtains related data through radio signals without requiring manual intervention.

a. RFID system

An RFID system is composed of RFID electronic tags (RFID tags for short) and RFID readers, and connected to an information processing system.

   i. RFID tag: consists of a chip and a tag antenna or coil. The tag has built-in identification information.

   ii. RFID reader: reads and writes tag information. Read-only RFID readers are often referred to as card readers.

Figure 4.18 shows an RFID system. In such a system, an RFID tag communicates with an RFID reader through inductive coupling or electromagnetic reflection. The RFID reader then reads information from the RFID tag and sends it to the information processing system over a network for storage and unified management.

b. Operating frequency

An RFID system typically operates on a low frequency (LF) band (120–134 kHz), high frequency (HF) band (13.56 MHz), ultra-high frequency (UHF) band (433 MHz, 865–868 MHz, 902–928 MHz, and 2.45 GHz), or super-high frequency (SHF) band (5.8 GHz), as shown in Figure 4.19.

RFID tags are cost-effective and energy-efficient because they only store a small amount of data and support a relatively short reading distance. The

FIGURE 4.18   RFID system.

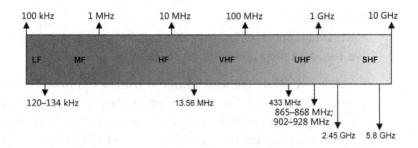

FIGURE 4.19 Typical operating frequency bands supported by RFID systems.

following describes how the characteristics of RFID tags vary depending on operating frequency:

i. LF RFID tag: characterized by mature technology, small amount of data, slow transmission, short read/write distance (less than 10 cm), strong media penetration, no special control, and low cost. RFID tags available on the market most often operate on the LF frequency band and are typically used for access control and attendance card swiping.

ii. HF RFID tag: characterized by mature technology, fast data transmission, short read/write distance (less than 1 m), strong media penetration, no special control, and high cost. HF RFID tags can store a large amount of data and therefore are most suitable for identifying objects moving at high speeds. In actual application, they are used for train number recognition and electronic toll collection (ETC) on highways.

iii. UHF RFID tag: supports fast data transmission and long read/write distance (3–100 m) but leads to high costs. UHF RFID tags are susceptible to interference due to mist, which greatly impacts radio waves. As one of the fastest-growing RFID products, UHF RFID tags are mainly used for supply chain management, logistics management, and production line automation.

iv. SHF RFID tag: requires line-of-sight transmission because signal transmission is easily reflected by water or metal. SHF RFID tags are mostly used in long-distance recognition and fast moving object recognition scenarios, such as logistics detection and highway ETC systems.

c. Power supply

RFID tags are classified into either active, semi-active, or passive, based on their power supply mode.

i. An active RFID tag uses a built-in battery to provide all or part of the energy for a microchip, without the need for an RFID reader to provide energy for startup. Active RFID tags support a long identification distance (up to 10 m), but have a limited service life (3–10 years) and high cost.

ii. A semi-active RFID tag features a built-in battery that supplies power to the internal circuits, but does not actively transmit signals. Before a semi-active RFID tag starts to work, it remains in dormant state, which means it is equivalent to a passive RFID tag. However, the battery inside the tag does not consume much energy and therefore has a long service life and low cost. When the tag enters the read/write area, the tag starts to work once it receives an instruction sent by the RFID reader. The energy sent by the reader supports information exchange between the reader and the tag.

iii. A passive RFID tag does not have an embedded battery. This type of tag stays in passive state when it is out of the reading range of an RFID reader. When it enters the reading range of an RFID reader, the tag is powered by the radio energy emitted by the reader to work at a distance ranging from 10 cm to several meters. Passive RFID tags are the most widely used RFID tags as they are lightweight, small, and have a long service life.

d. Read/write attributes of RFID tags

RFID tags are classified into either read-only or read-write tags, depending on whether or not stored information can be modified. Information on a read-only RFID tag is written when its integrated circuit is produced; however, it cannot be modified afterward and can only be read by a dedicated device. In contrast, a read-write RFID tag writes the stored information into its internal storage area and can be rewritten by either a dedicated programming or writing device.

On a campus network, RFID-based applications are applied in asset management and device locating scenarios. Active RFID tags, with their read/write attributes, can be used to implement flexible, long-distance asset management. Passive RFID tags, however, are cost-effective and widely used in scenarios such as access control and attendance card swiping.

3. Bluetooth

First conceived by Ericsson in 1994, Bluetooth is now one of the most widely used wireless IoT technologies for exchanging voice and data over short distances. In 1998, Ericsson, Intel, Nokia, Toshiba, and IBM jointly established the Bluetooth Special Interest Group (SIG) to oversee the development of Bluetooth standards. The following describes the Bluetooth standard versions that have been released by the Bluetooth SIG since 2010:

a. Bluetooth 4.0: Released in July 2010, it includes Classic Bluetooth, Bluetooth High Speed, and Bluetooth Low Energy (BLE) technologies. The Bluetooth 4.0 standard supports more device types, and features improved battery life and energy saving capabilities.

b. BLE 5.0: Released in 2016, it provides faster transmission, longer coverage, and lower power consumption. Note that Bluetooth versions released after Bluetooth 4.0 are referred to as BLE.

Leveraging various cutting-edge technologies, such as frequency hopping spread spectrum (FHSS), time division multiple access (TDMA), and code division multiple access (CDMA), BLE 5.0 connects devices in a small-scale information transmission system. As shown in Figure 4.20, BLE

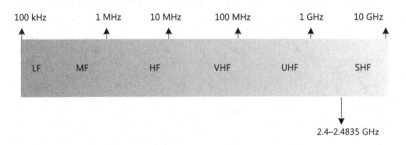

FIGURE 4.20   BLE's operating frequency band.

operates at the frequency band 2.4–2.4835 GHz and occupies 40 channels with the per capita bandwidth of 2 MHz. Among these 40 channels, there are 3 fixed broadcast channels and 37 frequency hopping data channels.

With the emergence of IoT devices such as intelligent wearable devices, smart household appliances, and Internet of Vehicles, Bluetooth, as a short-range communications technology, has become increasingly valued by developers. As such, a large number of Bluetooth products have emerged, such as Bluetooth headsets, Bluetooth speakers, smart bands, and smart appliances. The Bluetooth-based location solution has also been gradually promoted on campus networks. That is, users can implement location services by deploying Bluetooth tags or using the Bluetooth function on their smartphones or smart bands.

4. ZigBee

ZigBee is a short-range and low-energy wireless communication technology that is characterized by its short-distance transmission at a low speed, self-networking, low power consumption, and low cost. This technology is applicable to various devices in the automatic and remote control fields. The ZigBee Alliance periodically releases the ZigBee standard, whose PHY and MAC layer are designed based on the IEEE 802.15.4 standard. ZigBee enables vendors to develop stable, cost-effective, low-power, and wirelessly networked monitoring and control products based on globally open standards.

ZigBee wireless communication mainly operates on three frequency bands, namely, the 868 MHz frequency band in Europe, 915 MHz frequency band in the United States, and universal 2.4 GHz frequency band, as shown in Figure 4.21. These three bands provide

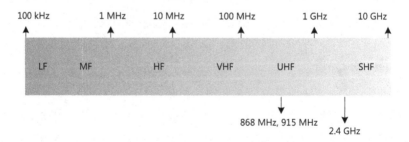

FIGURE 4.21   ZigBee's operating frequency bands.

1, 10, and 16 channels, with the per-channel bandwidths of 0.6, 2, and 5 MHz, respectively. ZigBee nodes leverage carrier sense multiple access with collision avoidance (CSMA/CA) technology to communicate with each other at the low data transmission rate of 250 kbit/s on the 2.4 GHz frequency band, 40 kbit/s on the 915 MHz frequency band, and 20 kbit/s on the 868 MHz frequency band.

ZigBee supports star, tree, and mesh modes, as shown in Figure 4.22. Three ZigBee node roles are available:

a. ZigBee Coordinator (ZC): responsible for building and managing the entire network. Once the network is built, the ZC also serves as a ZigBee Router (ZR).

b. ZR: provides routing information and determines whether or not to allow other devices to join the network.

c. ZigBee End Device (ZED): collects data.

ZigBee technology is used on campus networks in a variety of applications, such as energy efficiency management, intelligent lighting, and access control management. Based on its self-forming capabilities, ZigBee reduces gateway deployment and improves network reliability through inter-node communication. For these reasons, along with its low power consumption, ZigBee has become a widely used IoT solution.

5. Comparison between short-range wireless communication protocols
   Table 4.7 compares the parameters of the above mentioned short-range wireless communication protocols.

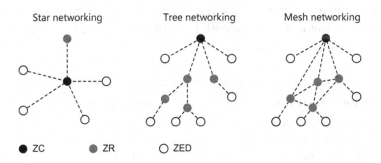

FIGURE 4.22   ZigBee network topologies.

TABLE 4.7 Comparison between Short-Range Wireless Communication Protocols

Wireless Technology Specifications	Wi-Fi	RFID	BLE	ZigBee
Operating frequency	2.4 and 5 GHz	120–134 kHz, 13.56 MHz, 433 MHz, 865–868 MHz, 902–928 MHz, 2.45 GHz, and 5.8 GHz	2.4–2.4835 GHz	868 MHz, 915 MHz, and 2.4 GHz
Transmission rate	<10 Gbit/s	≤2 Mbit/s	≤25 Mbit/s	≤250 kbit/s
Power consumption	≤15 W	≤1 W	≤100 mW	Hibernation: 1.5–3 μW Operating: 100 mW
Coverage	15 m	3–100 m	10 m	10 m
Networking	Point-to-point, star, tree, and mesh	Point-to-point and star	Point-to-point and star	Star, tree, and mesh

4.2.3 IoT Convergence Deployment Solution

When different wireless devices are deployed, the distance between them varies depending on the coverage they provide. If the different devices provide similar coverage, they can be deployed on the same base station where various wireless communication components or chips are integrated. Table 4.8 lists the deployment spacings supported by various wireless base stations. Wi-Fi and RFID base stations support a similar spacing, which is larger than that supported by Bluetooth and ZigBee base stations. In an ideal environment (with few obstacles and low penetration loss), Wi-Fi and RFID base stations can be deployed in the same location. In nonideal environments (with severe blocking and high penetration loss), we need to deploy more Bluetooth or ZigBee base stations in addition to Wi-Fi and RFID base stations.

TABLE 4.8 Deployment Spacing between Different Wireless Base Stations

Wireless Base Station	Deployment Spacing (m)
Wi-Fi base station (AP)	10–15
Bluetooth base station	8
RFID base station	10–15
ZigBee base station	10

1. IoT AP-based converged access

According to base station deployment analysis, APs on a campus network can integrate multiple wireless communication technologies for IoT terminal access, thereby enabling IoT and Wi-Fi access from the same base station. As a result, the IoT AP-based deployment solution can converge the IoT and campus networks, implementing unified cabling and power supply, and offering an effective option in terms of investment cost and installation convenience.

IoT APs integrate other wireless communication technologies through built-in integration or external expansion cards. Specifically, IoT APs have Wi-Fi and other radio modules integrated, or provide external interfaces such as mini PCIe or USB interfaces, all of which are used to accommodate IoT modules.

In terms of built-in integration, IoT APs offer both Wi-Fi and IoT functionality using a core chip, while independent IoT chips can also be utilized to deliver IoT functionality. The IoT module and AP system are deeply coupled, and devices are shipped with the ability to connect to IoT hardware without requiring additional IoT cards. The preinstalled system on IoT APs enables embedded IoT processing software to upload collected data to the IoT application system through standard interfaces. In addition, third-party software can be installed in container mode, which can then collect data and communicate with the application system through the AP.

Thanks to the built-in integration solution, APs offer IoT hardware with radio functionality and are therefore not dependent on external chips. The APs have IoT functionality integrated through customized software, which is both cost-effective and flexible. Now, more and more APs in the industry have built-in BLE modules.

IoT APs achieve protocol communication with IoT cards through mini PCIe or USB interfaces, as shown in Figure 4.23. The IoT cards provide access to IoT devices, collect data from them, and deliver this data to the IoT application system through the APs, which can then exchange data with IoT cards through serial or Ethernet interfaces. While this mode is a loose coupling solution for APs, third-party vendors provide IoT cards and corresponding IoT module drivers. The APs are responsible for power supply, simple configuration, card status maintenance, and data forwarding for the IoT cards.

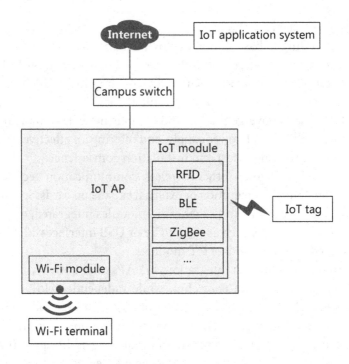

FIGURE 4.23   IoT AP with a built-in IoT module.

During the initial phase of IoT convergence access, APs do not have built-in IoT chips, and instead rely on external expansion cards to implement fast network convergence. Third-party vendors can offer mature IoT modular devices and drivers that can be easily integrated into APs, enabling network vendors to quickly build IoT solutions through cooperative win-win development.

2. IoT AP design optimization

The operating frequency band of Wi-Fi signals is similar to that of other short-range wireless communication signals, which will lead to interference. As such, optimization is required for IoT APs to reduce mutual interference between Wi-Fi and IoT radio signals (Figure 4.24).

a. Active channel avoidance

Wi-Fi signals are transmitted on unlicensed frequency bands, which may also be used by IoT device radio signals. When an IoT device and the connected AP operate on the same frequency band, mutual interference occurs. To avoid such a scenario,

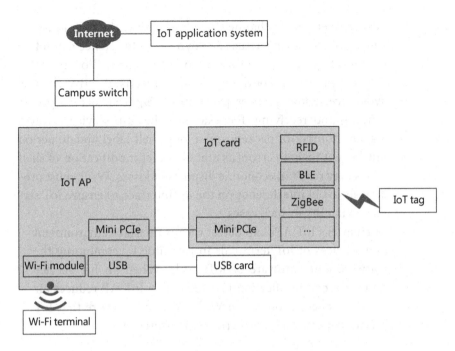

FIGURE 4.24   Communication between an IoT AP and an IoT card.

consider using different operating channels for the IoT device and the AP.

IoT devices may communicate with each other over a mesh network when carrying out certain services, such as electronic shelf label (ESL) and energy efficiency management. To properly support these services, the IoT system can no longer change channels randomly to prevent such issues as service interruptions and terminals failing to go online. To address these issues, we can configure APs to actively avoid the operating channels of IoT devices when setting or switching channels.

Huawei's IoT APs communicate with IoT devices in real time to detect their center frequency and bandwidth. The APs then mark the detected frequency channel as an IoT channel, and will proactively avoid the channel during subsequent calibration and Wi-Fi channel planning in order to prevent conflicts.

b. Interference avoidance

The 2.4 GHz frequency band supports a small frequency bandwidth, and only channels 1, 6, and 11 do not overlap with one another. Other channels may interfere with one another to some extent. During actual deployment, interference cannot be completely avoided, even if the operating channel of the AP is staggered from that of the IoT card by means of active channel avoidance. Some services place strict requirements on packet sending and receiving. For example, the ESL service involves a small number of packets and a long shelf label update period but cannot tolerate errors as any incorrect modification of shelf labels may result in economic disputes or losses. Wi-Fi must proactively avoid interference on the air interface to ensure IoT service reliability in such cases.

Huawei's IoT APs can detect the air interface environment and services of IoT terminals in real time by combining the software and hardware of IoT cards. When an IoT service requiring communication is detected, an IoT AP reduces air interface occupation of the Wi-Fi service to ensure that IoT service packets do not experience interference.

4.2.4 High-Power and Long-Distance PoE Power Supply

Various types of terminals are connected on a campus network, and managing their power supply is a major challenge. Terminals such as IP phones, cameras, and data collectors all require DC power supply, and such devices are usually installed in corridors or on high ceilings where power sockets are unavailable. On many large-scale LANs, administrators must deal with the complicated task of managing multiple terminals and access devices that require unified power supply and management.

PoE, a wired Ethernet power supply technology that is widely applied on campus networks, can address this problem. It provides DC power for IP-based terminals while transmitting data signals and, when compared with traditional power supply modes, offers the following advantages:

- Low-cost: The power cabling cost is significantly reduced, and it is easy to lay out power cables.

- Reliable: Multiple powered devices (PDs) are powered in a unified manner, facilitating power backup.

- Easy to deploy: Network terminals can be powered over Ethernet cables, without the need of external power sources.

- Standard-compliant: PoE complies with IEEE 802.3af and IEEE 802.3at, and all PoE devices use uniform power interfaces and can be connected to PDs of different vendors.

Figure 4.25 shows a typical PoE system, which consists of power sourcing equipment (PSE), PDs, and PoE module (built in a PoE switch).

- PSE: refers to a PoE device that supplies power to PDs through Ethernet. The PSE also provides functions such as detection, analysis, and intelligent power management.

- PD: refers to a powered device, such as AP, portable device charger, card reader, and IP camera. PDs are classified into standard and nonstandard PDs, depending on whether they conform to IEEE standards.

- PoE module: provides power to a PoE system. The number of PDs connected to a PSE is limited by the power output of a PoE power module. PoE power modules are classified into built-in and external power modules, depending on whether they are pluggable.

1. PoE power supply principle

   The following uses Huawei S series switches as an example to describe the PoE power supply principle:

   Step 1 PD detection: A PSE periodically transmits a low voltage with limited current through its ports to detect PDs (at 2.7–10.1 V,

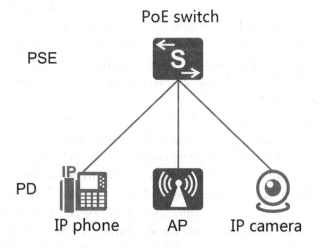

FIGURE 4.25   Components in a PoE system.

a detection period of 2 seconds). If a resistor with a specific resistance (19–26.5 kΩ) is detected, the PSE considers the cable terminal to be connected to a PD.

Step 2 Power supply capability negotiation: The PSE classifies PDs and negotiates the power supply capability by resolving detected resistors or using the Link Layer Discovery Protocol (LLDP).

Step 3 Power supply startup: During the startup period (generally within 15 μs), the PSE gradually increases the voltage to 48 V DC in order to supply power to PDs.

Step 4 Normal power supply: When the voltage reaches 48 V, the PSE provides stable and reliable 48 V DC power for PDs. The power of a PD cannot exceed the maximum output power of the PSE.

Step 5 Power supply disconnection: The PSE constantly detects the input current of PDs while providing power to them. If the current of a PD falls below the minimum value or increases sharply, the PSE stops providing power to this PD and repeats PD detection. This situation occurs when a PD is disconnected from the PSE, encounters a power overload or short circuit, or its power consumption exceeds the power supply capacity of the PSE.

2. PoE standards compliance

IEEE 802.3af was the earliest standard for PoE power supply and effectively provides centralized power supply for terminals such as IP phones, APs, portable chargers, card readers, and cameras. Subsequently, the IEEE 802.3at standard further proposes PoE+, which can supply power to devices installed with high-power applications such as dual-band access, videotelephony, and pan-tilt-zoom video surveillance.

Today, more advanced service types and terminals continue to emerge, requiring even higher PoE input power. To satisfy such demands, Huawei has actively engaged in the formulation of the IEEE 802.3bt standard (also known as PoE++). In compliance with IEEE 802.3bt (draft), Huawei has developed and launched many PoE++ switches capable of providing up to 60 W power. In addition, Huawei has unveiled the next-generation Universal Power Over Ethernet Plus (UPoE+) switches based on the new IEEE 802.3bt standard. The

UPoE+ function provides up to 90 W power, meeting the needs of more terminals. Table 4.9 provides detailed performance parameters of PoE, PoE+, PoE++, and UPoE+.

3. Perpetual PoE and fast PoE

Perpetual PoE technology delivers an uninterruptible power supply to PDs when a PoE device is rebooted or its software is upgraded. This technology ensures that PDs are not powered off during the reboot of the PoE device, eliminating any interruptions that may be triggered by a power failure.

When a PoE device reboots due to a power failure, it continues to supply power to the PDs immediately after being powered on without waiting for the reboot to complete. Compared with common PoE switches that typically require up to 3 minutes to begin supplying power to PDs, Huawei switches are capable of supplying power within 10 seconds of being rebooted, greatly reducing the service interruption time caused by power supply interruption.

4. Long-distance PoE power supply

Typically, the distance between a PoE device and a PD can be anywhere up to 100 m. Following the proliferation of wireless terminals, APs are now deployed across various scenarios, including outdoor spaces (such as campus playgrounds) that are inconvenient for

TABLE 4.9   Performance Parameters of PoE, PoE+, PoE++, and UPoE+

Category	PoE	PoE+	PoE++	UPoE+
Standard	IEEE 802.3af	IEEE 802.3at	IEEE 802.3bt (draft)	IEEE 802.3bt (draft)
Power supply distance (m)	100	100	100	100
Maximum current (mA)	350	720	720	960
PSE output voltage (V DC)	44–57	50–57	50–57	50–57
PSE output power (W)	≤15.4	≤30	≤60	≤90
PD input voltage (V DC)	36–57	42.5–57	42.5–57	42.5–57
Maximum input power available at PDs (W)	12.95	25.5	54	81.6

cabling, and where power supply problems become more apparent. PoE switches that provide PoE power supply are generally deployed in extra-low voltage rooms. As a result, in wireless scenarios, the PoE power supply distance must be increased without compromising the AP uplink bandwidth.

The signal-to-noise ratio (SNR) is the most important indicator in electrical interface transmission. To support a longer transmission distance, we must reduce the SNR loss of the entire link. Multiple types of components and media exist on the transmission link of electrical interfaces, including physical chips, cards, interface connectors, and network cables on both ends. Based on these components and media, SNR parameters can be optimized to enable multi-GE interfaces of Huawei PoE devices in order to support a maximum transmission distance of 200 m when connected to specific APs. Optimization methods include:

a. Huawei switches and APs have built-in customized physical chips that support a maximum transmission distance of 200 m. In addition, Huawei-developed algorithms are used to improve the driver software, ensuring that SNR parameters can be optimized for long-distance transmission.

b. Huawei uses high-quality connectors and STPs to reduce SNR loss.

4.3 BUILDING AN ULTRA-BROADBAND PHYSICAL NETWORK ARCHITECTURE

The physical network architecture is the foundation of a campus network. As such, any changes to the physical network architecture pose grave risks to networks, as well as incurring huge costs. Most campus networks have devices distributed across different buildings or floors, making it very difficult to lay and adjust cables between these devices. To overcome such challenges, it is paramount that we plan a proper physical network architecture before network construction. To this end, this section describes how to build ultra-broadband wired and wireless networks.

4.3.1 Building an Ultra-Broadband Wired Network

A physical campus network typically uses a tree topology, which is a hierarchical and modular networking architecture. Such an architecture is relatively stable and easy to expand and maintain. Network reliability can be further improved by planning a topology with device and port redundancy. Of these, device redundancy can be achieved by stacking, which is typically used to virtualize multiple switches into a single logical switch. In this way, management is simplified, and the forwarding, control, and management planes of the switches are unified. To provide port redundancy, port aggregation is typically used. With this technology enabled, multiple physical ports are bundled into a logical port, with these physical ports working in load balancing mode. In this way, link utilization is increased while ensuring reliability. What's more, as technologies such as Software-Defined Networking (SDN) develop, campus networks are gradually evolving from traditional three-layer tree networking toward simplified two-layer tree networking. The following is an illustration of the two types of tree networking.

1. Traditional three-layer tree networking architecture

   This networking architecture consists of three layers: core layer, aggregation layer, and access layer.

   Of these layers, the core layer forms the backbone of a campus network. As such, it is responsible for campus data exchange and provides high-speed interconnectivity for various parts of the campus network, such as the DC, aggregation layer, and egress zone. As this layer is so critical to the overall performance of the network, high-performance core switches need to be deployed here, thereby achieving high bandwidth utilization and fast network convergence when faults occur.

   The aggregation layer connects the access layer and the core layer of a campus network. At this layer, aggregation devices forward east-west traffic between users and north-south traffic to the core layer. The aggregation layer can also function as the switching core for a department or zone and connect to dedicated servers for that department or zone.

   The access layer is the first network layer to which terminals connect. As such, it needs to provide a wide range of access modes for users. This layer is usually composed of access switches, with Layer 2 switches being the most common. What's more, these access switches tend to be large in number and sparsely distributed throughout the network.

   Figure 4.26 shows a typical three-layer tree networking architecture. Such an architecture shortens the optical fiber cabling distance

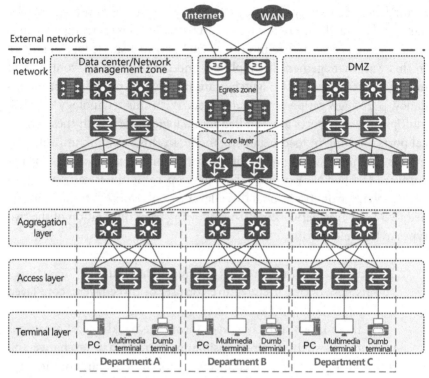

DMZ refers to demilitarized zone, which is a semi-trusted zone.

FIGURE 4.26 Traditional three-layer tree networking architecture.

from the core layer to the access layer, supports a large network scale, and is relatively easy to reconstruct. However, the downside is that it requires more optical modules and devices, leading to high networking cost.

2. Simplified two-layer tree networking architecture

This networking architecture consists of only the core layer and access layer, which are directly connected, as shown in Figure 4.27.

The simplified two-layer tree networking architecture has the following advantages:

a. Low network deployment cost

In this architecture, there is no aggregation layer, meaning that fewer devices and optical modules are required and the overall

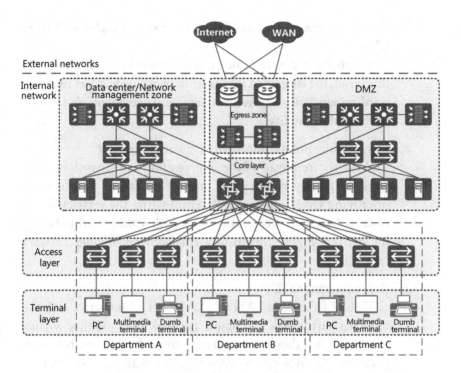

FIGURE 4.27   Simplified two-layer tree networking architecture.

network deployment cost is decreased. The following makes a comparison between three-layer networking and two-layer networking, assuming the campus network is Wi-Fi 6-enabled and has 10 access switches, each delivering 25GE uplinks: If two aggregation switches are deployed in three-layer networking, with access switches connected to the core layer through 100GE uplinks, we need ten pairs of 25GE optical modules (used for connecting access switches to aggregation switches through Eth-Trunks) and two pairs of 100GE optical modules (used for connecting aggregation switches to core switches through Eth-Trunks). If we use two-layer networking, however, only ten pairs of 25GE optical modules are required to connect access switches to ten pairs of 25GE downlink ports of core switches. With two-layer networking, there is no need for 100GE optical modules and aggregation switches, significantly lowering the network deployment cost.

b. Flattened networking, making the network simpler and more efficient

Simplicity is the trend of campus networks. In keeping with this trend, the simplified two-layer networking features high forwarding efficiency, convenient service deployment, and strong horizontal scalability.

c. Fewer failure points, improving network reliability

With the removal of the aggregation layer, two-layer networking greatly reduces the number of failure points while improving O&M efficiency. By eliminating the restrictions of intermediate oversubscription ratios, this simplified networking delivers higher forwarding efficiency and higher network reliability.

d. Better catering to SDN

Virtual Extensible LAN (VXLAN) is an innovative technology that virtualizes a physical network into multiple virtual networks. This then allows services to be flexibly controlled. In addition, VXLAN can be used together with an SDN controller to provide a wide range of network functions. SDN requires a simplified physical network to facilitate automated service orchestration and intelligent O&M on the SDN controller, as well as a network with strong plug-and-play capabilities. Such requirements are fully met by two-layer networking.

In short, such advantages of a simplified two-layer tree networking architecture make it a better choice for newly constructed networks.

3. Deployment suggestions for ultra-broadband wired networks

Figures 4.28 and 4.29 show the recommended tree networking architectures for campus networks.

The typical downlink rates of access switches are 2.5 and 5 Gbit/s. Regardless of whether the network is being newly constructed or expanded, access switches with multi-GE ports (1GE, 2.5GE, 5GE, and 10GE) are recommended. Category 6 or higher twisted pair cables are used for connections of these access switches. Category 5 twisted pair cables used on legacy networks can support a maximum transmission rate of only 5 Gbit/s.

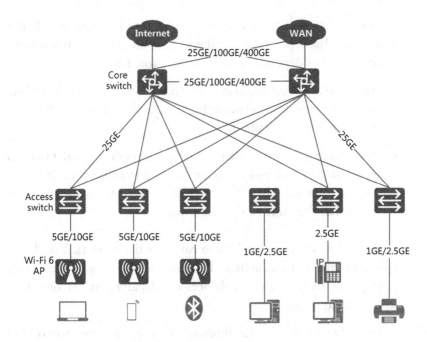

FIGURE 4.28 Recommended simplified two-layer networking.

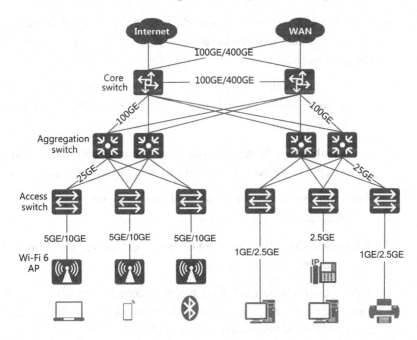

FIGURE 4.29 Recommended traditional three-layer networking.

The mainstream transmission rates of core switches are 25 and 100 Gbit/s. The type of optical fibers recommended for connections of core switches depends on the transmission distance:

a. Long transmission distance: Use single-mode 25GE optical fibers for interconnection between the access and core layers, which are far apart from each other.

b. Short transmission distance: Use multimode optical fibers to connect core switches to each other and to connect them to the egress zone and the DC, which have devices that are usually located in the same equipment room.

The following describes how to calculate the network model. For this calculation, the simplified two-layer networking is used as an example, in which two core devices are deployed in 1+1 dual-link load balancing mode.

a. Number of APs=total number of users on the network/30 (assuming that each AP connects to 30 users on average.)

b. Number of ports on access devices: One third of the downlink ports on access devices need to be reserved for future capacity expansion; therefore:

   i. Number of downlink ports=(total number of users on the network/average number of access users associated with an AP)×(1+1/3)

   ii. Number of uplink ports=number of downlink ports×downlink port rate/access oversubscription ratio/uplink port rate

c. Total number of downlink ports at the core layer=total number of uplink ports at the access layer

d. Number of downlink ports on each core device = total number of downlink ports at the core layer/2

e. Interconnection bandwidth required by core devices=bandwidth required for interconnection with the network egress+bandwidth required for interconnection with a DC

Table 4.10 lists the device port models calculated for different user scales based on the preceding formulas.

4.3.2 Building an Ultra-Broadband Wireless Network

With every new generation of Wi-Fi, the data rate of the air interface on a wireless network greatly increases. For example, with Wi-Fi 6, the transmission rate of a single AP increases to nearly 10 Gbit/s. Such APs can be used to deploy an ultra-broadband wireless network, which can be built using either of the following networking architectures:

- Wireless access controller (WAC)+Fit AP: also known as centralized networking. This architecture reduces AP management and maintenance costs and implements large-scale wireless network deployment.

- Wired and wireless convergence: also known as the integrated WAC networking. This architecture improves management, forwarding, and maintenance efficiency and makes service policies more flexible.

1. WAC+Fit AP networking architecture

In this networking architecture, a WAC centrally manages Fit APs through the Control And Provisioning of Wireless Access Points (CAPWAP) protocol, achieving unified wireless service

TABLE 4.10   Device port models for different user scales

User Scale (Indicated by N)	Access-Layer Port Model		Port Model of a Single Core Switch (1+1 Protection)	
$N \leq 10000$	Downlink: 445×multi-GE ports	Uplink: 30×25GE ports	Downlink: 15×25GE ports	Uplink: 5×25GE ports
$10000 < N \leq 50000$	Downlink: 2223×multi-GE ports	Uplink: 149×25GE ports	Downlink: 75×25GE ports	Uplink: 6×100GE ports
$50000 < N \leq 100000$	Downlink: 4445×multi-GE ports	Uplink: 297×25GE ports	Downlink: 149×25GE ports	Uplink: 12×100GE ports
$100000 < N \leq 400000$	Downlink: 17778×multi-GE ports	Uplink: 1185×25GE ports	Downlink: 593×25GE ports	Uplink: 45×100GE ports

management. Fit APs provide radio signals for STAs to access a wireless network, and provide almost no management or control capabilities at all.

The WAC+Fit AP networking architecture is applicable to large- and medium-sized campus networks. On such a network, a WAC can be deployed at either the aggregation or core layer, depending on the wireless network capacity and locations of Fit APs at the access layer. For reliability purposes, deploying the WAC at the core layer is considered best practice. The WAC is responsible for service control such as configuration delivery and upgrade management for all Fit APs, which are also plug-and-play, greatly reducing WLAN management, control, and maintenance costs.

WACs provide the same features in both in-path and off-path deployments. Figure 4.30 shows the latter deployment mode for

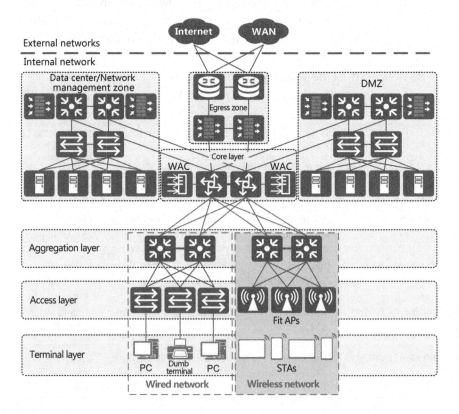

FIGURE 4.30 WAC+Fit AP networking architecture with off-path WACs.

WACs connecting to core switches. WACs can also connect in off-path mode to aggregation switches.

In off-path networking, WACs manage APs over CAPWAP tunnels, whereas data flows can either travel through the WACs over CAPWAP tunnels or bypass the WACs by being directly forwarded by core switches to the upper-layer network. This networking mode facilitates the deployment of new WACs on live networks. These WACs are connected to idle ports on core or aggregation switches, without affecting the original physical connections or services. In addition, the off-path networking applies when we need to centrally deploy WACs to manage sparsely distributed APs that are within the management scope of switches connected to the WACs.

Figure 4.31 shows the WAC in-path deployment, where WACs are located in the data forwarding path between downstream APs and upstream devices such as core switches or egress routers.

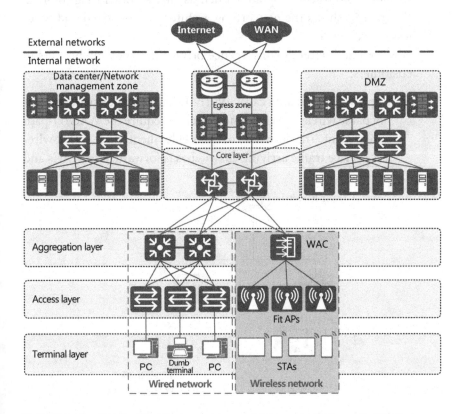

FIGURE 4.31 WAC+Fit AP networking architecture with in-path WACs.

In the in-path networking mode, WACs also act as switches to process all data packets, facilitating centralized management of user data. However, if a large amount of user data is involved in this networking mode, the data processing capability of the WACs will be affected. Therefore, it is considered best practice to deploy WACs at or below the core layer. Such characteristics make in-path networking applicable to small- and medium-sized campus networks where APs are densely deployed. Due to restrictions of the in-path deployment locations, off-path networking is widely used on live networks.

2. Wired and wireless convergence networking architecture

Huawei's agile switches are able to implement wired and wireless convergence. That is, in addition to processing wired packets, the agile switches can identify and process CAPWAP packets. They can centrally manage wired and wireless service traffic in a wired and wireless convergence topology, such as that shown in Figure 4.32. Figure 4.33 shows the networking architecture for wired and wireless convergence.

Wired and wireless convergence brings the following benefits to customers:

a. Improved forwarding capacity: Traditional campus switches cannot parse wireless packets and therefore require the WAC off-path deployment. The downside of this deployment mode is that wireless service traffic arriving at switches is forwarded to WACs and

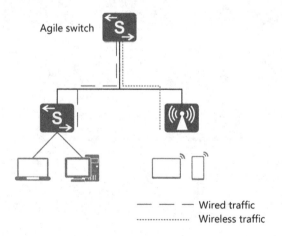

FIGURE 4.32   Wired and wireless convergence solution.

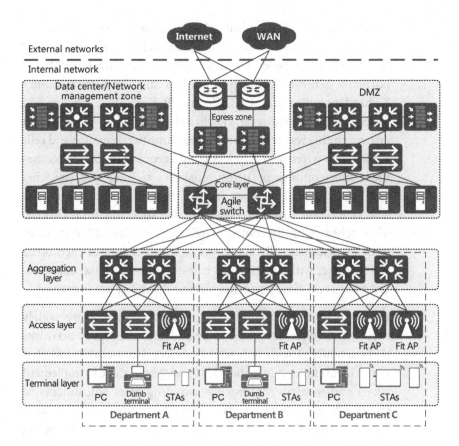

FIGURE 4.33   Wired and wireless convergence networking architecture.

then forwarded back to the switches, causing a longer delay. In addition, the overall forwarding capacity for wireless service traffic depends on the forwarding performance of WACs. With wired and wireless convergence technology, Ethernet Network Processor (ENP) cards on agile switches are able to decapsulate and forward wireless packets together with wired packets. This simplifies the forwarding path and eliminates forwarding bottlenecks.

b. Unified management of devices and user policies: The management plane of WACs is independent from that of switches, complicating network management and maintenance. Wired and wireless convergence technology centralizes management and control points on an agile switch, allowing for unified management of wired and wireless users.

c. High reliability: In a traditional solution where two standalone WACs work in 1+1 backup mode, an additional channel must be established between them for data synchronization. This is normally achieved using technologies such as Virtual Router Redundancy Protocol (VRRP) and Bidirectional Forwarding Detection (BFD). However, synchronizing data between different devices in this way compromises real-time performance and reliability. The wired and wireless convergence solution overcomes these shortcomings by using reliability technologies (stacking and Eth-Trunk) of switches to implement device-level and link-level redundancy. In this way, the main processing units (MPUs) of switches centrally control wireless data, while ENP cards of switches automatically synchronize wireless data to each other in real time, without the need to establish additional channels. Therefore, real-time performance and reliability are improved.

d. Flexible capacity expansion: As wireless terminals are widely used, wireless services need to be added to the legacy networks that currently provide only wired services. Huawei agile switches are an ideal choice due to their ENP cards being able to implement wired and wireless convergence, avoiding great adjustments or changes to the physical network. Additionally, the ENP cards can be easily expanded to meet requirements of the increasing wireless users and wireless service volume.

e. Fast wireless roaming: When a switch with ENP cards functions as a WAC, APs connected to its different ENP cards are centrally managed and controlled by the switch MPUs. So, when STAs roam between these APs, they actually roam on the same WAC. In contrast, if STAs roam between APs connected to different standalone WACs, they roam between WACs. In this way, the integrated WAC solution provides faster wireless roaming and shorter forwarding paths.

3. Deployment suggestions for ultra-broadband wireless networks

The air interface technology is critical to ultra-broadband wireless networks. And next-generation Wi-Fi 6 APs can be employed to improve wireless coverage and throughput capabilities, so as to implement ultra-broadband forwarding. For example, a Wi-Fi 6 AP can provide a maximum rate of 9.6 Gbit/s on a single 5 GHz radio. When higher bandwidth is required, APs with dual 5 GHz radios can be deployed, with rates of up to 19.2 Gbit/s delivered by a single AP. The APs use the Eth-Trunk technology to bundle two 10GE wired uplinks into a 20GE uplink. Additionally, 10GE access switches are deployed to implement ultra-broadband networking of the wireless access layer. Different AP models are better suited to different scenarios, for example:

a. In high-density stadiums, APs with directional antennas are preferred, as they provide centralized coverage and reduce interference.

b. In classrooms, APs with triple radios are preferred, as they provide higher access capacity.

c. In hotels, agile distributed APs are preferred. An independent remote unit is deployed in each room, as they provide ubiquitous coverage and avoid mutual interference.

On a campus wireless network, ultra-broadband forwarding must be achieved by not only a single AP but also the overall wireless network. For example, AP position planning and intelligent radio calibration need to be performed to maximize the coverage while minimizing interference.

For wireless networking, the wired and wireless convergence architecture is the best choice, with core switches implementing integrated WAC functions. This prevents traffic from being forwarded to standalone WACs. The advantage of standalone WAC networking is that it features a strong control capability. However, it delivers a relatively low forwarding capability (only 40 or 100 Gbit/s). Therefore, the standalone WAC networking is applicable to scenarios that do not require high wireless network forwarding capabilities. However, there is a growing demand for network bandwidth as the Wi-Fi 6 standard is popularized, and bandwidth-hungry services, such as campus augmented reality, VR, and 8K high definition video, are widely being put into commercial use. Huawei core switches integrated with WAC functions are suitable for future ultra-broadband campus networks due to their wireless management capabilities and ultra-broadband forwarding capacities of 50 Tbit/s or higher.

CHAPTER 5

Building Virtual Networks for an Intent-Driven Campus Network

O N AN INTENT-DRIVEN CAMPUS network, independent virtual networks (VNs) carry services. In this way, users' service requirements are decoupled from the physical network. Each VN can be separately deployed and managed, thereby simplifying network management. VNs can enable the pooling of network resources, which then can be flexibly allocated to users to improve resource utilization efficiency. More importantly, a VN has a standard and fixed topology, based on which programming can be easily applied during application deployment, without considering varying physical network topologies. This significantly simplifies network automation.

5.1 INTRODUCTION TO NETWORK VIRTUALIZATION TECHNOLOGIES

Commonly used traditional network virtualization technologies include virtual local area network (VLAN), virtual private network (VPN), stacking, clustering, and Super Virtual Fabric (SVF). Among these technologies, stacking, clustering, and SVF all virtualize multiple physical switches into one logical switch, integrating the control plane and achieving unified management. Strictly speaking, they are device-level virtualization technologies and cannot be used as independent protocols on campus networks. VLAN and VPN technologies, on the other hand, cannot meet the network virtualization requirements of intent-driven campus networks.

5.1.1 VLAN and VPN

1. VLAN

On traditional LANs, virtualization is used mainly for isolation between different organizations. The most well-known virtualization technology is VLAN, which divides users on a physical network into multiple VNs. This promotes convenient and flexible network construction and maintenance, because users on the same VLAN can be physically distant from each other. VLANs effectively limit the scope of Layer 2 bridge domains (BDs) and therefore enhance LAN security.

Figure 5.1 shows the application of VLANs. Switch A and its directly connected PC 1 and PC 3 are located on the tenth floor of an office building; and Switch B and its directly connected PC 2 and PC 4 are located on the 11th floor of the office building. PC 1 and PC 2 belong to company M, and PC 3 and PC 4 belong to company N. The PCs of different companies can be added to different VLANs to only allow intracompany communication. In this example, PC 1 and PC 2 are added to VLAN 2, and PC 3 and PC 4 are added to VLAN 3.

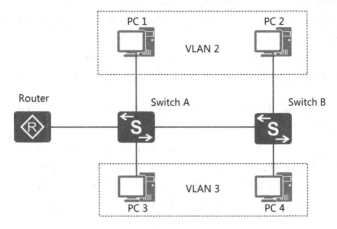

FIGURE 5.1   Application of VLANs.

VLAN is a Layer 2 network virtualization technology that cannot be used to build cross-region VNs. When VLANs are used to build VNs, VLAN information needs to be added on each network device one by one, resulting in tight coupling between services and the physical topology. VLAN configuration and maintenance on a large-scale network is complex due to the wide variety of services and large number of terminals. Therefore, VLANs are more suited to small-scale and simple networks.

2. VPN

VPN technology is not a protocol or standard; rather, it is a technology that extends a private network across a wide area network (WAN) and enables users to communicate across the WAN. A VPN is an isolated VN built on a physical network and can be considered a dedicated channel used exclusively by specified users. Security settings and network management can be performed separately for these channels.

VPNs use various tunneling technologies to encapsulate packets in tunnels and transparently transmit them through dedicated channels established over public networks. A tunneling technology uses one protocol to encapsulate the packets of another protocol. Packets of an encapsulation protocol can also be encapsulated or carried by another encapsulation protocol.

Based on the tunneling protocol's layer, VPN technologies are classified as follows:

a. Layer 2 VPN technologies: Point-to-Point Tunneling Protocol (PPTP) and Layer 2 Tunneling Protocol (L2TP), among others.

b. Layer 3 VPN technologies: Multiprotocol Label Switching (MPLS, most widely used), Internet Protocol Security (IPsec), Secure Sockets Layer (SSL), and Generic Routing Encapsulation (GRE), among others.

On a campus network, VPNs can be used to create isolated VNs for independent service resources, and implement cross-WAN service access within a VN. As shown in Figure 5.2, a large enterprise uses VPNs to build R&D and non-R&D networks over the same physical network. In such cases, the two networks are isolated,

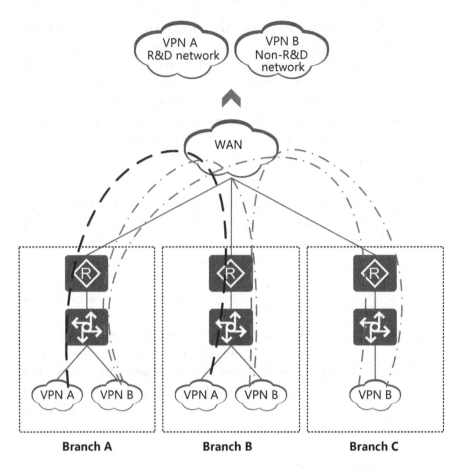

FIGURE 5.2 Application of VPNs.

and resources in different branches on the respective network can access each other through the network's dedicated channels across the WAN.

However, deploying and maintaining VPNs requires meticulous planning. When provisioning a new service, network engineers usually configure the VLAN and VPN parameters on each device individually, which is both time-consuming and error-prone. Additionally, it is difficult to locate incorrect configurations on VPNs. What's more, Internet-based VPNs cannot be directly managed by enterprises, and their normal running relies on Internet service providers (ISPs).

5.1.2 Outstanding NVO3 Technology — VXLAN

VLAN and VPN used on traditional campus networks are hardware-centric technologies that tightly couple services with physical networks. In contrast, Network Virtualization over Layer 3 (NVO3) technology — an overlay network technology originally proposed by IT vendors — is independent of the traditional physical network architecture.

NVO3 is a network virtualization technology developed amid cloud computing that was first proposed for data center network virtualization to support large-scale tenant networks. NVO3 builds an overlay network through tunnels over an IP-based Layer 3 underlay network. An ingress Network Virtualization Edge (NVE) node of a tunnel encapsulates the original packets sent by terminals with the egress NVE node information and the destination address. The egress NVE node of the tunnel decapsulates the received packets to obtain the original packets and forwards the original packets to the destination end users. Regarding the transmission over the tunnel, IP devices (such as routers and switches) forward the encapsulated packets on the transport network based on routing tables. In this sense, NVO3 is similar to a traditional Layer 3 tunneling technology.

NVO3 adds a logical network that is independent of the physical network to the traditional IP network. Physical devices are unaware of this logical network that uses the same IP forwarding mechanism as an IP network. This significantly reduces the threshold to use NVO3, making it to become widely used on data center networks over the course of just a few years.

On a campus network, multiple isolated service networks are deployed and managed independently, thereby making network deployment and maintenance very difficult. To address this issue, NVO3 can be used to pool physical resources on a campus network and create multiple VPNs on the physical network to reduce customers' investment and management costs.

The Internet Engineering Task Force (IETF) proposes three NVO3 solutions: Virtual Extensible LAN (VXLAN), Network Virtualization using Generic Routing Encapsulation (NVGRE), and Stateless Transport Tunneling (STT). These solutions use different MAC-in-IP encapsulation technologies to build VNs over an IP network:

- VXLAN uses Media Access Control (MAC)-in-UDP encapsulation. That is, Layer 2 packets are encapsulated into User Datagram Protocol (UDP) packets.

- NVGRE uses MAC-in-GRE encapsulation. That is, Layer 2 packets are encapsulated into GRE packets.

- STT uses MAC-in-Transmission Control Protocol (TCP) encapsulation. That is, Layer 2 packets are encapsulated into TCP packets.

Table 5.1 compares the mainstream NVO3 solutions.

Among the mainstream NVO3 solutions, VXLAN is the most popular. It has the following advantages over NVGRE and STT:

- VXLAN does not require the existing network to be reconstructed, whereas NVGRE requires that network devices support GRE.

TABLE 5.1   Comparison of Mainstream NVO3 Solutions

Solution	Encapsulation Method	Technical Implementation	Leading Vendor
VXLAN	MAC in UDP	VXLAN encapsulates Ethernet packets into UDP packets for tunnel transmission. The UDP destination port number is a known port number VXLAN performs load balancing based on the source port number in the outer UDP header. This load balancing mode relies on the standard 5-tuple information, making it easy to implement on an IP network	VMware, Cisco, Arista, Broadcom, Citrix, Red Hat, and Huawei
NVGRE	MAC in GRE	NVGRE complies with the GRE standards defined in RFC 2784 and RFC 2890, and encapsulates Ethernet packets into GRE packets for tunnel transmission. The main difference between NVGRE and VXLAN is that NVGRE uses the GRE extension field **FlowID** to load balance traffic. This mode requires the physical network to identify the GRE extension information	Microsoft, Arista, Intel, Dell, Hewlett-Packard, and Broadcom
STT	MAC in TCP	STT encapsulates Ethernet packets into TCP packets for tunnel transmission. It uses stateless TCP, which differs greatly from VXLAN and NVGRE	Nicira

- VXLAN uses the standard UDP protocol to transmit traffic without modifying the transport layer. In contrast, STT requires the traditional TCP to be modified.

- Unlike NVGRE and STT, VXLAN is supported by most commercial network chips.

For these reasons, VXLAN is ideal for virtualization of campus networks. VXLAN's technical details will be expanded upon in the following sections.

5.2 ARCHITECTURE OF VNS ON AN INTENT-DRIVEN CAMPUS NETWORK

On an intent-driven campus network, services need to be decoupled from the network. In this way, a multipurpose network is achieved and fast, flexible service deployments are implemented without changing the basic network. As such, a new VN architecture, which differs from the traditional VN architecture, is proposed. This section describes VN architecture built using VXLAN on an intent-driven campus network.

5.2.1 VN Architecture

VXLAN-based VNs can decouple service networks from the physical network, irrespective of that network's complexity. When service networks need to be adjusted, the physical network topology does not need to be changed. As shown in Figure 5.3, the VN architecture of an intent-driven campus network features two layers, consisting of underlay and overlay networks.

The underlay network is the physical infrastructure consisting of various physical devices, such as access switches, aggregation switches, core switches, routers, and firewalls.

The overlay network is completely decoupled from the underlay network. It is a fully connected logical fabric topology built on top of the physical topology using VXLAN technology. In the logical fabric topology, resources such as user IP addresses, VLANs, and access points are pooled in a unified manner and allocated to VNs on demand. Through the logical fabric topology, users can create multiple VN instances based on service requirements to achieve a multi-purpose network, service isolation, and fast service deployment.

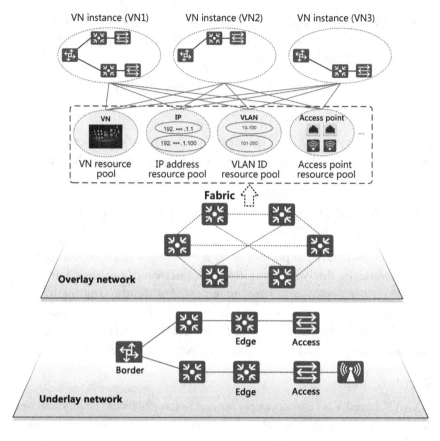

FIGURE 5.3   VN architecture on a campus network.

1. Fabric: an overlay network resource pool abstracted from the underlay network, including the following resources:

   a. VN resource pool, which mainly includes VNs that can be created on the overlay network

   b. IP address resource pool used for client access

   c. VLAN ID resource pool for client access

   d. Client access point resource pool (switch ports or SSIDs)

2. VN instance: One or more VNs can be created, with one VN corresponding to one isolated network (service network), such as the R&D private network. Each VN has all network functions.

a. Each VN has access points. The physical port on the access switch is the wired client access point, and the Service Set Identifier (SSID) is the wireless client access point.

b. Each VN has one or more Layer 2 BDs. A VN can be divided into multiple Layer 2 BDs, which is similar to Layer 2 isolation and subnet division on the R&D private network based on service requirements.

c. Each VN has a Layer 3 routing domain. As a result, the VN can communicate with external networks and other VNs.

VNs are constructed on a campus network in order to achieve the following two goals:

- Creating basic conditions for network automation. By building VNs, physical network resources form a network resource pool that can be invoked by the service layer. In this way, the Software-Defined Networking (SDN) controller can implement automatic allocation of network resources. Ultimately, this automation simplifies service provisioning and network deployment, implementing a real software-defined network.

- Providing network-level service isolation to achieve a multi-purpose network. Multiple VNs can be created on a campus network based on service requirements, and network-level isolation and interworking can be implemented between VNs. For example, a traditional office network, a scientific research network, and a video surveillance network may be three independent physical networks, wasting resources and complicating network O&M. To address this issue, three VNs can be created on one physical network to isolate services, implementing one network for multiple purposes.

5.2.2 Roles of a VN

On a VN, four roles are defined: border node, edge node, access node, and transparent node. The entities of all four roles are physical devices on the physical network. The border node and edge node are assigned new functions on the VN. Figure 5.4 shows the VN roles on a campus network.

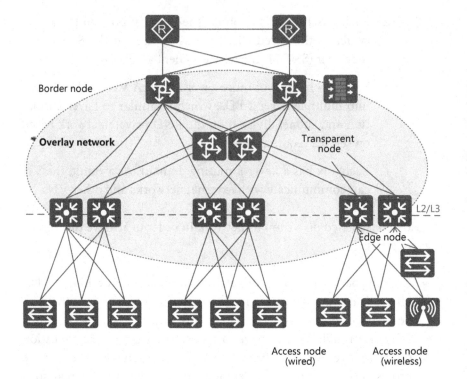

FIGURE 5.4 VN roles on a campus network.

- Border node: a border gateway on a VN, which provides a data forwarding channel between the VN and an external network. Generally, core switches that support VXLAN are used as border nodes.

- Edge node: an edge node on a VN, from which user traffic enters the VN. Generally, VXLAN-capable access switches or aggregation switches are used as edge nodes.

- Access node: includes both wired and wireless access nodes, which are usually access switches and Access Points (APs). An access node can be integrated with an edge node. If an access node is deployed independently, VXLAN is not necessarily required.

- Transparent node: a transparent transmission node on a VN, which is VN-unaware and is not required to support VXLAN.

5.3 TYPICAL VIRTUALIZATION SCENARIOS

The VN technology applies to various types of campus networks. The following uses high-tech industrial campus, education campus, and commercial building scenarios as examples to describe specific use cases.

5.3.1 High-Tech Industrial Campus: One Network for Multiple Purposes Achieved through Network Virtualization

In the following example, a new high-tech R&D enterprise enters an economic development zone. The enterprise has ten independent multi-story office buildings, where multiple service networks such as office, IoT, and security surveillance networks must be constructed.

As shown in Figure 5.5, the enterprise needs to deploy only one physical network. Then, by using network virtualization technology and SDN controller, the enterprise constructs three VNs that are isolated from one another. These VNs can then be managed in a unified manner, reducing overall network construction costs.

On the underlay network, redundancy protection schemes such as Cluster Switch System (CSS) and Virtual Router Redundancy Protocol (VRRP) are used to ensure the reliability of all service networks, as well as the normal use of user services after network virtualization. On the overlay

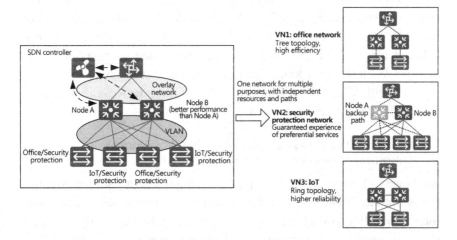

FIGURE 5.5 Multipurpose network implemented through network virtualization on a high-tech industrial campus.

network, the office network uses a tree structure, the security network uses an active/standby structure, and the IoT network uses a ring structure. The three types of networks are logically independent of each other and can flexibly construct a VN topology based on service characteristics.

5.3.2 Education Campus: Integrating Server Resources through Network Virtualization

In the following example, a university/college has more than ten departments catering to 20000 students. Due to historical reasons, the university/college does not have a unified data center. Server resources are distributed across each department and cannot be aggregated in a centralized manner. While the server resources of some departments are insufficient, the overall server resource usage throughout the university/college is lower than 30%. The campus network uses devices from multiple vendors, making network reconstruction difficult. Deployment of VNs on the campus network is a viable solution, enabling server resources across the entire campus to be integrated and shared among all departments. As such, server resources can be integrated without changing the original network, and resource utilization is improved.

As shown in Figure 5.6, the SDN controller is used to deploy a VXLAN-based large Layer 2 server network, so that Virtual Machines (VMs) (one server is virtualized into multiple VMs) can be dynamically migrated on the large Layer 2 network. This technique enables server resources distributed across each department to be managed and used in a centralized manner. With this solution, only aggregation and core devices need to be replaced with VXLAN-capable alternatives. Other devices can be reused, effectively reducing customers' investment costs.

5.3.3 Commercial Building: Quick Service Provisioning through Network Virtualization

As shown in Figure 5.7, a commercial building is a high-rise structure that leases out administrative or office resources to enterprises. On the commercial building's existing network, one access switch is deployed on each floor and one aggregation switch is deployed for every three floors. Each aggregation switch is dual-homed to two core switches, and traffic is routed out through firewalls. Tenants in the building can share access switches (for example, tenants 1 and 2 share one access switch) or aggregation

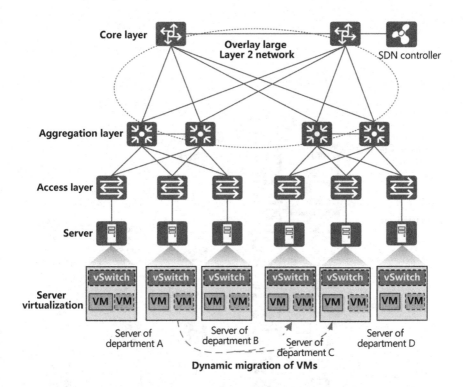

FIGURE 5.6 Server resource integration through network virtualization on the education campus.

switch (for example, tenants 4–7 share one aggregation switch). Large- and medium-sized enterprises generally rent multiple floors of the building. As a result, small-sized data centers need to be deployed. However, small-sized enterprises rent only a portion of a single floor and just require access to the Internet. In addition, tenant networks must be isolated from one another. Traditionally, each time a new tenant arrives, the network would need to be re-planned and commissioned again based on the tenant's scale and requirements. This process is inefficient and features slow service provisioning.

In this case, network virtualization technology is a feasible way. Network virtualization does not require reconstruction or complex configuration of the existing network. Instead, the SDN controller can be used to quickly create VNs based on various service requirements to provide services for new tenants. Tenants can manage their own VNs.

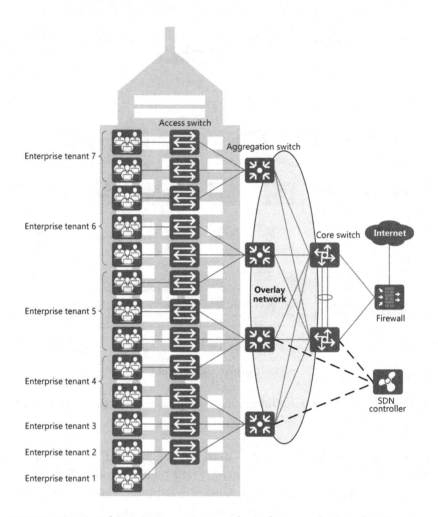

FIGURE 5.7 Rapid service provisioning through network virtualization in a commercial building.

5.4 VXLAN TECHNOLOGY BASICS

VXLAN is a standard NVO3 technology defined by the IETF. VXLAN is essentially a tunneling technology that extends Layer 2 networks across Layer 3 infrastructure by encapsulating Layer 2 packets into UDP packets using MAC-in-UDP encapsulation.

5.4.1 Basic Concepts of VXLAN

Figure 5.8 shows the basic structure of a VN constructed on top of an IP network through VXLAN tunnels.

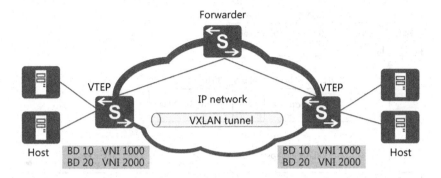

FIGURE 5.8   VXLAN model.

VXLAN involves the following elements that do not exist on traditional campus networks:

- VXLAN Tunnel Endpoints (VTEPs): VXLAN edge devices of a VXLAN tunnel that encapsulate and decapsulate VXLAN packets. In a VXLAN packet, the source IP address is the local VTEP address, and the destination IP address is the remote VTEP address. One pair of such VTEP addresses identifies a VXLAN tunnel. The border node and edge node described in Section 5.2.2 are both VTEPs, and they play an important role in VXLAN.

- VXLAN Network Identifier (VNI): In a similar way to VLAN IDs on a traditional network, VNIs differentiate subnets in a VN. Layer 2 communication is not allowed between users on different subnets. A VNI is 24 bits long, which means it supports approximately 16 million subnets.

- BD: Each VNI is mapped to a BD, and users in the same BD communicate with each other at Layer 2.

5.4.2 VXLAN Packet Format

In the VXLAN encapsulation process, after receiving a packet from a host, the source VTEP adds a VXLAN header, UDP header, IP header, and Ethernet header to the original packet in sequence. This new packet is known as a VXLAN packet. Upon receipt of the VXLAN packet, the destination VTEP decapsulates the packet to obtain the original packet and forwards the original packet to the destination host. Figure 5.9 shows the VXLAN packet format.

Table 5.2 describes the fields in a VXLAN packet.

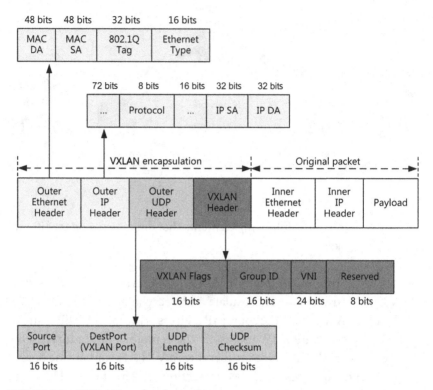

FIGURE 5.9   VXLAN packet format.

TABLE 5.2   Description of Fields in a VXLAN Packet

Field	Description
VXLAN header	VXLAN Flags (16 bits) Group ID (16 bits): specifies a user group ID. When the first bit of **VXLAN Flags** is 1, the value is a group ID; when the first bit of **VXLAN Flags** is 0, the value is all 0s VNI: specifies a VXLAN network identifier Reserved: this 8-bit field is reserved and set to 0
Outer UDP header	Source port (16 bits): specifies the source UDP port number, which is calculated based on the inner Ethernet packet header using the hash algorithm DestPort (16 bits): specifies the destination UDP port number. The value is 4789 UDP length (16 bits): specifies the length of a UDP packet, which is total length of the UDP header and UDP data UDP checksum (16 bits): used for error-checking of the UDP header and UDP data

(Continued)

TABLE 5.2 (*Continued*)   Description of Fields in a VXLAN Packet

Field	Description
Outer IP Header	IP SA (32 bits): specifies the source IP address, which is the IP address of the source VTEP of a VXLAN tunnel IP DA (32 bits): specifies the destination IP address, which is the IP address of the destination VTEP of a VXLAN tunnel Protocol (8 bits): specifies the next protocol following the datagram
Outer Ethernet header	MAC DA (48 bits): specifies the destination MAC address, which is the MAC address of the next-hop device on the path to the destination VTEP MAC SA (48 bits): specifies the source MAC address, which is the MAC address of the source VTEP that sends the packet 802.1Q tag (32 bits): specifies the VLAN tag in the packet. This field is optional Ethernet type (16 bits): specifies the type of the Ethernet frame. The value is 0x0800 in an IP packet

As described, the VXLAN network model and packet format convey the following VXLAN features:

- Compared with Layer 2 isolation using 12-bit VLAN IDs, VXLAN uses 24-bit VNIs to support the isolation of up to 16 million VXLAN segments, meeting the requirements of a large number of tenants.

- A VXLAN header includes a VNI that can be flexibly associated with other services, such as Layer 2 and Layer 3 VPNs.

- On a VXLAN network, only edge devices need to identify the MAC addresses of hosts.

- VXLAN uses MAC-in-UDP encapsulation to extend Layer 2 networks, decoupling VNs from the physical network. In addition, tenants can plan VNs without the need to consider IP addresses and BDs on the physical network, greatly simplifying network management.

- The VXLAN-encapsulated UDP source port number is calculated based on the inner Ethernet packet header using the hash algorithm. Load balancing can be performed on the physical network without parsing inner packets, increasing the network throughput.

5.5 VXLAN CONTROL PLANE

In the initial VXLAN solution (RFC 7348), the control plane is not defined. Instead, VXLAN tunnels require manual configuration and host MAC addresses need to be learned through traffic flooding. Although the flood-and-learn approach is much simpler, it causes a large amount of flooded traffic on the network and makes the network difficult to expand.

To address these problems, Ethernet Virtual Private Network (EVPN) is introduced as the VXLAN control plane. EVPN relies on the Border Gateway Protocol (BGP)/MPLS VPN mechanism. By extending BGP, EVPN defines three new types of BGP EVPN routes to implement VTEP autodiscovery and host MAC address learning. Using EVPN as the VXLAN control plane has the following advantages:

- VTEPs are discovered automatically and VXLAN tunnels are established automatically, simplifying network deployment and expansion.

- EVPN can advertise both Layer 2 MAC addresses and Layer 3 routing information.

- Flooded traffic on the network is significantly decreased.

5.5.1 Understanding BGP EVPN

EVPN defines three new types of BGP EVPN routes to transmit VTEP addresses and host information by extending BGP. As such, the applications of EVPN on VXLAN move VTEP autodiscovery and host MAC address learning from the data plane to the control plane. The functions of the control-plane routes are as follows:

- Type 2 route (MAC/IP route): used to advertise host MAC addresses, host Address Resolution Protocol (ARP) entries, and host route information.

- Type 3 route (inclusive multicast route): used to automatically discover VTEPs and dynamically establish VXLAN tunnels.

- Type 5 route (IP prefix route): used to advertise the imported external routes and host route information.

1. Type 2 route: MAC/IP route

     Figure 5.10 shows the format of a MAC/IP route.

     Table 5.3 describes the fields in a MAC/IP route.

     MAC/IP routes function on the VXLAN control plane as follows:

  a. Advertising host MAC addresses

       To implement Layer 2 communication between intrasubnet hosts, the local and remote VTEPs of a VXLAN tunnel need to learn the host MAC addresses from each other. To achieve this, the VTEPs function as BGP EVPN peers to exchange MAC/IP routes.

  b. Advertising host ARP entries

       A MAC/IP route can carry both the MAC and IP addresses of a host, and therefore can be used to advertise ARP entries between

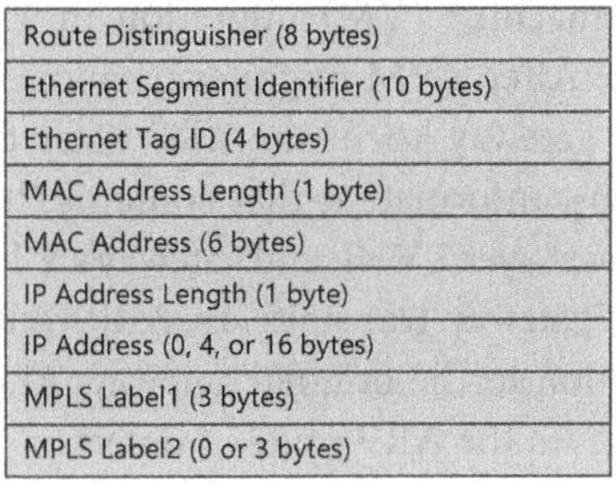

Route Distinguisher (8 bytes)
Ethernet Segment Identifier (10 bytes)
Ethernet Tag ID (4 bytes)
MAC Address Length (1 byte)
MAC Address (6 bytes)
IP Address Length (1 byte)
IP Address (0, 4, or 16 bytes)
MPLS Label1 (3 bytes)
MPLS Label2 (0 or 3 bytes)

FIGURE 5.10   Format of a MAC/IP route.

TABLE 5.3   Fields in a MAC/IP Route

Field	Description
Route distinguisher	RD value of an EVPN instance
Ethernet segment identifier	Unique identifier of the connection between the local and remote devices
Ethernet tag ID	VLAN ID configured on the local device
MAC address length	Length of the host MAC address carried in the route
MAC address	Host MAC address carried in the route
IP address length	Mask length of the host IP address carried in the route
IP address	Host IP address carried in the route
MPLS Label1	Layer 2 VNI carried in the route
MPLS Label2	Layer 3 VNI carried in the route

VTEPs. This type of MAC/IP route is also called the ARP route. ARP entry advertisement applies to the following scenarios:

i. ARP broadcast suppression: After a Layer 3 gateway learns the ARP entries of hosts, it generates host information that contains the host IP and MAC addresses, Layer 2 VNI, and gateway's VTEP IP address. The Layer 3 gateway then transmits an ARP route carrying the host information to a Layer 2 gateway. Upon receiving an ARP request, the Layer 2 gateway checks whether it includes the host information corresponding to the destination IP address of the packet. If such host information exists, the Layer 2 gateway replaces the broadcast MAC address in the ARP request with the destination unicast MAC address and unicasts the packet. This implementation suppresses ARP broadcast packets.

ii. Virtual machine (VM) migration in a distributed gateway scenario: After a VM migrates from one gateway to another, the new gateway learns the ARP entry of the VM, and generates host information that contains the host IP and MAC addresses, Layer 2 VNI, and gateway's VTEP IP address. Then, the new gateway transmits an ARP route carrying the host information to the original gateway. After the original gateway receives the ARP route, it detects a VM location change and triggers ARP probe. If ARP probe fails, the original gateway withdraws the ARP entry and host route of the VM.

c. Advertising host IP routes

In a distributed VXLAN gateway scenario, to implement Layer 3 communication between intersubnet hosts, the local and remote VTEPs that function as Layer 3 gateways need to learn host IP routes from each other. To achieve this, the VTEPs function as BGP EVPN peers to exchange MAC/IP routes. This type of MAC/IP route is also called the Integrated Routing and Bridging (IRB) route.

d. Advertising neighbor discovery (ND) entries

A MAC/IP route can carry both the MAC and IPv6 addresses of a host. This means that this type of route can be used to transmit ND entries between VTEPs and implement ND entry advertisement. The MAC/IP route is also called an ND route. ND entry flooding applies to the following scenarios:

i. Neighbor Solicitation (NS) multicast suppression: After a VXLAN gateway collects information about a local IPv6 host, it generates an NS multicast suppression entry and transmits the entry through a MAC/IP route. After receiving the MAC/IP route, other VXLAN gateways (BGP EVPN peers) each generate a local NS multicast suppression entry. In this way, when a VXLAN gateway receives an NS message, it searches the local NS multicast suppression table. If a matching entry is found, the VXLAN gateway performs multicast-to-unicast processing to reduce or suppress NS message flooding.

ii. IPv6 VM migration in a distributed gateway scenario: After an IPv6 VM is migrated from one gateway to another, the VM sends a gratuitous Neighbor Advertisement (NA) message. After receiving this message, the new gateway generates an ND entry and transmits it to the original gateway through a MAC/IP route. Upon receipt of the entry, the original gateway detects that the location of the IPv6 VM changes and triggers neighbor unreachability detection (NUD). If the original gateway cannot detect the IPv6 VM in the original location, it deletes the corresponding local ND entry and uses an MAC/IP route to instruct the new gateway to delete the old ND entry for the IPv6 VM.

e. Advertising host IPv6 routes

In a distributed VXLAN gateway scenario, to implement Layer 3 communication between intersubnet IPv6 hosts, the VTEPs that function as Layer 3 gateways need to learn host IPv6 routes from each other. To achieve this, the VTEPs function as BGP EVPN peers to exchange MAC/IP routes. In this case, MAC/IP routes are also called IRBv6 routes.

2. Type 3 route: inclusive multicast route

An inclusive multicast route encompasses a prefix and a P-Multicast Service Interface (PMSI) attribute, as shown in Figure 5.11.

Table 5.4 describes the fields in an inclusive multicast route.

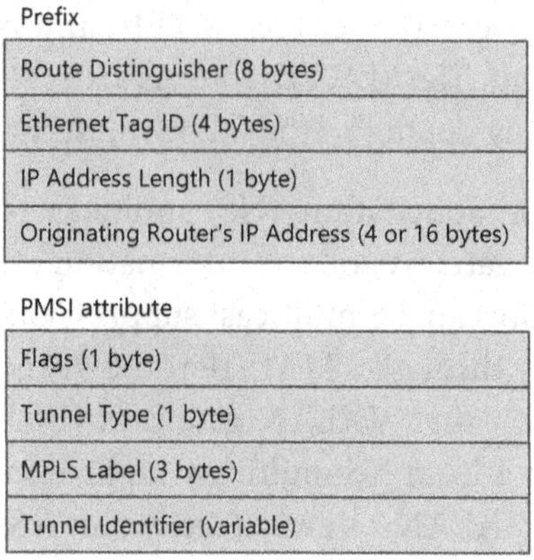

Prefix

Route Distinguisher (8 bytes)
Ethernet Tag ID (4 bytes)
IP Address Length (1 byte)
Originating Router's IP Address (4 or 16 bytes)

PMSI attribute

Flags (1 byte)
Tunnel Type (1 byte)
MPLS Label (3 bytes)
Tunnel Identifier (variable)

FIGURE 5.11   Format of an inclusive multicast route.

The inclusive multicast route is used on the VXLAN control plane for VTEP autodiscovery and dynamic VXLAN tunnel establishment. VTEPs function as BGP EVPN peers to exchange inclusive multicast routes so that they can learn Layer 2 VNIs and VTEPs' IP addresses from each other. If the remote VTEP's IP address is reachable at Layer 3, the local VTEP establishes a VXLAN tunnel

TABLE 5.4   Fields in an Inclusive Multicast Route

Field	Description
Route distinguisher	RD value of an EVPN instance
Ethernet tag ID	VLAN ID configured on the local device The value is all 0s in an inclusive multicast route
IP address length	Mask length of the local VTEP's IP address carried in the route
Originating router's IP address	Local VTEP's IP address carried in the route
Flags	Flags indicate whether or not leaf node information is required for the tunnel This field is meaningless in VXLAN scenarios
Tunnel type	Tunnel type carried in the route The value can only be 6, representing ingress replication in VXLAN scenarios. It is used to forward broadcast, unknown unicast, and multicast (BUM) packets
MPLS label	Layer 2 VNI carried in the route
Tunnel identifier	Tunnel identifier carried in the route This field is the local VTEP's IP address in VXLAN scenarios

with the remote VTEP. If the remote VNI is the same as the local VNI, an ingress replication list is created for subsequent BUM packet forwarding.

3. Type 5 route: IP prefix route

Figure 5.12 shows the format of an IP prefix route.

Table 5.5 describes the fields.

The **IP Prefix Length** and **IP Prefix** fields can identify a host IP address or network segment.

a. If the **IP Prefix Length** and **IP Prefix** fields identify a host IP address, the route is used for IP route advertisement in distributed VXLAN gateway scenarios. In such cases, the route functions the same as an IRB route on the VXLAN control plane.

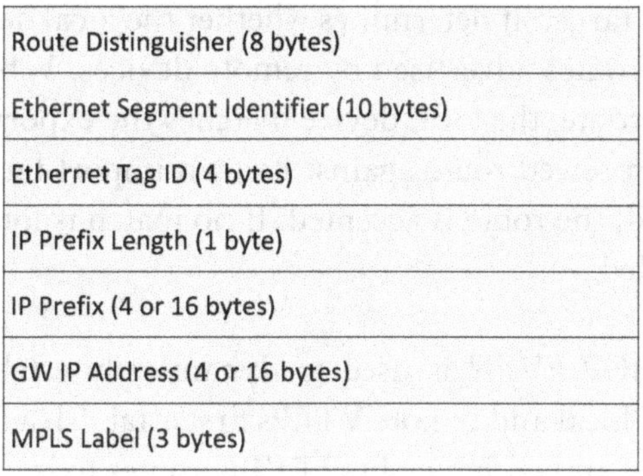

Route Distinguisher (8 bytes)
Ethernet Segment Identifier (10 bytes)
Ethernet Tag ID (4 bytes)
IP Prefix Length (1 byte)
IP Prefix (4 or 16 bytes)
GW IP Address (4 or 16 bytes)
MPLS Label (3 bytes)

FIGURE 5.12   Format of an IP prefix route.

TABLE 5.5   Fields in an IP Prefix Route

Field	Description
Route distinguisher	RD value of an EVPN instance
Ethernet segment identifier	Unique identifier of the connection between the local and remote devices
Ethernet tag ID	VLAN ID configured on the local device
IP prefix length	Length of the IP prefix carried in the route
IP prefix	IP prefix carried in the route
GW IP address	Default gateway address
This field is meaningless in VXLAN scenarios
MPLS label	Layer 3 VNI carried in the route

b. If the **IP Prefix Length** and **IP Prefix** fields in an IP prefix route identify a network segment, the route enables access to external networks.

Advertised EVPN routes carry RDs and VPN targets (also known as route targets).

RDs are used to identify different VXLAN EVPN routes. In addition, VPN targets are BGP extended community attributes used to control the export and import of EVPN routes.

A VPN target is either an export target or an import target.

a. Export target: It is carried in the EVPN routes advertised by the local device and defines which remote devices can accept the EVPN routes.

b. Import target: It determines whether the local device accepts the EVPN routes advertised by remote devices. When receiving an EVPN route, the local device matches the export targets carried in the received route against its own import targets. If a match is found, the route is accepted. If no match is found, the route is discarded.

When BGP EVPN is used to dynamically establish a VXLAN tunnel, the local and remote VTEPs first establish a BGP EVPN peer relationship and exchange BGP EVPN routes to learn the VNIs and VTEP IP addresses from each other. This approach is applicable to both centralized and distributed VXLAN gateway scenarios. The following uses the centralized VXLAN gateway scenario to describe the process of VXLAN tunnel establishment.

5.5.2 VXLAN Tunnel Establishment

A VXLAN tunnel is identified by a pair of VTEP IP addresses. During VXLAN tunnel establishment, the local and remote VTEPs attempt to obtain IP addresses from each other. A VXLAN tunnel can only be established if the obtained VTEP IP addresses can reach each other at Layer 3. When BGP EVPN is used to dynamically establish a VXLAN tunnel, the local and remote VTEPs first establish a BGP EVPN peer relationship before exchanging BGP EVPN routes to learn the VNIs and VTEP IP addresses from each other.

On the network shown in Figure 5.13, Host 1 and Host 3 are attached to VTEP 2, Host 2 is attached to VTEP 3, and a Layer 3 gateway is deployed on VTEP 1. To allow Host 3 and Host 2, which are on the same subnet, to communicate with each other, a VXLAN tunnel needs to be established between VTEP 2 and VTEP 3. To allow Host 1 and Host 2 on different subnets to communicate with each other, VXLAN tunnels need to be established between VTEP 2 and VTEP 1 and between VTEP 1 and VTEP 3. Although Host 1 and Host 3 are both attached to VTEP 2, they belong to different subnets and must communicate through the Layer 3 gateway (VTEP 1). For this reason, a VXLAN tunnel is also required between VTEP 2 and VTEP 1.

The following example illustrates how to use BGP EVPN to dynamically establish a VXLAN tunnel between VTEP 2 and VTEP 3, as shown in Figure 5.14.

1. VTEP 2 and VTEP 3 first establish a BGP EVPN peer relationship. Then, local EVPN instances are created on VTEP 2 and VTEP 3, and a route distinguisher (RD), export VPN target (ERT), and import

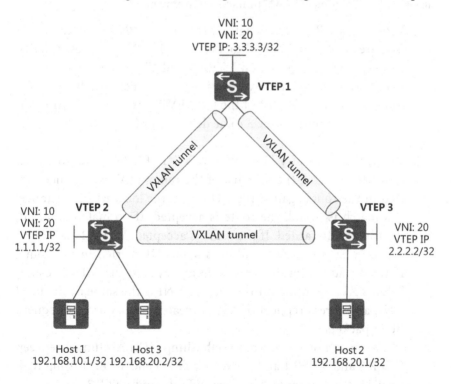

FIGURE 5.13 VXLAN tunnel establishment.

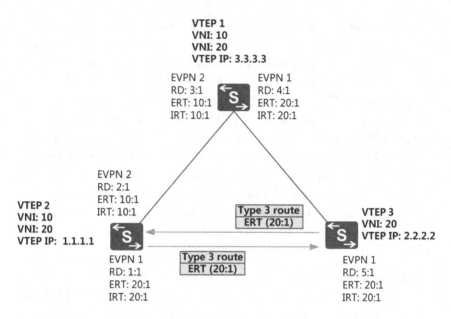

FIGURE 5.14 Dynamic VXLAN tunnel establishment.

VPN target (IRT) are configured for each EVPN instance. Layer 2 BDs are created and bound to VNIs and EVPN instances. After IP addresses are configured on VTEP 2 and VTEP 3, they generate a BGP EVPN route and advertise it to each other. The BGP EVPN route carries the ERT list of the local EVPN instance and an inclusive multicast route (Type 3 route defined in BGP EVPN).

2. When VTEP 2 and VTEP 3 receive a BGP EVPN route from each other, they match the ERT list of the remote EVPN instance carried in the route against the IRT list of the local EVPN instance. If a match is found, the route is accepted. If no match is found, the route is discarded. If the route is accepted, VTEP 2 and VTEP 3 obtain each other's IP address and VNI carried in the route. If the IP addresses are reachable at Layer 3, the VTEPs establish a VXLAN tunnel. If the remote VNI is the same as the local VNI, an ingress replication list is created to forward subsequent BUM packets.

The process of dynamically establishing a VXLAN tunnel between VTEP 2 and VTEP 1 and between VTEP 3 and VTEP 1 using BGP EVPN is the same as that between VTEP 2 and VTEP 3.

5.5.3 Dynamic MAC Address Learning

VXLAN uses dynamic MAC address learning to facilitate communication between end users. MAC address entries are dynamically created and therefore do not require manual maintenance, greatly reducing the maintenance workload. Figure 5.15 illustrates how intrasubnet hosts dynamically learn each other's MAC address.

1. When Host 3 communicates with VTEP 2 for the first time, VTEP 2 learns the mapping between Host 3's MAC address, BD ID, and inbound interface (Port 1) that has received the ARP packet, and generates a MAC address entry for Host 3, with the outbound interface set to Port 1. In addition, VTEP 2 generates a BGP EVPN route based on the ARP entry of Host 3 and advertises the route to VTEP 3. The BGP EVPN route carries the ERT list of VTEP 2's EVPN instance, next-hop attribute (VTEP 2's IP address), and MAC/IP route (Type 2 route defined in BGP EVPN). Figure 5.16 shows the format of a MAC/IP route. In this example, the **MAC Address Length** and **MAC Address** fields identify the MAC address of Host 3, and the **MPLS Label1** field identifies the Layer 2 VNI.

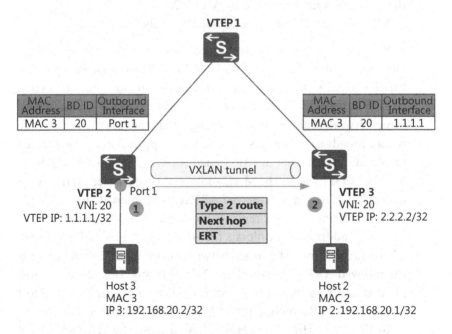

FIGURE 5.15   Dynamic MAC address learning.

Route Distinguisher (8 bytes)
Ethernet Segment Identifier (10 bytes)
Ethernet Tag ID (4 bytes)
MAC Address Length (1 byte)
MAC Address (6 bytes)
IP Address Length (1 byte)
IP Address (0, 4, or 16 bytes)
MPLS Label1 (3 bytes)
MPLS Label2 (0 or 3 bytes)

FIGURE 5.16   MAC/IP route.

2. When receiving the BGP EVPN route from VTEP 2, VTEP 3 matches the ERT list of the EVPN instance carried in the route against the IRT list of the local EVPN instance. If a match is found, the route is accepted. If no match is found, the route is discarded. If the route is accepted, VTEP 3 obtains the mapping between Host 3's MAC address, BD ID, and VTEP 2's IP address (next-hop attribute), and generates a MAC address entry for Host 3. Based on the next-hop attribute, the MAC address entry's outbound interface is recursed to the VXLAN tunnel destined for VTEP 2.

   VTEP 2 learns Host 2's MAC address in the same way.

3. When Host 3 attempts to communicate with Host 2 for the first time, Host 3 sends an ARP request for Host 2's MAC address, with the destination MAC address set to all Fs and the destination IP address set to IP 2. By default, VTEP 2 broadcasts the ARP request to devices on the same network segment as the interface that receives the request. To reduce broadcast packets, ARP broadcast suppression can be enabled on VTEP 2. With this function enabled, VTEP 2 searches the local MAC address table for the MAC address of Host 2 based on the destination IP address in the received ARP request. Then, if Host 2's MAC address is found, VTEP 2 replaces the destination MAC address with this MAC address, and unicasts the ARP request to VTEP 3 through the VXLAN tunnel established between them. VTEP 3 then forwards the received ARP request to Host 2. In this way, Host 2 learns Host 3's MAC address and responds with a unicast ARP reply. After Host 3 receives the ARP reply, it learns Host 2's MAC address.

   By this stage, Host 3 and Host 2 have learned the MAC address of each other, and they can communicate in unicast mode.

5.6 VXLAN DATA PLANE

The VXLAN data plane is responsible for forwarding data over VXLAN tunnels after address mappings are learned on the control plane. A source VTEP adds a UDP header to the original data frame, which is later removed by the destination VTEP. The intermediate network devices forward the packet based on the destination address in the outer IP header. Data forwarding scenarios include intrasubnet known unicast packet forwarding, intrasubnet BUM packet forwarding, and intersubnet packet forwarding.

5.6.1 Intrasubnet Packet Forwarding

1. Intrasubnet forwarding of known unicast packets

   Known unicast packets for intrasubnet communication are forwarded only between Layer 2 VXLAN gateways. Layer 3 VXLAN gateways are unaware of this process. Figure 5.17 shows the packet forwarding process.

   a. When receiving a packet from Host 3, VTEP 2 determines the Layer 2 BD of the packet based on the access interface and VLAN information in the packet, and searches for the outbound interface and encapsulation information in the Layer 2 BD.

   b. VTEP 2 performs VXLAN encapsulation based on the encapsulation information and forwards the packet through the outbound interface.

   c. Upon receipt of the VXLAN packet, VTEP 3 verifies the VXLAN packet based on the UDP destination port number, source and destination IP addresses, and VNI. Then, VTEP 3 obtains the Layer 2 BD based on the VNI and performs VXLAN decapsulation to obtain the inner Layer 2 packet.

   d. Once VTEP 3 obtains the destination MAC address of the inner Layer 2 packet, it processes VLAN tags for the packet based on the outbound interface and encapsulation information in the local MAC address table, and forwards the packet to Host 2.

   Host 2 sends packets to Host 3 in the same manner.

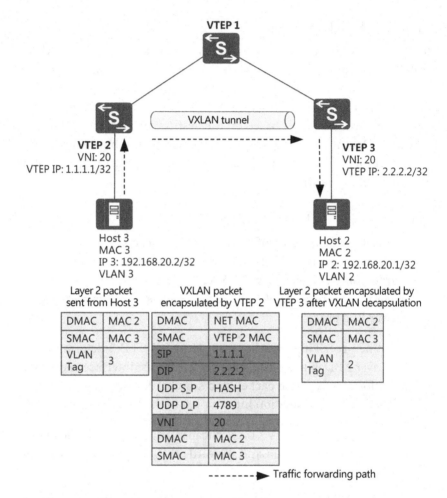

FIGURE 5.17   Intrasubnet forwarding of known unicast packets.

2. Intrasubnet forwarding of BUM packets

Intra-subnet BUM packets are forwarded in ingress replication mode between Layer 2 VXLAN gateways. Layer 3 VXLAN gateways are unaware of this process.

In ingress replication mode, after a BUM packet enters a VXLAN tunnel, the ingress VTEP performs VXLAN encapsulation based on the ingress replication list and forwards the packet to all egress VTEPs in the list. Then, when the BUM packet leaves the VXLAN tunnel, the egress VTEPs decapsulate the BUM packet. Figure 5.18 shows the forwarding process of a BUM packet in ingress replication mode.

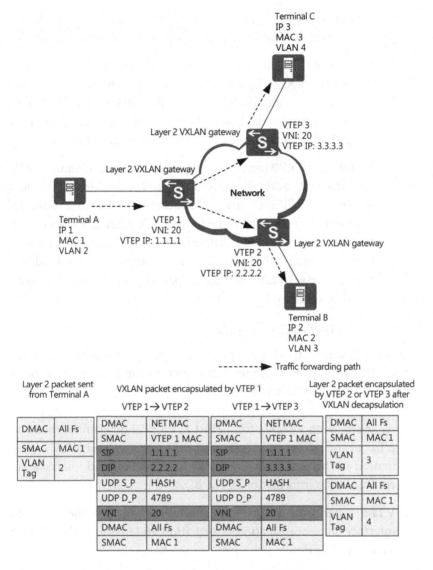

FIGURE 5.18 Intrasubnet forwarding process of a BUM packet in ingress replication mode.

a. After receiving a packet from Terminal A, VTEP 1 determines the Layer 2 BD of the packet based on the access interface and VLAN information in the packet.

b. VTEP 1 obtains the ingress replication list for the VNI based on the Layer 2 BD, replicates the packet based on the list, and performs VXLAN encapsulation on the packet. VTEP 1 then forwards the VXLAN packet through the outbound interface.

c. Upon receipt of the VXLAN packet, either VTEP 2 or VTEP 3 verifies the VXLAN packet based on the UDP destination port number, source and destination IP addresses, and VNI. VTEP 2 or VTEP 3 obtains the Layer 2 BD based on the VNI and perform VXLAN decapsulation to obtain the inner Layer 2 packet.

d. VTEP 2 or VTEP 3 then checks the destination MAC address of the inner Layer 2 packet and finds it a BUM MAC address. Therefore, VTEP 2 or VTEP 3 broadcasts the packet on the network connected to terminals (user side) in the Layer 2 BD. Specifically, VTEP 2 or VTEP 3 finds the outbound interfaces and encapsulation information not related to the VXLAN tunnel from the local MAC address table, processes VLAN tags for the packet, and forwards the packet to either Terminal B or Terminal C.

5.6.2 Intersubnet Packet Forwarding

Intersubnet packets must be forwarded through a Layer 3 gateway. Figure 5.19 shows the intersubnet packet forwarding process in a centralized VXLAN gateway scenario.

1. After receiving a packet from Host 1, VTEP 2 determines the Layer 2 BD of the packet based on the access interface and VLAN information in the packet, and searches for the outbound interface and encapsulation information in the BD.

2. VTEP 2 performs VXLAN encapsulation based on the outbound interface and encapsulation information and forwards the VXLAN packet to VTEP 1.

3. After receiving the VXLAN packet, VTEP 1 decapsulates the packet and finds that the destination MAC address of the inner packet is the MAC address (MAC 3) of the Layer 3 gateway interface (VBDIF 10). In this case, the packet must be forwarded at Layer 3.

4. VTEP 1 removes the inner Ethernet header, parses the destination IP address, and searches the routing table for the next-hop address. Then, VTEP 1 searches the ARP table based on the next-hop address

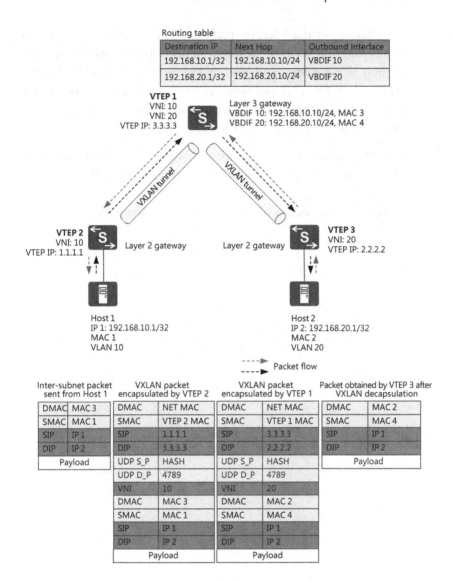

FIGURE 5.19 Intersubnet packet forwarding.

to obtain the destination MAC address, VXLAN tunnel's outbound interface, and VNI.

5. VTEP 1 performs VXLAN encapsulation on the inner packet again and forwards the VXLAN packet to VTEP 3, with the source MAC address in the inner Ethernet header being the MAC address (MAC 4) of the Layer 3 gateway interface (VBDIF 20).

6. Upon receipt of the VXLAN packet, VTEP 3 verifies the VXLAN packet based on the UDP destination port number, source and destination IP addresses, and VNI. VTEP 3 then obtains the Layer 2 BD based on the VNI and removes the outer headers to obtain the inner Layer 2 packet. It then searches for the outbound interface and encapsulation information in the Layer 2 BD.

7. VTEP 3 processes VLAN tags for the packet based on the outbound interface and encapsulation information and forwards the packet to Host 2.

Host 2 sends packets to Host 1 in the same manner.

CHAPTER 6

Automated Service Deployment on an Intent-Driven Campus Network

A s INTERNET TECHNOLOGIES CONTINUE to develop, the amount of application software also increases, meaning that software will soon replace hardware to dominate the Information and Communications Technology (ICT) development. As such, the intent-driven campus network is gradually transferring network complexity from the network to software by leveraging Software-Defined Networking (SDN) technology, thereby implementing automated deployment of network services.

6.1 OVERVIEW OF INTENT-DRIVEN CAMPUS NETWORK AUTOMATION

Varying in scale, campus networks can be classified into small-sized campus networks and medium- and large-sized campus networks, which have different automated service deployment focuses.

- A small-sized campus network is usually a simple-structured network that consists of only a few devices and carries basic services, often with no full-time network administrator. It usually covers multiple branches, to which the same configurations need to be delivered quickly.

On a small-sized campus network, device installation, deployment, and service commissioning can be completed quickly. If this network has multiple scattered branches, it would be difficult to perform repeated commissioning.

For small-sized campus networks, the recommended automated service deployment involves two phases: pre-configuring services and bringing devices online in plug-and-play mode. Specifically, the network administrator pre-configures services on the SDN controller, completes commissioning for a single site, and duplicates the configuration of this site to new sites to quickly complete service configuration. During onsite deployment, field engineers only need to power on devices, use a mobile app to scan devices, add the devices to corresponding sites, wait for the devices to register with the SDN controller, and test network connectivity. This deployment approach is highly efficient and does not pose high skill requirements on the field engineers.

- A medium- and large-sized campus network is typically a complex-structured network that consists of many network devices and carries a variety of services, and is usually managed by a full-time network administrator.

  With regard to automated service deployment, a medium- and large-sized campus network uses the SDN controller to simplify network management through hierarchical decoupling and network abstraction. To be specific, the intent-driven campus network architecture abstracts a network into a network model with an underlay network, overlay network, and service layer, creates service models for each layer, and pools network resources. Additionally, the SDN technology is used to implement automated service deployment for the physical network, virtual networks, and service policies, in that order.

  Medium- and large-sized campus networks use the plug-and-play function to improve network deployment efficiency. There are two plug-and-play scenarios: "plan-and-deploy" and "deploy-and-verify". If device installation is centralized within a specific period of time, the network administrator needs to plan topology connections and add devices on the SDN controller. Installation engineers then connect and power on devices according to the network plan, and use a mobile app to verify device connections. This scenario is referred to as "plan-and-deploy". In contrast, if device installation is spread out over a long period of time, installation engineers can connect and power on devices before any network plan is made. The network administrator can then verify topology connections and add devices to the SDN controller. This plug-and-play scenario is referred to as "deploy-and-verify". In both scenarios, the plug-and-play function enables devices to automatically register with the SDN controller. In addition, to construct virtual networks using resource pools on a physical network, the network administrator needs to configure IP address segments used for generating routes between devices on the SDN controller. The SDN controller then automatically orchestrates and distributes IP address segments to related devices.

Each virtual network requires an independent IP address segment to avoid conflicts with the physical network. In this case, the administrator needs to specify an IP address segment based on the number of terminals on a virtual network, create the virtual network, and allocate resources to the virtual network, ultimately achieving automated deployment of the virtual network.

A virtual network can be accessed at any location through either wired or wireless connections. As such, each employee needs to be assigned an account. The administrator can create accounts and specify account identities (such as R&D or marketing) on the SDN controller. What's more, based on identity management policies, the administrator can configure the resource groups accessible to different accounts, such as R&D or marketing accounts, to achieve automated user access policy deployment.

6.2 PHYSICAL NETWORK AUTOMATION

The automated deployment of physical networks can lower skill requirements for installation personnel and implement fast deployment of campus networks. The following sections describe the technologies used to achieve automated deployment of physical networks.

6.2.1 Device Plug-and-Play

As network technologies develop rapidly and enterprise networks continue to expand, enterprise customers need to manage and maintain hundreds or even thousands of devices. The time spent in network planning and deployment (such as device installation, initial configuration, and device upgrade) accounts for one-third or even longer of the entire network management and O&M period, as shown in Figure 6.1, and most of that time is wasted on completing simple and repetitive tasks. To address this, customers are in urgent need of simplified network device installation, deployment, and software upgrade.

Traditional networks also include unconfigured device deployment solutions. The typical deployment process is shown in Figure 6.2.

Step 1 An administrator creates a general-purpose configuration script and device-specific configuration files using a text editor (for example, Notepad), and uploads them to a file server.

Step 2 Engineering personnel obtain devices and notify the administrator of the equipment serial numbers (ESNs), Media Access Control (MAC) addresses, and deployment locations of the devices.

Step 3 The administrator records the devices' ESNs, MAC addresses, and deployment locations, and specifies configuration files for the devices.

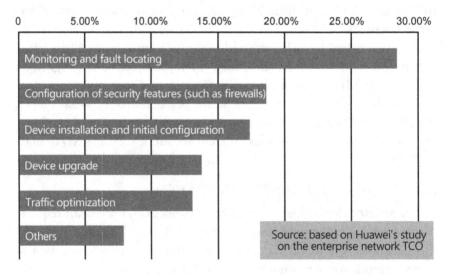

TCO refers to total cost of ownership.

FIGURE 6.1 Work-time distribution of network management personnel.

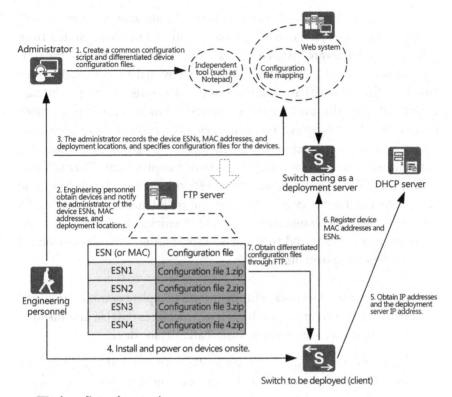

FIGURE 6.2 Traditional plug-and-play solution involving manually recorded device MAC addresses and ESNs.

Step 4 Engineering personnel install devices onsite, connect them through cables, and power on these devices.

Step 5 After the devices are started, they obtain their IP addresses and the IP address of a switch acting as the deployment server through the Dynamic Host Configuration Protocol (DHCP).

Step 6 The devices communicate with the deployment server through a protocol deployed by the administrator to register their MAC addresses and ESNs.

Step 7 The devices download corresponding configuration files from the file server. Device deployment is now complete.

However, the traditional plug-and-play solution also has many disadvantages. For example, network planning and design are separated from software and hardware installation and cable connection. That is, the configuration files need to be manually associated with devices to be deployed based on the device MAC addresses and ESNs, often leading to human errors. Additionally, deployment efficiency is low because services need to be manually configured on each device. To address these issues, the intent-driven campus network introduces the next-generation automated plug-and-play solution to simplify network deployment. This solution uses the SDN controller to automate network service deployment and seamlessly combine network planning and maintenance, thereby greatly improving network management and O&M efficiency while also significantly reducing labor and time costs. This solution provides customers with the following benefits:

- Visualization: Network administrators complete configuration and network planning on graphical user interfaces (GUIs), and installation engineers use an app to install and deploy devices.

- High efficiency: The SDN controller automatically delivers the preconfigured services to devices during network deployment, reducing the end-to-end deployment duration from several days to several hours.

- Few errors: GUI-based configuration on the SDN controller ensures fewer configuration errors than CLI-based configuration. The SDN controller can also detect connection errors in real time to help quickly rectify faults.

Figure 6.3 shows an example of the device plug-and-play process for a small- and medium-sized intent-driven campus network.

Step 1 The network administrator preconfigures services on the SDN controller, including creating tenants, binding licenses to tenants, creating sites, importing device information, and configuring services for network devices by site. For example, the network administrator needs to configure security zones, VLANIF 1, DHCP server, domain name service (DNS), and network address translation for

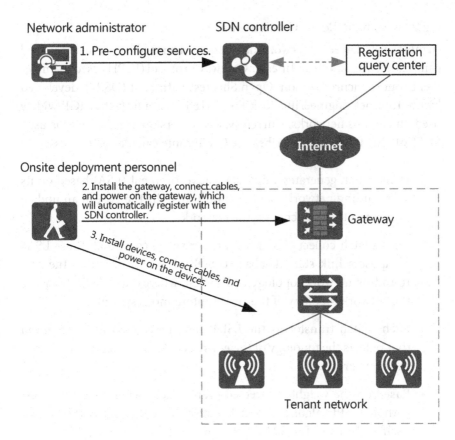

FIGURE 6.3   Device plug-and-play process for an intent-driven campus network.

the egress gateway, and configure service set identifiers (SSIDs) for wireless access devices.

Step 2 Deployment personnel install hardware, connect cables, and power on the egress gateway onsite, and manually configure the gateway to connect to the Internet. The gateway then registers with the SDN controller for management, and the SDN controller delivers service configurations to the gateway.

Step 3 Deployment personnel install devices attached to the egress gateway, connect cables, and power on the devices onsite. The devices then obtain their IP addresses and the DNS server IP address from the egress gateway functioning as the DHCP server, and register with the SDN controller to obtain service configurations.

6.2.2 Automatic Route Configuration

In most cases, Layer 3 interworking is required on medium- and large-sized campus networks, and an Interior Gateway Protocol (IGP) is used to implement route synchronization. Open Shortest Path First (OSPF), developed by the Internet Engineering Task Force (IETF), is a link-state IGP widely used on campus networks. Currently, OSPF Version 2 (RFC 2328) is used for IPv4. The following describes the OSPF route calculation process:

- Each switch generates a link-state advertisement (LSA) based on its surrounding network topology and transmits the LSA in an update packet to other switches on the network.

- Each switch collects LSAs from other switches, and all these LSAs compose a link state database (LSDB). An LSA describes the surrounding network topology of a switch, whereas an LSDB describes the network topology of the entire autonomous system (AS).

- Each switch transforms the LSDB into a weighted directed graph that reflects the topology of the entire AS. All switches in an AS have the same graph.

- Based on the weighted directed graph, each switch uses a shortest path first (SPF) algorithm to calculate a shortest path tree (SPT) with itself as the root. The SPT displays routes to nodes in the AS.

For example, if all switches on a medium- and large-sized campus network run OSPF and the number of switches keeps increasing, each switch will have a large LSDB, which will ultimately occupy a large amount of storage space and complicate the operating of the SPF algorithm, leading to switch overload. In addition, network expansion increases the possibility of network topology changes and causes frequent route flapping. Consequently, a large number of OSPF packets are transmitted on the network, wasting further bandwidth resources. More seriously, each network change causes all switches on the network to recalculate routes.

To address this issue, OSPF partitions an AS into different areas, each regarded as a logical group that is identified by a unique area ID. At the border of an area resides a switch instead of a link, and a network segment or a link belongs to only one area. Therefore, the area to which each OSPF interface belongs needs to be specified.

When an OSPF network is divided, not all areas are equal. The area with ID 0 is known as the backbone area, which is responsible for transmitting routes between nonbackbone areas. Therefore, OSPF requires that all nonbackbone areas be connected to the backbone area, and the backbone area's devices must all be connected to each other.

On the network shown in Figure 6.4, all switches in an AS run OSPF, and the AS is divided into three areas. Switch A and Switch B function as area border routers (ABRs) to forward interarea routes. Then, after basic OSPF functions are configured, each switch learns the routes to all network segments in the AS, including the Virtual Local Area Network (VLAN) ID of each interface and the IP address of each VLANIF interface.

On the preceding network, traditional route deployment poses many problems. First, it is time-consuming to log in to each device to configure IP addresses for nearly 20 OSPF-enabled interfaces and OSPF-dependent interfaces. Second, configuring each device using commands can lead to errors and faults that are difficult to locate. Lastly, OSPF convergence upon network changes is slow, causing long periods of service interruption.

The automatic route configuration solution improves on traditional route deployment. In this solution, the network administrator only needs

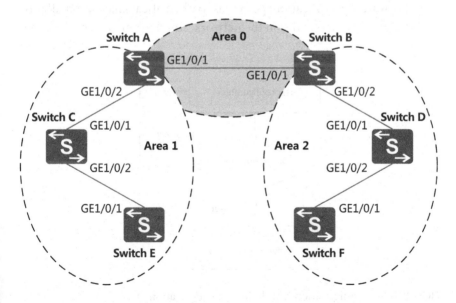

FIGURE 6.4 OSPF area partitioning.

to plan the network topology on the SDN controller. The SDN controller will automatically divide OSPF areas and generate configurations based on the network topology, and deliver the configurations after verifying they match with the network service requirements. Therefore, this solution improves the efficiency and correctness of the route configuration, and reduces the service interruption period.

The configuration simulation and verification technology is used to check whether or not the end-to-end network behaviors (determined by the configuration and status) meet configuration expectations. After the administrator configures services, the SDN controller performs config-uration simulation and verification, and delivers only the verified con-figurations to devices, as shown in Figure 6.5. Additionally, the SDN controller notifies the administrator of incorrect configurations, and con-tinues simulation and verification after the administrator modifies these configurations.

The configuration simulation and verification technology verifies ser-vice configurations and network forwarding.

- Configuration verification: verifies configurations before network deployment, for example, checking for IP address conflicts.

- Forwarding verification: performs mathematical analysis on all pos-sible end-to-end paths on the network, simulates all data packets that

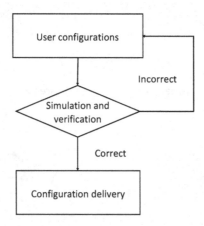

FIGURE 6.5 Configuration simulation and verification.

may pass through the network, and compares the traffic model with the configuration objective to verify the configurations. For example, the network verification module checks whether there are reachable routes between any two nodes on the network and whether the maximum transmission unit (MTU) is correct.

An intent-driven campus network uses the SDN controller to support configuration simulation and verification and change the network IT model from passive response to proactive response. It analyzes the network's current design automatically to eliminate misoperations and configuration errors. Such automation helps IT engineers quickly diagnose faults, record network requirements, and verify configurations.

6.2.3 Security Guarantee during Automation

In the automation solution, the SDN controller acting as the network center sends various control instructions to devices to control device behaviors. If an attacker impersonates the SDN controller to deliver attack instructions to devices, network faults may occur. In this case, a security solution is required to protect the confidentiality and integrity of control instructions and data by performing strict authentication and authorization between the controlling and controlled parties.

To ensure secure and reliable communication between the SDN controller and devices, identity authentication is commonly used to verify both parties' identities. That is, the SDN controller verifies that the messages received from devices are authentic, and the devices also verify that the instructions received from the SDN controller are authentic. In the network communications field, digital certificates similar to electronic copies of driving licenses or passports are typically used between communication parties to confirm each other's identity.

A digital certificate is an electronic document issued and signed by a certificate authority (CA) to prove the ownership of a public key. Digital certificates can be used to achieve the following:

- Data encryption: After a key is negotiated between two communication parties using a handshake protocol, all the transmitted messages are encrypted using a single-key encryption algorithm, such as Advanced Encryption Standard (AES).

- Identity authentication: The identities of both communication parties are signed using public key encryption algorithms such as Rivest-Shamir-Adleman (RSA) and Data Security Standard (DSS) to prevent spoofing.

- Data integrity: All messages transmitted during communication contain digital signatures to ensure message integrity.

The basic architecture for managing digital certificates is the public key infrastructure (PKI), which uses a pair of keys (a private key and a public key) for encryption and decryption. A private key is mainly used for signature and decryption and is user-defined and known only by the key generator. A public key, however, is used for signature verification and encryption and can be shared by multiple users. The following describes how digital certificates work:

- Public key encryption is also known as asymmetric key encryption. When a confidential file is sent, the sender uses the public key of the receiver to encrypt the file, and the receiver uses its own private key to decrypt the file, as shown in Figure 6.6.

- Digital signature: A sender can also use a private key to generate a digital signature in a message to be sent. The digital signature uniquely identifies the sender. The receiver can then determine whether the message is tampered with by checking the digital signature. Figure 6.7 shows the encryption and decryption process using digital signatures.

Table 6.1 describes four types of digital certificates.

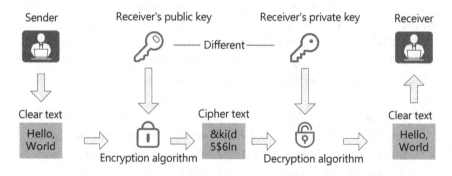

FIGURE 6.6   Public key encryption.

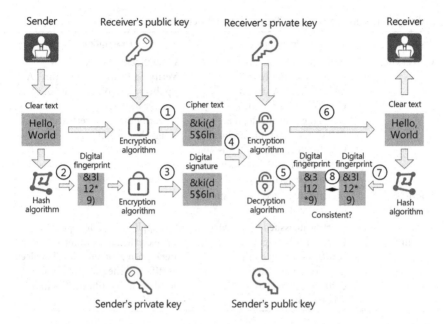

FIGURE 6.7   Encryption and decryption using digital signatures.

Figure 6.8 shows the automatic authentication process between the SDN controller and a device based on the digital certificate mechanism.

Step 1 A device is preconfigured with a device certificate and a private key pair before delivery. The device certificate and SDN controller certificate are sourced from the same root certificate.

TABLE 6.1   Four Types of Digital Certificates

Type	Definition	Description
Self-signed certificate	A self-signed certificate, also known as a root certificate, issued by an entity to itself. In this certificate, the issuer name and subject name are the same	If an applicant cannot apply for a local certificate from the CA, the device can generate a self-signed certificate. The process of issuing a self-signed certificate is simple A device does not support lifecycle management (such as certificate update and revocation) over its self-signed certificate. In this case, to ensure the security of the device and certificate, it is recommended that the self-signed certificate be replaced with a local certificate

(Continued)

TABLE 6.1 (*Continued*) Four Types of Digital Certificates

Type	Definition	Description
CA certificate	CA's own certificate. If a PKI system does not have a hierarchical CA structure, the CA certificate is the self-signed certificate. If a PKI system has a hierarchical CA structure, the top CA is the root CA, which owns a self-signed certificate	An applicant trusts a CA by verifying its digital signature. Any applicant can obtain the CA's certificate (including a public key) to verify the local certificate issued by the CA
Local certificate	A certificate issued by a CA to an applicant	-
Local device certificate	A certificate issued by a device to itself according to the certificate issued by the CA The issuer name in the certificate is the CA server's name	If an applicant cannot apply for a local certificate from the CA, the device can generate a local device certificate. The process of issuing a local device certificate is simple

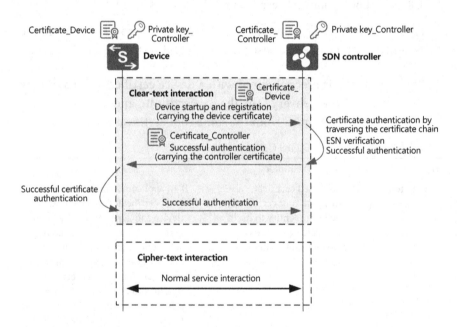

FIGURE 6.8 Certificate authentication process between the SDN controller and device.

Step 2 After being powered on, the device obtains an IP address and sends a registration request carrying the device certificate to the SDN controller.

Step 3 After obtaining the device certificate, the SDN controller traverses the certificate chain to verify the device certificate and verifies the device ESN.

Step 4 If authentication is successful, the SDN controller sends an authentication success message carrying the SDN controller certificate to the device.

Step 5 Upon receipt of the SDN controller certificate, the device performs certificate authentication and sends an authentication success message to the SDN controller if authentication is successful.

Step 6 After verifying each other's identity, the SDN controller and device start to use the encrypted channel for service delivery.

The SDN controller communicates with devices and systems using the Secure Sockets Layer (SSL) or Hypertext Transfer Protocol Secure (HTTPS) encryption protocol. The SDN controller is preconfigured with devices' SSL certificates for temporary communication with devices after the devices are installed. After temporary communication is complete, the pre-configured certificates need to be updated. Users can obtain the replacement certificates of the devices and SDN controller using either of the following methods:

- Method 1: If the carrier or enterprise has its own CA, apply for SSL client and SSL server certificates from the CA and obtain the corresponding trust certificates.

- Method 2: Use tools such as OpenSSL or XCA to make digital certificates.

6.3 VIRTUAL NETWORK AUTOMATION

As described in previous sections, physical network automation achieves Layer 3 connectivity between any two devices on the network. At this stage, the network administrator can start to construct virtual networks (VNs). To create a VN, they just need to perform two steps on the SDN controller.

First, specify resources including the physical device roles, IP address segment, and VN access location for the VN. This step is also called creating a fabric on an intent-driven network. The fabric virtualizes and pools all resources on the network and is presented in the form of VNs to carry services.

Second, create the VN based on service requirements, including specifying the VN name, available IP address segment, and access interfaces.

Throughout the entire process, the network administrator does not need to consider the specific network implementation. This significantly reduces the degree to which service requirements and network implementation are coupled, and improves network planning efficiency.

This two-step operational simplicity can be partially credited to the orchestration by the SDN controller. The following illustrates what happens to the SDN controller and network devices in the two steps performed by the network administrator.

6.3.1 Mapping between VNs and Resources

The SDN controller implements Framework as a Service (FaaS) to support the mapping from a physical network to a fabric by virtualizing a campus network, pooling network resources, and abstracting network services. When VNs are being created, FaaS resources are instantiated based on rules, as shown in Figure 6.9.

When a fabric is created, the SDN controller abstracts fabric resources into a virtual router pool, access port pool, subnet pool, and network egress pool. The following describes the functions of each pool:

- Virtual router pool: A virtual router uses a Virtual Private Network (VPN) to create an independent Layer 3 routing domain that provides the same functions as a physical router. Each VN occupies one VPN resource.

- Access port pool: Access port pool is a collection of ports that can be used for accessing VNs. The ports can be wired ports or wireless ports. Wired ports refer to ports on all access switches, and wireless ports refer to SSIDs. User access automation enables multiple VNs to use the same physical port.

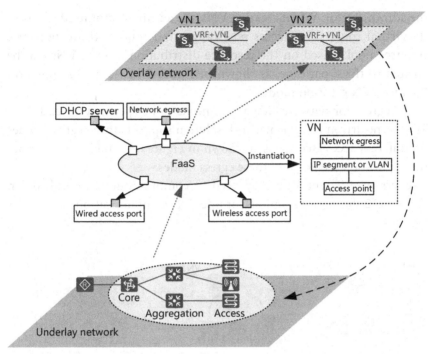

VRF refers to virtual routing and forwarding.

FIGURE 6.9   Resource instantiation.

- Subnet pool: Subnet pool contains IP address segments that can be assigned to VNs. A large network segment (for example, a class B network segment) assigned to a fabric is divided into subnets, with a subnet assigned to one VN.

- Network egress pool: Network egress pool connects VNs to external resources such as the Internet and other VNs.

6.3.2 Automated VN Deployment

After physical network resources are pooled, the administrator can create VNs based on service requirements. In real-world situations, a VN is an independent service network that is typically created for an independent department. For example, if a company has a marketing department, a finance department, and an R&D department, a VN can be created for each of the three departments. Automated deployment of these VNs involves resource pool instantiation and VN creation by the administrator. When creating VNs, the administrator needs to specify virtual routers, an access port range, and subnets from the fabric resource pool. Based on the resource distribution modes, VNs can be created in three patterns, as shown in Figure 6.10. Table 6.2 compares the three VN creation modes.

An egress for network-side access needs to be specified for each VN. The intent-driven campus network solution supports three egress modes for different access scenarios, as shown in Figure 6.11. Table 6.3 describes the application scenarios of these egress modes.

Users in different VNs may need to communicate with each other. Figure 6.12 shows two solutions for inter-VN communication.

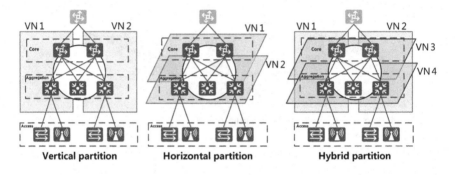

FIGURE 6.10   VN creation modes.

TABLE 6.2   Comparison of VN Creation Modes

Creation Mode	Application Scenario	Resource Distribution
Vertical partition	Users statically access VNs by using invariably authorized VLANs	Resources of specified physical switches are exclusively used by a VN
Horizontal partition	Users dynamically access VNs by using flexibly authorized VLANs	Resources of specified physical switches are shared by VNs
Hybrid partition	There are both users who access VNs statically and those who access VNs dynamically (applicable to network reconstruction and migration scenarios)	Resources of specified physical switches are shared by the VNs created in horizontal partition mode. When creating VNs in vertical partition mode, specify the physical switches to be used by the VNs

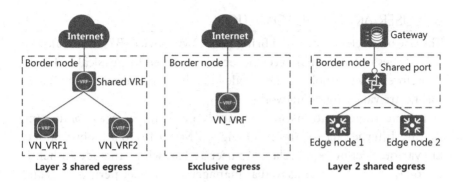

FIGURE 6.11   VN egress modes.

TABLE 6.3   Application Scenarios of Different VN Egress Modes

VN Egress Mode	Application Scenario
Layer 3 shared egress	Multiple VNs need to access the Internet or a DC Multiple VNs use the same security policy
Exclusive egress	Multiple VNs need to access the Internet or DC Each VN uses a customized security policy
Layer 2 shared egress	The border node does not function as a user gateway

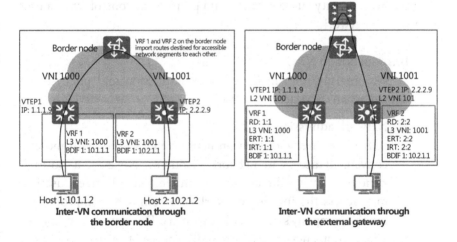

FIGURE 6.12   VN mutual access.

- Inter-VN communication through a border node: This mode is applicable when application-level policy control is not required.

- Inter-VN communication through an external gateway: This mode is applicable when application-level policy control is required.

6.4 USER ACCESS AUTOMATION

The rise of wireless access and Bring Your Own Device (BYOD) breaks the boundaries of traditional network security typified by physical isolation. Therefore, new technologies are needed to rebuild the security boundaries for the arrival of digital networks.

Behind the operations we commonly perform, such as conducting dial-up Internet access and inserting a SIM card into a mobile phone, are various technologies related to user access, which include authentication of user identities as well as management of user permissions and user accounts. However, these technologies are not suitable for campus networks as it is unlikely that each employee uses Point-to-Point Protocol over Ethernet (PPPoE) dial-up software on their tablets or that each desktop computer has a SIM card. Therefore, user access technologies need to be optimized to cater to differentiated requirements on campus networks.

6.4.1 User Access Technology

Campus networks typically use network access control (NAC) technology to implement identity authentication and permission control when a user accesses the network.

1. Introduction to NAC

NAC mainly implements network admission and policy-based control when users access a network.

a. Network admission

An open network environment provides users with more convenient access to network resources; however, it poses various security threats. For example, unauthorized users may access the internal network of a company, compromising the company's information security. Diversified terminals on a campus network are the main source of security threats as user activities on the network are difficult to manage and control. For security purposes, the campus network needs to authenticate users based on their identity and terminal status, and grant access permission only to those who pass the authentication. Figure 6.13 shows an example of network admission for various terminals.

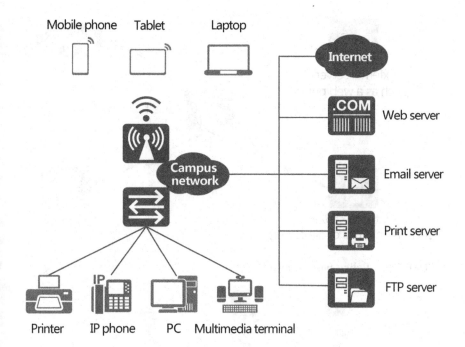

FIGURE 6.13   Network admission.

Network access authentication is like a gate of a courtyard, with the courtyard representing the network and the gate being typically an access device on the network. Terminals can pass through the gate to access the network only when they meet certain conditions such as time, location, and identity; in other words: when they are allowed access, from what device, and who can access it.

b. Policy-based control

Even if a terminal is authenticated and accesses the network, the terminal may not have access to all resources on the network. Rather, policies are used to grant terminals different network access permissions based on user identities. This mechanism is called policy-based control. On the campus network shown in Figure 6.14, guests can access external servers and the Internet, but not internal servers and the data center (DC) on the campus network.

On the network shown in Figure 6.15, different terminals in a same physical location may need to access different VNs.

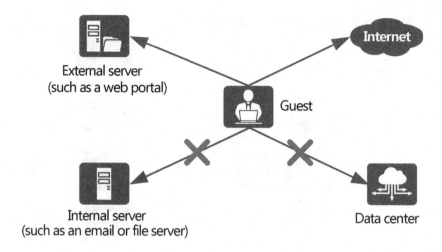

FIGURE 6.14   Policy-based control.

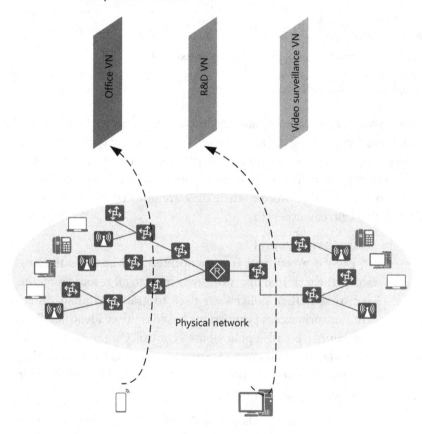

FIGURE 6.15   Different terminals accessing different VNs.

TABLE 6.4   Examples of Network Access Control Requirements of Different Terminals

Terminal Type	Access Control Requirement
Computer (staff access)	These terminals can access the internal company printer, file server, email server, and public resources on the network, with bandwidth limited to 100 Mbit/s
Mobile phone (staff access)	These terminals can access only the Internet, with bandwidth limited to 10 Mbit/s
Computer (guest access)	These terminals can access internal public resources and the Internet, but not internal servers such as the email server and file server. The bandwidth is limited to 100 Mbit/s
Mobile phone (guest access)	These terminals can access only the Internet, with bandwidth limited to 5 Mbit/s

After a terminal enters the courtyard from the gate, policies are used to control which rooms the terminal can enter. The requirements for NAC vary according to different types of terminals, as listed in Table 6.4.

2. Working mechanism of NAC

An NAC system is composed of terminals, admission devices, and admission servers, which work together to implement network admission and policy-based control, as shown in Figure 6.16.

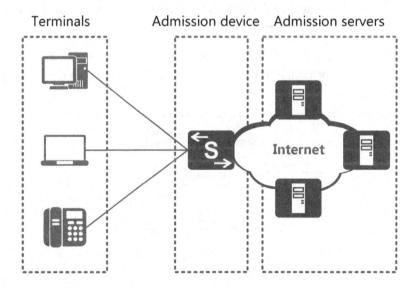

FIGURE 6.16   NAC system architecture.

a. Terminals: devices such as PCs, mobile phones, and printers that access the network

b. Admission devices: authentication control points that execute network security policies for network admission (for example, allowing or denying the access of users) on terminals. Admission devices can be switches, routers, APs, VPN gateways, or security devices.

c. Admission servers: devices that authenticate and authorize users. Admission servers verify the user identity of terminals that attempt to access the network, and grant network access permission to authenticated terminals. Admission servers usually include an authentication server such as a Remote Authentication Dial-In User Service (RADIUS) server or a user data source server for storing information on user identities.

The NAC process involves three phases: user identity authentication, user identity verification, and user policy authorization, as shown in Figure 6.17.

a. User identity authentication: When a terminal accesses a network, its identity needs to be authenticated first. The terminal sends its identity credentials to the admission device, and then to the admission server for identity authentication.

b. User identity verification: The admission server stores user identity information and manages users. After receiving the identity

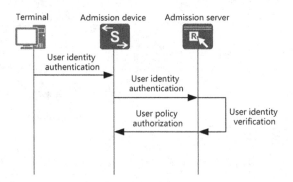

FIGURE 6.17   NAC process.

credentials of a terminal, the admission server determines whether the identity of the terminal is valid and sends the verification result and corresponding policy to the admission device.

c. User policy authorization: The admission device executes different policies on terminals based on the authorization result sent by the admission server. Policy-based control essentially allows or denies access from terminals, or performs more complex control such as increasing or decreasing the forwarding priority, limiting the network access rate, restricting access to some server resources, and allowing terminals to access a specific VLAN or VN.

3. Key NAC technology: user authentication

MAC address authentication, 802.1X authentication, and Portal authentication are widely used to authenticate users on campus networks.

a. MAC address authentication

Through this type of authentication, terminals are authenticated based on their MAC addresses. When a terminal accesses a network, an admission device obtains the terminal's MAC address and uses it as the user name and password of the terminal for authentication.

MAC address authentication is less secure because MAC addresses can easily be forged. In addition, terminals' MAC addresses need to be configured on the admission server for authentication purposes and are complex to manage. Therefore, MAC address authentication is used when end users do not wish to or are unable to enter user account information for authentication in some special cases. For example, some privileged terminals want to directly access the network without authentication, and user account information cannot be entered on dumb terminals such as printers and IP phones.

For network security purposes, MAC address authentication is usually implemented together with other security technologies. For example, static access control lists (ACLs) can be configured for the network segment of dumb terminals; or terminals can be bound to access interfaces after they are authenticated for the first time.

b. 802.1X authentication

Extensible Authentication Protocol (EAP) is used to exchange information between terminals, admission devices, and admission servers. EAP can run without an IP address over various bottom layers, including the data link layer and upper-layer protocols such as User Datagram Protocol (UDP) and Transmission Control Protocol (TCP). This offers much flexibility to 802.1X authentication.

i. EAP packets transmitted between terminals and admission devices are encapsulated in the EAP over LAN (EAPoL) format and transmitted across the Local Area Network (LAN).

ii. We can determine to use either of the following authentication modes between the admission device and admission server based on the support of the admission device and based on network security requirements.

- EAP termination mode: The admission device terminates EAP packets and encapsulates them into RADIUS packets. The admission server then uses the standard RADIUS protocol to implement authentication, authorization, and accounting.

- EAP relay mode: The admission device directly encapsulates the received EAP packets into RADIUS packets and transmits the packets over a complex network to the admission server.

In 802.1X authentication, an authentication client is required on a terminal for users to enter account information. The terminal (authentication client), admission device, and admission server exchange protocol packets to complete user identity authentication.

802.1X authentication provides the highest security and is applicable to scenarios where new networks are added, users are centralized, and stringent requirements on information security are imposed. It is recommended that 802.1X authentication be deployed on access devices.

c. Portal authentication

Portal authentication is also known as web authentication. Generally, a Portal authentication website is referred to as a web portal. When a user accesses the Internet, the user must be first authenticated on the web portal. If the authentication fails, the user can access only specified network resources. The user can access more network resources only after being successfully authenticated.

The Portal server pushes a web authentication page to the terminal, on which the user needs to enter account information, as shown in Figure 6.18. The terminal, admission device, and admission server exchange protocol packets to implement

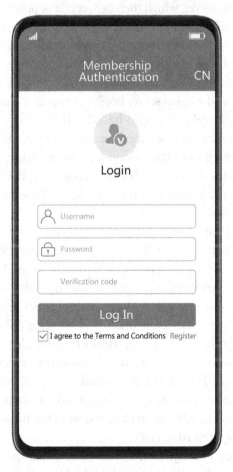

FIGURE 6.18   Customized Portal authentication page.

user identity authentication. In addition, Portal authentication can work with third-party servers such as Facebook for
authentication using third-party accounts.

Portal authentication does not require any client software
on a terminal; it is flexible and ideal for guests and users on
business trips. Additionally, the authentication page can be
customized.

4. Key NAC technology: policy-based control
To achieve policy-based control, user policies need to be defined on
the admission server, which then delivers these policies to an admission
device. The admission device performs actions defined in the policies
on users. Huawei's SDN controller can function as an admission server
to implement user identity authentication and policy-based authorization. The following uses the Huawei SDN controller as an example to
show how the admission server implements policy-based authorization.

Access Control List (ACL)-based authorization is typically used
when the SDN controller functions as an admission server. In this
authorization mode, the admission server delivers different ACL
numbers to the admission device according to the type of user. The
admission device then controls user access based on the preconfigured ACL rules.

Figure 6.19 shows the process of ACL-based authorization. Based
on requirements for user access control, the network administrator
preconfigures user access policies (including ACL numbers, ACL
rules, and corresponding policies) on the admission device and
defines user authorization policies (mapping between users and ACL
numbers) on the SDN controller. When a terminal accesses the network, the admission device transmits the user identity credentials to
the admission server. After the authentication succeeds, the admission server sends back the ACL number corresponding to the user
type to the admission device according to the configured authorization policies. The admission device then performs actions defined in
the preconfigured ACL rules.

ACL-based management of user policies depends on the network
topology, as well as IP address and VLAN plans. It is usually used on
small- and medium-sized campus networks with simple networking.

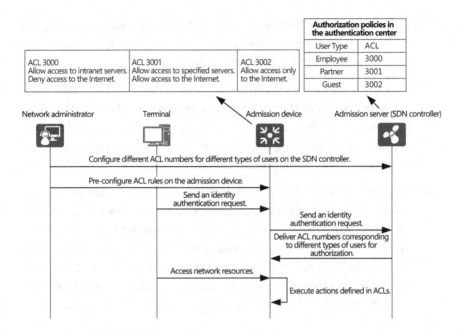

FIGURE 6.19 Process of ACL-based authorization.

6.4.2 User Access Challenges and Solutions

On a campus network, access technologies are required to meet the following requirements:

- Simple NAC configuration: To control user access, the network administrator needs to specify an admission server, and configure interconnection interfaces, admission server address, and authentication mode on each admission device. On a campus network, access devices are typically used as admission devices. Therefore, heavy configuration workload can occur with a large number of access devices, manual configuration is error-prone, and service adjustment is complex. The network administrator urgently requires a simplified process for network access configuration and automated delivery of configurations.

- Convenient account management system: The network administrator usually deploys the user name and password authentication for users who access the network dynamically, and therefore needs to obtain MAC addresses of dumb terminals in advance for MAC address authentication, which brings huge challenges to managing accounts. Managing accounts easily and conveniently during access control is a concern for network administrators.

- User-unaware identity identification: For network security purposes, users need to enter their user names and passwords for authentication each time they access the network, reducing convenience for the user and degrading user experience. As such, users require an identity authentication process that is unnoticeable.

- Flexible and simplified policy definition: After authenticated users access a campus network, the network administrator needs to manage and control these users by varying policies, for example, assigning different priorities and bandwidths to users, and configuring the specific VNs and resources that users can access. Therefore, the network administrator wants to manage campus network policies flexibly and efficiently while meeting various service requirements.

The user access automation solution used on an intent-driven campus network addresses the preceding challenges by providing the following functions:

- Automatic access configuration: The SDN controller implements centralized user access control, simplifying network access configuration.

- Automatic account management: The SDN controller provides northbound application programming interfaces (APIs) for interconnection with third-party account management systems and allows guest self-registration, facilitating account management.

- Automatic user identity identification: MAC address–prioritized authentication, terminal identification, and certificate authentication are supported to implement user-unaware access.

- Automatic user policy management: Security groups are used to flexibly manage user policies.

6.4.3 Automatic Access Configuration

In an NAC system, authentication channels need to be established between admission devices and the SDN controller in advance. Automatic access configuration is implemented on an intent-driven campus network, greatly simplifying user operations. Specifically, devices automatically register with the SDN controller after being powered on, and obtain the authentication and interconnection configurations automatically generated by the SDN controller. Figure 6.20 shows the automatic access configuration process.

Step 1 Using the plug-and-play function, an admission device automatically registers with the SDN controller after being powered on.

Step 2 The network administrator enables authentication for sites in a batch or for a single site on the network management page of the SDN controller.

Step 3 The SDN controller with the integrated admission server function automatically generates authentication and interconnection configurations based on the registration information of a device and delivers the configurations to the admission device. In this way, automatic interconnection is implemented between the SDN controller and admission device, without the need to manually configure their IP addresses.

Step 4 Upon receipt of the configurations, the admission device automatically establishes an authentication channel with the SDN controller.

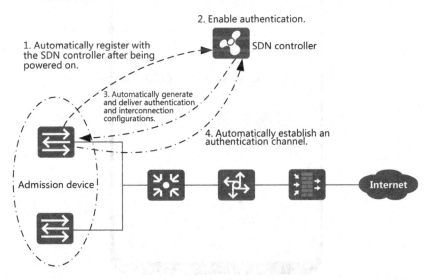

FIGURE 6.20   Automatic access configuration process.

6.4.4 Automatic Account Management

The intent-driven campus network uses the following methods to implement automatic account management in the NAC system, to simplify management of employees' and guests' accounts:

1. Self-service guest account registration

The SDN controller provides the self-service account registration function for guests. When a guest accesses a campus network, the SDN controller automatically pushes an account registration page to the guest, as shown in Figure 6.21. For network security purposes,

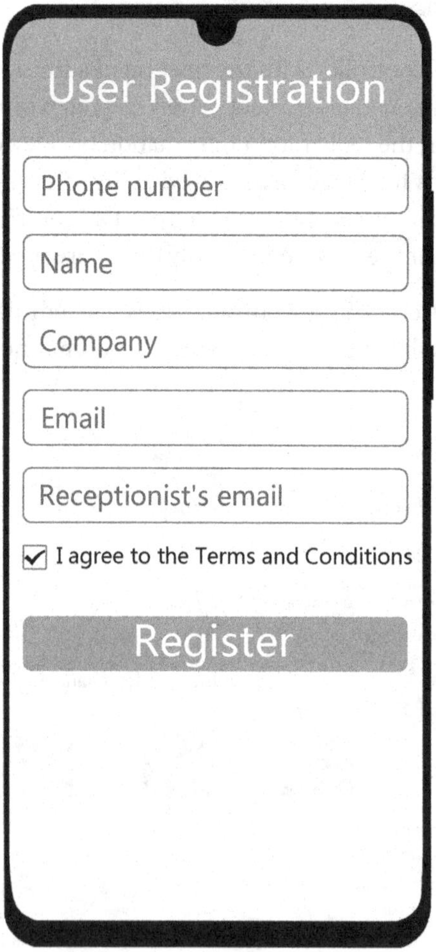

FIGURE 6.21 User account registration page.

the account approval function is also supported. An account is successfully created only after it is approved. Figure 6.22 shows the page of registration success, waiting for further approval.

2. Social media system interconnection

The SDN controller can interconnect with third-party social media systems to enable guests to directly access a campus network through social media accounts. Figure 6.23 shows an example of authentication using a social media account.

FIGURE 6.22   Registration success page.

FIGURE 6.23   Social media account authentication.

3. Interconnection with a third-party account management system

   In terms of enterprise employee accounts, the SDN controller can interconnect with a third-party account management system, for example, a Lightweight Directory Access Protocol (LDAP) or Active Directory (AD) system, to provide more powerful account management capabilities (such as registration, distribution, and approval). During user authentication, the user account and password information is automatically verified by the third-party account management system, eliminating the need to manage account and password information on the SDN controller.

6.4.5 Automatic User Identity Identification

As described above, user access permission control is mainly implemented by an NAC system. When a user accesses the network, the NAC system first needs to identify the user's identity. The intent-driven campus network solution implements automatic user identity identification using the following methods:

1. MAC address–prioritized authentication

   Portal or 802.1X authentication can be used to authenticate a user when the user accesses the network for the first time. If the user enters the correct user name and password, the user is successfully authenticated. The SDN controller then records the user's MAC address and binds it to the user identity. When the user accesses the network next time, the user MAC address is preferentially used for authentication. This process is MAC address–prioritized authentication, which implements user-unaware identity identification, as shown in Figure 6.24.

2. Terminal identification for automatic network access

   To enable dumb terminals such as printers and IP phones to access the network, the network administrator needs to obtain their MAC addresses in advance and record them in the account management

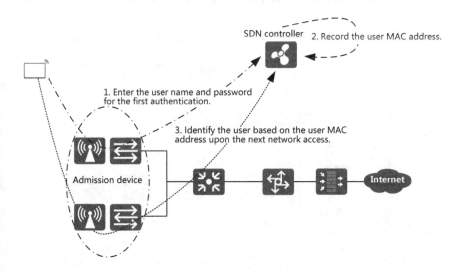

FIGURE 6.24   MAC address–prioritized authentication.

system of the SDN controller, which is tedious and error-prone. To address this issue, terminal identification technology is introduced to automatically identify dumb terminals. With this technology, the network administrator only needs to preconfigure an authentication-free access policy for dumb terminals on the SDN controller, as shown in Figure 6.25.

3. Certificate authentication

If a network has higher security requirements, certificate authentication can be used for terminal access control. Clients can automatically apply for and install user certificates, configure access parameters, and connect to the network using the certificates. This achieves user-unaware network access and simplifies user operations, while ensuring high security. Figure 6.26 shows the typical certificate authentication process.

6.4.6 Automatic User Policy Management

Traditional campus networks mainly use ACLs as user access control policies. However, ACL-based policy configuration relies on the network topology, IP address plan, and VLAN plan. ACL rules need to be changed if the network topology, IP addresses, VLANs, or location of user changes. As a result, the user policy configuration cannot be decoupled from the physical network, leading to poor flexibility and

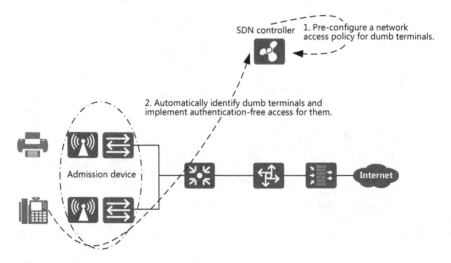

FIGURE 6.25   Terminal identification for automatic network access.

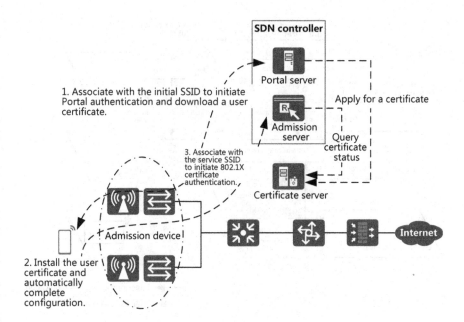

FIGURE 6.26   Typical certificate authentication process.

maintainability. With the popularization of mobile office, users hope for a consistent experience when they use different terminals to access a network at different locations.

1. Policy automation

Traditionally, policy-based control was focused on IP addresses. This conventional approach is not suitable for an intent-driven campus network. Ideally, policy-based control is centered on user identities, which means it is decoupled from the physical network topology, IP address plan, and VLAN plan. Huawei's free mobility technology perfectly meets these requirements. It focuses on services, users, and experience, uses the SDN controller for centralized management, and uses service languages and global user groups to maintain consistency in user policies regardless of changes in user locations and IP addresses.

Free mobility divides a campus network into multiple logical layers, as shown in Figure 6.27.

a. User terminal layer: provides man–machine interfaces for users to complete authentication and access resources on servers.

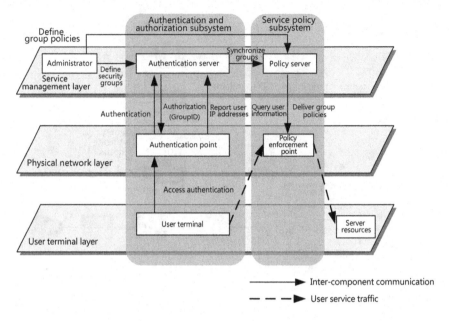

FIGURE 6.27   Logical architecture of free mobility.

b. Physical network layer: Network devices at this layer respond to authentication requests initiated by user terminals, report user information to an authentication server, and control users' service traffic based on policies. Both authentication and policy enforcement points reside at the physical network layer.

   i. An authentication point responds to authentication requests from clients and interacts with an authentication server to complete user authentication.

   ii. A policy enforcement point enforces service policies for users.

c. Service management layer: provides UIs for the network administrator to determine global authentication and authorization rules as well as service policies. The SDN controller at this layer provides an authentication and authorization subsystem as well as a service policy subsystem, and works with network devices to implement user authentication and policy management, achieving decoupling between service policies and IP addresses.

The authentication and authorization subsystem completes user authentication, as well as collecting and managing information about network-wide online users. The service policy subsystem synchronizes service policies between the policy server and policy enforcement points to implement unified policy deployment, as well as controlling user traffic based on security groups.

2. Free mobility

The free mobility solution abstracts IP-based policies into user language–based policies that are implemented based on security groups. As shown in Figure 6.28, network objects of the same type and with the same permission are added to the same security group. The network objects can be PCs, mobile phones, printers, and servers. For example, users in the R&D department in an enterprise have the same access permissions on network resources; therefore, hosts in the R&D department can be added to an R&D host group, and servers accessible to users in the R&D department can be added to an R&D server group. Another example is that all printers in the enterprise can be added to a printer group.

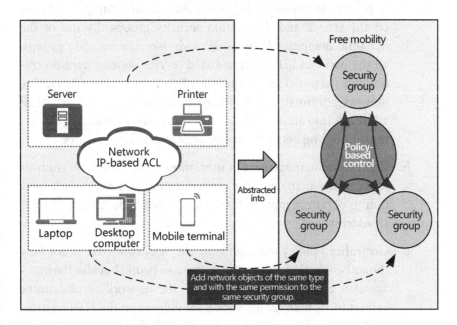

FIGURE 6.28   Security group division.

The free mobility solution classifies network objects by security group and defines the network services that a security group can access through security group policies. On the SDN controller, the network administrator centrally plans network services available to security groups in the policy matrix to achieve access permission and application control.

The free mobility solution uses security groups to decouple policies from IP addresses, enabling the network administrator to implement policy-based control between security groups without the need to know the users' IP addresses. Intergroup policies are defined on the SDN controller, enabling quick policy creation and automated policy provisioning. The free mobility solution resolves issues that traditional campus networks face in the following ways:

a. Decoupling of service policies from IP addresses: Using the SDN controller, the network administrator can divide network-wide users and resources into different security groups from different dimensions. A device first determines the source and destination security groups based on the source and destination IP addresses of packets and then finds the matching intergroup policy based on the source and destination security groups. By use of the dynamic mappings between IP addresses and security groups, all the user- and IP address–based service policies used on traditional networks can be migrated to intergroup policies. The network administrator no longer needs to take users' actual IP addresses into account when predefining service policies, achieving decoupling between service policies and IP addresses.

b. Centralized management of user information: The SDN controller centrally manages user authentication and online information to obtain mappings between network-wide users and their IP addresses.

c. Centralized policy management: The SDN controller is not only the authentication center on a campus network, but also the management center of service policies. The network administrator can centrally manage network-wide policies on the SDN controller. Once service policies are configured, they are automatically delivered to policy enforcement devices on the network.

Table 6.5 compares free mobility with a traditional access control solution in terms of network experience consistency as well as deployment and maintenance efficiency. The comparison result shows the noticeable advantages in free mobility.

3. Automatic policy delivery by the SDN controller

Figure 6.29 shows how free mobility is jointly implemented by the SDN controller and network devices.

Step 1 The network administrator creates security groups and intergroup policies on the SDN controller.

Each security group is a collection of users with the same network access permission. Intergroup policies define the network services that users can access.

TABLE 6.5   Comparison between Free Mobility and Traditional Access Control Solution

Solution	Network Experience Consistency	Deployment and Maintenance Efficiency
Free mobility	Security groups are used to decouple policies from IP addresses, eliminating the need to take into account the network topology, VLANs, and host IP addresses. Consistent user permissions and experience are delivered based on user identities regardless of the terminal types and user locations	Only security groups and inter-group policies need to be defined based on user identities, simplifying network planning and deployment. The network administrator centrally plans security groups and manages service policies on the SDN controller, facilitating management and maintenance
Traditional access control	ACL-based policies are tightly coupled with the network topology and IP addresses, and service policies need to be reconfigured if user locations or the network topology changes. In mobility scenarios, the planning for configuration is complex, the configuration workload is heavy, and it is difficult to ensure consistent network access permissions for users	A large number of VLANs, IP addresses, and ACL rules need to be planned in the early phase of network design. User policies are mapped to ACL rules based on the planned VLANs and IP addresses, which require complex configurations. The network administrator needs to manually configure devices one by one, complicating management and maintenance

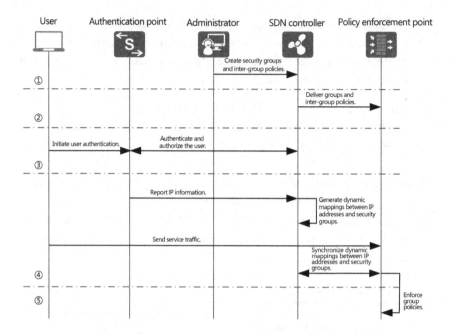

FIGURE 6.29   Implementation of the free mobility solution.

Step 2 The SDN controller delivers security groups and intergroup policies to the policy enforcement point.

Security groups and intergroup policies take effect only after they are deployed on network devices.

Step 3 A user is authenticated and authorized based on the security group to which the user belongs.

The user can be authenticated using MAC address, Portal, or 802.1X authentication. The authentication process is the same as that on a traditional campus network. The SDN controller verifies the user identity by checking the user name and password, and associates the user with the corresponding security group based on preconfigured authorization rules. If the authentication succeeds, the SDN controller sends the authorization result containing the security group to which the user belongs to the authentication point.

Step 4 The SDN controller generates dynamic mappings between IP addresses and security groups and synchronizes them to the policy enforcement point.

After the user is authenticated, the authentication point can obtain the user IP address from a user packet, such as an ARP packet, and sends the user IP address to the SDN controller. The SDN controller then generates a dynamic mapping between the user IP address and the corresponding security group based on the security group contained in the authorization result. If the user IP address changes due to a terminal location change or re-authentication, the SDN controller automatically updates the dynamic mapping through packet exchange. When the user initiates a service flow, the policy enforcement point synchronizes the dynamic mappings between IP addresses and security groups from the SDN controller.

Step 5 Execute intergroup policies.

The policy enforcement point obtains the source and destination IP addresses of the service flow that pass through it, queries the dynamic mapping table for the security group corresponding to the user's IP address, and then executes the corresponding inter-group policy.

Note: In actual applications, a policy enforcement point may not be an authentication point. For example, if users are authenticated on the core switch while traffic control is performed on the firewall, the firewall is the policy enforcement point and synchronizes the IP address-to-security group mappings from the SDN controller. In contrast, if a device serves as both the policy enforcement point and authentication point, such synchronization is not required because the device can obtain user IP addresses and generate dynamic mappings by acting as the authentication point.

For policy-based control on east-west traffic, for example, denying mutual access between security groups, it is recommended that switches be used as policy enforcement points. For policy-based control on north-south traffic, for example, controlling access from a security group to the Internet, it is recommended that firewalls be used as policy enforcement points.

CHAPTER 7

Intelligent O&M on an Intent-Driven Campus Network

TRADITIONALLY, NETWORK OPERATIONS AND maintenance (O&M) is usually device-centric. That is, O&M personnel have to log in to devices one by one or use a network management system (NMS) to centrally manage multiple devices. However, the NMS can only display data immediately after obtaining it or display it after basic processing. Data analysis still needs to be carried out manually, which is time-consuming and labor-intensive, as well as incurring high costs and requiring operators to be highly trained. For example, in traditional O&M scenarios, when customers or employees complain about network problems such as video freezing, authentication failure, and Internet access failure, O&M personnel need to immediately log in to network devices to check logs. If a problem has persisted for a long time, they also need to download large amounts of historical logs and search for desired data. Although the NMS generates logs to offer some statistical data, this is often not enough to help administrators locate all problems, which may require further data analysis. As such, O&M personnel need to be familiar with the system processing procedures. In addition, with the capacity expansion of campus networks, large quantities of branch networks must be managed in a unified manner. In most cases, branch networks have no dedicated O&M

personnel and are instead centrally managed by O&M professionals at the headquarters. If the network is large in scale and has many branches, it is almost impossible for O&M personnel to effectively maintain the entire network manually.

What's more, as more campus networks are going wireless, a larger number of wireless terminals access campus networks, such as mobile phones, tablets, printers, and electronic whiteboards. This therefore brings new O&M considerations to campus networks, such as the mobility and air interface interference of wireless networks. Fast fault locating, quick service restoration, and user experience guarantee in the wireless environment are new demands for O&M of campus networks. In the Internet of Everything (IoE) network model, large amounts of devices and data need to be operated and maintained. With all of these changes to campus network, the lack of automation and intelligence offered by traditional O&M means is far from able to meet customers' O&M requirements. Instead, overcoming these challenges requires Artificial Intelligence (AI) technology to be applied to the network O&M field, thereby implementing automatic analysis and intelligent processing of network data.

7.1 GETTING STARTED WITH INTELLIGENT O&M

The intelligent O&M solution radically changes the resource status-centric monitoring mode in traditional O&M solutions and applies AI to the network O&M field. Based on existing O&M data (such as device performance indicators and terminal logs), the intelligent O&M solution uses big data analytics, AI algorithms, and more advanced analysis technologies to digitize user experience on the network as well as visualize the network operational status. These capabilities enable rapid detection of network issues, prediction of network faults, correct network operations, and improved user experience.

7.1.1 Experience-Centric Proactive O&M

As shown in Figure 7.1, in traditional network O&M, problems can be located and resolved only after service exceptions occur or users complain. This reactive, postevent network O&M mode is becoming increasingly inadequate on large and midsize campus networks, failing to meet their requirements for fast and efficient O&M.

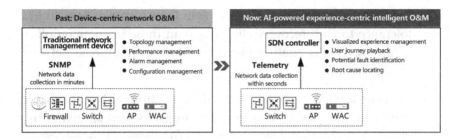

FIGURE 7.1 Transformation of network O&M from device-centric to experience-centric.

Traditional O&M faces the following problems:

- Passive response to faults: Anomalies or faults on traditional networks are mainly detected based on device alarms or user complaints. In most cases, network faults can only be handled after they occur. Therefore, network O&M personnel have to passively respond to customer needs. Especially during major holidays or big events, they have to always be prepared for possible network faults to occur. In addition, with the convergence of wired networks, wireless networks, and the Internet of Things (IoT), the boundary of campus networks extends and the number of Network Elements (NEs) increases. Once a fault occurs, a large number of alarms are generated on the network. This overloads O&M personnel and leads to more user complaints, ultimately forming a vicious cycle.

- Slow fault recovery: In traditional O&M, when a service exception occurs, O&M personnel first need to check the network topology before logging in to devices through the Command Line Interface (CLI) to locate the fault. Using this method, more than 60% of the faults need to be rectified on site. If the symptom of a fault disappears, O&M personnel have to wait until the symptom recurs or reproduce the fault. What's worse, wireless reconstruction further increases the troubleshooting difficulty. Because of the complexity of the wireless environment, more than 90% of faults need to be located on site. However, a growing number of digital businesses are oriented to production and customer services such as automatic medication dispensing systems in medical scenarios, unattended payment systems in commercial scenarios, and Automated Guided Vehicles (AGVs) in warehousing scenarios. Customers in these scenarios have far less tolerance to fault recovery efficiency than those in common office scenarios. This is another major challenge to traditional network troubleshooting.

- Difficult service experience detection: Traditional NMSs provide functions like device management, topology management, and alarm configuration. Leveraging such functions, O&M personnel can monitor the network topology and alarms to discover network anomalies. However, even if the network devices are running normally, it does not necessarily mean that services are too, or that user experience is smooth. This is because more terminals are connecting to the Internet as mobility demands and IoT development surge, as well as terminals, operating systems, services, and traffic models becoming more complex and diverse. For example, when strong co-channel interference exists, wireless users connected to an access point (AP) have low Internet access quality even if the AP is running properly. And when there is no differentiated management for a latency-sensitive service, user experience of the service will be poor even if network devices are running properly.

Intelligent network O&M centers on user experience. That is, it uses a predictive and intelligent NMS, namely, Software-Defined Networking (SDN) controller, to improve user and service experiences, and dramatically transforms the traditional O&M mode that is based on manual and delayed data analysis.

The SDN controller provides an intelligent, automatic, proactive network analysis system that integrates big data analytics and AI computing capabilities. It can predict faults and adjust the network in advance to reduce the chance of them occurring. Intelligent network O&M has the following benefits over traditional network O&M:

1. Fault identification and proactive prediction
    a. Automatic fault identification: Big data and AI technologies are used to automatically identify issues regarding connection, air interface performance, roaming, devices, and applications, improving the identification rate of potential issues.

b. Proactive fault prediction: Historical data are learned through machine learning to dynamically generate a baseline, which is then compared and analyzed with real-time data to predict possible faults.

With the automatic fault identification and proactive prediction functions in intelligent O&M, the SDN controller can predict some faults on the network based on big data analytics and provide relevant warnings. For example, it can predict faults on optical modules, which, due to a large number of network devices communicating through optical links, must be used to connect optical fibers to devices. Predicting such faults is critical because optical modules are vulnerable to dust and electrostatic discharge, which leads to great link loss. Additionally, incorrect installation and removal of optical modules may cause module faults and affect services. This function can also be used to display the status of all optical modules on the entire network, as shown in Figure 7.2. With the help of big data

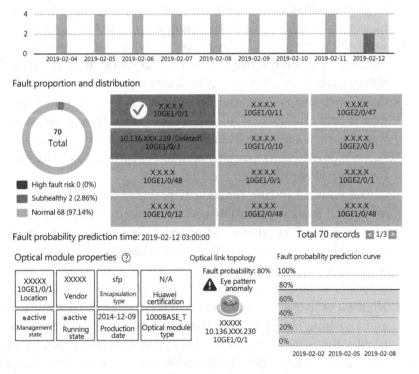

FIGURE 7.2　Optical link fault prediction in intelligent O&M.

analytics and machine learning algorithms, the SDN controller can detect and predict optical link faults before they affect services.

2. Fast fault locating and intelligent root cause analysis

   a. Fast fault locating: The fault type and impact range are intelligently identified using the network O&M expert system and various AI algorithms, helping administrators to better locate faults.

   b. Intelligent root cause analysis: Possible causes of faults are analyzed based on the big data platform and troubleshooting suggestions are provided.

These merits make life much easier for network O&M personnel, who need to ensure normal running of networks through routine maintenance, as well as locating and rectifying faults quickly. Traditionally, this would have to be done by manually analyzing large amounts of data and relying on personal experience, which is hugely challenging and time-consuming. For example, when network O&M personnel receive reports about user network access difficulties or failures, they would need to check all login logs and obtain packets. In addition, network O&M personnel may be unfamiliar with the complex user network access process, which varies greatly between different authentication modes. In such cases, they would have to seek help from professional engineers to locate user access failures. Faults may also need to be reproduced on site, making troubleshooting even more difficult.

The protocol tracing function is provided in intelligent O&M. With such function, the SDN controller can display each phase of user access and the corresponding result (success or failure) on the graphical user interface (GUI), as well as providing root causes of user access issues, helping O&M personnel quickly locate faults. Figure 7.3 details the protocol tracing function, which enables visualization of user access in three phases: association, authentication, and Dynamic Host Configuration Protocol (DHCP). The function enables the SDN controller to collect statistics on the results and time spent in each protocol interaction phase, and performs refined

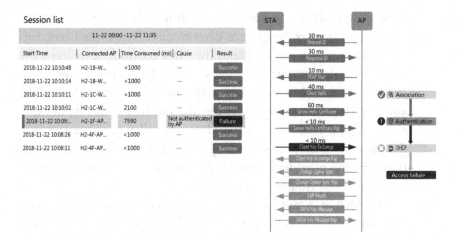

FIGURE 7.3   Protocol tracing.

analysis on the user access process to quickly detect anomalies, achieving precise fault locating.

When receiving an authentication failure report from a user, O&M personnel can search for the user's session records based on the terminal's Media Access Control (MAC) address, and then get a clear understanding of the number of authentication successes and failures. O&M personnel can also check details about the user access authentication process of an access failure record, determine the phase where a fault occurs, and rectify the fault based on the provided troubleshooting suggestions.

3. Experience visualization

The SDN controller displays the network operational status in real time, allowing O&M personnel to be aware of the actual network status. It uses Telemetry, a real-time hardware-based data collection technology, to collect data of specified nodes based on service requirements and at exact time points, so as to accurately display the network status. In this way, O&M personnel can learn about the network status including user experience and application running status, and then perform multidimensional O&M on the network.

In traditional network O&M, O&M personnel cannot proactively detect user experience deterioration or be aware of the actual operational status of the network, and are therefore unable to optimize the network in real time. In addition, it is difficult for them to locate the root causes of user experience deterioration. With intelligent network O&M, on the other hand, the SDN controller can learn the indicator deterioration thresholds using dynamic learning algorithms based on historical key performance indicator (KPI) data, and then determine and display the users with poor quality of experience (QoE). Additionally, the SDN controller analyzes the QoE data of each of them based on big data, identifies KPIs that impact the QoE, and provides causes for the QoE deterioration, facilitating both O&M and troubleshooting.

As shown in Figure 7.4, the SDN controller automatically analyzes QoE data of users on the entire network and provides detailed data about the poor-QoE users, including the user information, duration with poor QoE, and possible causes.

Figure 7.5 shows the cause analysis on poor QoE. Specifically, the network status when a user experiences poor network quality is checked and analyzed. In this example, it is determined that the user remained in weak-signal connection state for a long time, which ultimately led to the poor QoE. Such problems are generally caused by poor signal coverage, which can be solved by increasing the AP transmit power or deploying an additional AP in the coverage area.

7.1.2 Visualized Quality Evaluation System

As enterprises develop and may expand their businesses globally, worldwide deployment of enterprise networks is required to meet their service demands. Against this backdrop, traditional network O&M, which does not support unified and visualized network management, would surely overburden O&M personnel as they have to constantly pay attention to the network status of networks all around the world. To solve this issue, the SDN controller can be deployed to build a visualized network evaluation and monitoring system, which covers the entire network, branch networks, network devices, and even users and applications. In this way, the SDN controller can intuitively display the network operational status, as well as detecting faults and analyzing their root causes.

As the Chinese saying goes, even the cleverest housewife cannot cook a meal without rice. And this is also true with the big data-based intelligent fault analysis system. That is, data are the foundation for fault analysis, and only when there is sufficient valid data can the correct

Poor-QoE user

XXXXXX
XXXX

User journey

 According to analysis, user XXXXXX had poor QoE. The user was online for a total of 10°15'30", and experienced poor network quality for 3°30', accounting for 34.12% of the total online duration.

According to analysis, the user encountered the following issues:

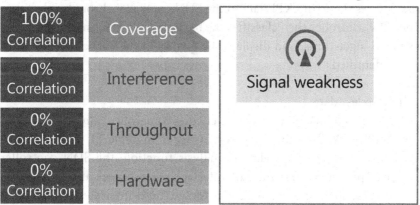

100% Correlation	Coverage
0% Correlation	Interference
0% Correlation	Throughput
0% Correlation	Hardware

Signal weakness

FIGURE 7.4 Data of a poor-QoE user.

fault analysis be obtained. This makes data collection especially important for fault analysis. The SDN controller uses Telemetry, a hardware-based data collection mechanism, to collect data through chips. This enables data collection to complete within microseconds. In addition,

The system performs correlation analysis based on big data to automatically identify KPIs that affect the QoE of users:

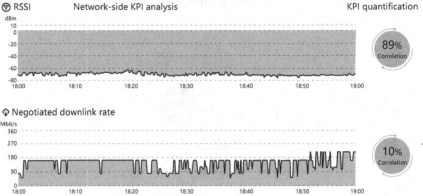

FIGURE 7.5   Cause analysis on QoE deterioration.

hardware-based data collection can be customized based on service requirements. All these advantages help build a big data support system that is optimal in terms of data timeliness and satisfaction, implementing intelligent O&M.

The intelligent O&M solution supports fault identification and root cause analysis for both wired and wireless networks, with wired and wireless devices reporting KPI data to the SDN controller through Telemetry. The SDN controller then classifies the big data and performs data analysis using AI algorithms, and displays the quality of the entire network with faults identified.

1. Wireless data

    The quality of a wireless network is mainly evaluated based on KPIs of APs, radios, and users, as listed in Table 7.1. With AI algorithms and the correlation analysis function, the SDN controller can proactively identify air interface performance and user access issues, such as weak signal coverage, high interference, and high channel usage.

2. Wired data

    Wired network devices use Telemetry to collect performance indicator data of devices, interfaces, and optical links, as listed in Table 7.2. Such devices also predict baselines of KPIs including the device CPU usage and memory usage using AI algorithms such as

TABLE 7.1   KPI Data Collected by Wireless Network Devices Using Telemetry

Measurement Object	KPIs	Applicable Device Type	Minimum Sampling Precision (Seconds)
AP	CPU usage, memory usage, and number of online users	AP	10
Radio	Number of online users, channel usage, noise, traffic, backpressure queue, interference rate, and power	AP	10
User	RSSI, negotiated rate, packet loss rate, latency, DHCP IP address obtaining, and Dot1x authentication data	AP	10

TABLE 7.2   KPI Data Collected by Wired Network Devices Using Telemetry

Measurement Object	KPIs	Applicable Device Type	Minimum Sampling Precision (Minutes)
Device/card	CPU usage	Switch and WAC	1
	Memory Usage	Switch and WAC	1
Interface	Number of received/sent packets, number of received/sent broadcast packets, number of received/sent multicast packets, number of received/sent unicast packets, number of discarded received/sent packets, number of received/sent error packets	Switch and WAC	1
Optical link	Rx/Tx power, bias current, voltage, and temperature	Switch	1

time series decomposition and aperiodic sequence Gaussian fitting. Then they compare the predicted baselines with the dynamic baselines to identify the deterioration of KPIs before services are affected. The SDN controller can detect fiber connection issues using the logistic regression and linear regression algorithms, and displays the status of network-wide optical modules (known faults, possible

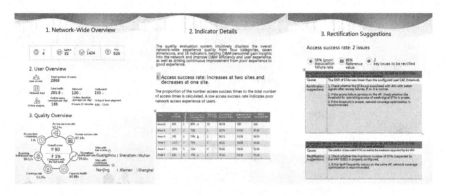

FIGURE 7.6   Quality evaluation report.

faults, and fault probability distribution), achieving proactive network monitoring, anomaly prediction, and fault prewarning.

3. Visualized quality evaluation system

Based on the data reported by devices, the SDN controller builds a visualized campus user experience quality evaluation system that covers the following perspectives: user access experience, roaming experience, throughput experience, and network availability. It also displays the overall network quality, helping O&M personnel gain insights into the network and improving both O&M efficiency and user experience. In addition, the SDN controller provides professional network evaluation reports on the network overview, indicator details, and rectification suggestions in real time or periodically. In this way, it provides quantifiable network services. Figure 7.6 shows an example of a quality evaluation report.

7.2 KEY TECHNOLOGIES USED FOR INTELLIGENT O&M

As outlined in the previous section, the intelligent O&M solution uses Telemetry to collect high-performance real-time data and adopts AI algorithms to analyze and present data, thereby implementing user experience visualization. In addition to these technologies, this solution leverages Enhanced Media Delivery Index (eMDI) technology to monitor audio and video services, as well as detect the quality of these services, ensuring user experience.

7.2.1 Architecture of Intelligent O&M

1. Logical architecture of intelligent O&M

    The SDN controller uses Telemetry and other mechanisms to collect information such as packet loss, traffic, status, and configuration of network devices. Combining Telemetry with AI algorithms (including dynamic baseline and Gaussian process regression) and the fault library that is built based on Huawei's years of O&M experience, the SDN controller automatically builds fault baselines through machine learning and uses the AI algorithms to improve O&M efficiency. Leveraging continuous scenario-based learning and expert experience, the SDN controller provides multilayer correlation analysis capabilities on service flows, forwarding paths, and network services, freeing O&M personnel from being pestered by constant alarms. The SDN controller displays application behaviors and network quality in a structured manner, making network O&M more intelligent and automatic, as well as enabling proactive evaluation of the network service status. Figure 7.7 shows the logical architecture of the intelligent analysis system of the SDN controller.

    a. Data collection: The intelligent analysis system of the SDN controller interconnects with network devices through the following types of southbound interfaces: Telemetry-based Hypertext Transfer Protocol version 2 (HTTP/2) + ProtoBuf (referring to Google Protocol Buffers, a mechanism for serializing structured data), Simple Network Management Protocol (SNMP), and Syslog. The following describe these interfaces in detail:

        i. HTTP/2 + ProtoBuf: The intelligent analysis system of the SDN controller uses Telemetry-based HTTP/2 and ProtoBuf protocols to collect device performance data. ProtoBuf is a mechanism for serializing structured data, similar to the Extensible Markup Language (XML), JavaScript Object Notation (JSON), and Hessian. It is widely applied in data storage and communication protocols. HTTP/2, on the other hand, uses the Secure Sockets Layer (SSL) and Transport Layer Security (TLS) protocols for authentication and encryption of communication channels.

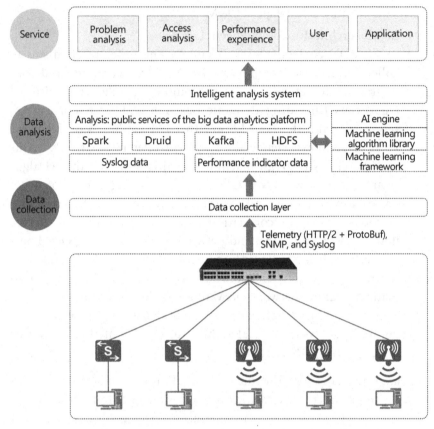

Note: Kafka is a high-throughput, distributed publish-subscribe messaging system.
Spark is a fast general-purpose computing engine designed for large-scale data processing.
Druid is a fast column-oriented distributed data store. It provides fast data aggregation and data query in sub-seconds, and can ingest millions of events per second.
The Hadoop Distributed File System (HDFS) provides high-throughput data access to application data and is suitable for applications that have large data sets.

FIGURE 7.7 Logical architecture of the intelligent analysis system of the SDN controller.

ii. SNMP: The intelligent analysis system of the SDN controller supports standard SNMPv2c and SNMPv3 interfaces, through which the SDN controller can interconnect with network devices. SNMP is an application-layer network management protocol based on the Transmission Control Protocol/Internet Protocol (TCP/IP) architecture. SNMP uses the User Datagram Protocol (UDP) as its transport-layer protocol and can be used to manage network devices that support proxy processes.

iii. Syslog: Syslog, a protocol for forwarding system log information on the IP network, has become a standard industrial protocol for recording device logs. It is used by the intelligent analysis system of the SDN controller to receive logs reported by network devices.

b. Data analysis: The big data analytics platform can collect and analyze millions of data flows per minute based on the distributed database, high-performance message distribution mechanism, and distributed file system. Of these, the distributed database provides distributed computing, aggregation, and storage of large amounts of real-time data, as well as supporting multidimensional data retrieval and statistics query in seconds. The machine learning algorithm library currently contains multiple network O&M analysis algorithms, providing AI services for upper-layer O&M applications. It can be constantly expanded.

c. Application service: The intelligent analysis system of the SDN controller provides a large number of application services for data analysis based on typical O&M and troubleshooting scenarios of campus networks. For example, it can intelligently detect connection, air interface performance, roaming, and device issues, analyze connection and performance issues, play back user journeys, analyze AP details, and detect the quality of audio and video services.

2. Data processing by the SDN controller

Processing data such as network performance data and logs are key for the SDN controller to implement digital and intelligent network O&M. Based on big data and AI analytics, the SDN controller displays the data processing result in a user-friendly and easy-to-understand way. Figure 7.8 shows the types of data collected by the SDN controller.

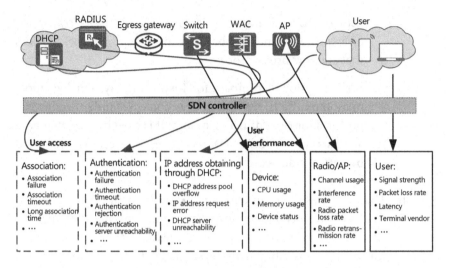

FIGURE 7.8   Types of data collected by the SDN controller.

Currently, the SDN controller collects two types of data:

a. User access data: refers to the access logs generated when users access networks. Such logs include those generated in the association, authentication, and DHCP phases, as well as those recording success and failure information in addition to access failure causes. Network devices use the Syslog protocol to report log data.

b. User performance data: refers to the performance data of terminals, radios, APs, wireless access controllers (WACs), and switches. The performance data includes the received signal strength indicator (RSSI) and packet loss rate of terminals, radio interference rate and channel usage of radios, as well as the CPU usage, memory usage, access user quantity and application quantity of APs, WACs, and switches. Devices use various protocols to report user performance data.

Figure 7.9 illustrates how the SDN controller processes data. Starting from data reporting by network devices and culminating in data display on the GUI, the data processing flow consists of five parts: subscription, collection, buffering/distribution, analysis/AI computing (filtering, combination, expert library-based analysis, and machine learning), and storage/display.

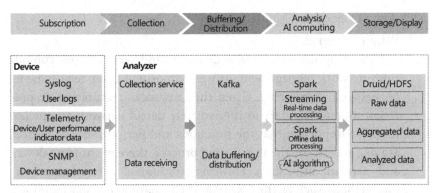

Note: Spark Streaming is an extension of the core Spark API that enables scalable, high-throughput, fault-tolerant processing of live data streams.

FIGURE 7.9 Data processing by the SDN controller.

1. Data subscription: To meet the varying requirements that different devices and O&M logics have regarding data, the SDN controller selectively subscribes to device data. Network devices report data to the SDN controller using different approaches, such as Syslog, Telemetry, and SNMP.

2. Data collection: Subscribed data are collected by the collection service module of the SDN controller. With the help of Telemetry, data collection takes just seconds.

3. Data buffering/distribution: The high-throughput distributed messaging system buffers a large amount of collected data to the SDN controller, and then distributes the data to the respective analysis and computing service modules for analysis.

4. Data analysis/AI computing: The SDN controller analyzes and processes the collected raw device data from multiple aspects. For example, it classifies access logs by phase and computes the counts of successful or failed user access attempts. Given this, the SDN controller can evaluate the Internet access quality of end users based on user performance, and implement offline identification and analysis of typical service issues based on raw user data using AI machine learning algorithms.

5. Data storage/display: After completing data analysis, the SDN controller saves the processed data to the fast column-oriented distributed data storage system, before finally displaying relevant data and functions on the GUI.

7.2.2 Data Collection Using Telemetry

1. Why Telemetry?

Increasing network service demands and service types complicates network management, as well as posing higher requirements on network monitoring. Given this, networks require higher precision monitoring data to immediately detect network anomalies. It is also important that the functions and performance of devices are not affected by the monitoring process, thereby ensuring device and network utilization. However, traditional network monitoring methods such as the SNMP and CLI have many shortcomings, resulting in low management efficiency and the failure to fulfill user requirements.

In traditional network monitoring, device monitoring data (such as interface traffic) is collected in pull mode, as shown in Figure 7.10. This is inadequate for monitoring a large number of network nodes. The SDN controller has higher precision requirements on network node data, yet the only way to improve the precision of obtained data in traditional network monitoring is to increase the data query frequency. However, doing so would cause high CPU usage of network nodes and thereby affect device functions. What's more, due to the network transmission delay, data of network nodes can only be obtained at an interval of minutes, whereas it needs to be obtained every few seconds or even more often. Although SNMP trap and Syslog use the push mode and can push data immediately, only traps or events can be pushed. Other data such as interface traffic cannot be pushed.

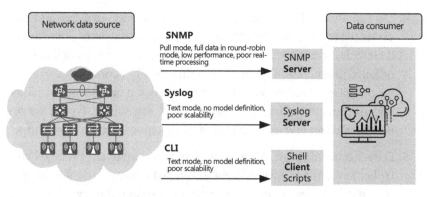

FIGURE 7.10 Limitations of traditional data collection methods.

All of the above challenges can be overcome using Telemetry, a technology that implements large-scale and high-performance network monitoring. Table 7.3 compares Telemetry with traditional data collection methods. Here, you can see that Telemetry supports large-scale data reporting in push mode and enables data reporting at an interval of seconds, meeting intelligent O&M requirements. The SDN controller on a campus network uses Telemetry to collect performance data of devices, interfaces, and queues. It also uses intelligent algorithms to analyze and present network data, thereby enabling proactive network monitoring, anomaly prediction, and intelligent network O&M.

TABLE 7.3   Comparison between Telemetry and Traditional Data Collection Methods

Data Collection Method	Description	Sampling Interval	Inter-Vendor Compatibility
SNMP/Syslog/CLI	Pull mode: devices respond only after receiving query requests. The collector queries NEs using SNMP in round-robin mode Push mode: devices proactively report data such as SNMP traps and Syslog files. The data format varies between vendors, making data analysis more difficult	Minutes	The data format is defined by each vendor. For example, SNMP traps need to be parsed based on the Management Information Base (MIB) tree. Character strings are unstructured and defined by each vendor, meaning that adaptation is required for each trap
Telemetry	Push mode: Data are proactively reported upon subscription. That is, data is continuously reported as scheduled with one subscription, avoiding the impact of query in round-robin mode on the collector and network traffic	Seconds	The unified data flow format (ProtoBuf) simplifies data analysis

2. Understanding Telemetry

Telemetry is a network monitoring technology developed to quickly collect performance data from physical or virtual devices remotely. Telemetry enables the SDN controller to manage a larger number of devices, laying a foundation for fast network fault locating and network optimization. It transforms network quality analysis into big data analytics, effectively supporting intelligent O&M. As shown in Figure 7.11, Telemetry enables network devices to push high-precision performance data to the collector in real time and at high speeds, improving the utilization of devices and networks during data collection.

Telemetry has the following advantages over traditional network monitoring technologies:

a. Sample data are uploaded in push mode, increasing the number of nodes to be monitored.

In the SNMP query process, the NMS and devices interact with each other by alternatively sending requests and responses. If 1000 SNMP query requests need to be sent in the first data query, the SNMP query requests are parsed 1000 times. In the second query, SNMP query requests are parsed another 1000 times. This process is subsequently repeated. In fact, the SNMP query requests are the same in the first and second queries, with the query requests parsed repeatedly in each subsequent query. Parsing these query requests consumes CPU resources of devices, and therefore the number of monitored nodes must be limited to ensure normal device running.

In the Telemetry process, the NMS and devices interact with each other in push mode. In the first subscription, the NMS sends 1000 subscription request packets to a device and the device parses these packets 1000 times. During the parsing process, the device records the subscription information. Then, in the subsequent sampling process, the NMS does not need to send subscription packets to the device again. Instead, the device automatically and continuously pushes the subscribed data to the NMS based on the recorded subscription information. Telemetry saves the packet parsing time and CPU resources of the device, and increases the device monitoring frequency.

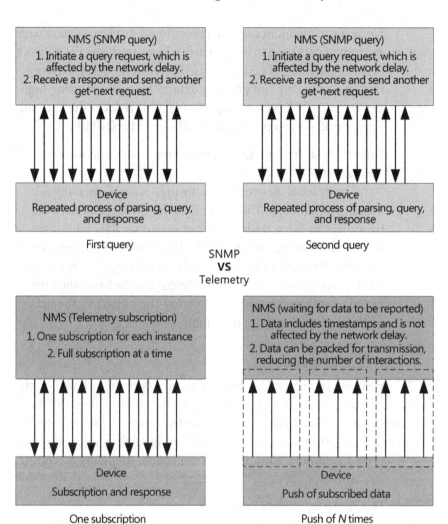

FIGURE 7.11 Comparison between the SNMP query process and Telemetry sampling process.

b. Sample data are uploaded after being packaged, improving the time precision of data collection.

In the Telemetry process, after the NMS sends subscription packets to a device, the device reports multiple pieces of sample data being packaged and sent to the NMS. This reduces the number of packet interactions between the NMS and device, and achieves a sampling precision of subseconds or even milliseconds.

However, to maintain such high precision of displayed data, a large amount of data need to be reported, which consumes the network egress bandwidth. Telemetry-based intelligent O&M overcomes this challenged by compressing data to be reported using high-performance compression algorithms. This reduces the data volume of packets and frees up network egress bandwidth.

c. Sample data contain timestamps, improving the accuracy of sample data.

In traditional network monitoring, sample data do not contain timestamps. This, as well as the existence of network transmission delay, leads to the network node data monitored by the NMS being inaccurate. In the Telemetry process, on the other hand, sample data contain timestamps. In this way, when parsing data, the NMS can determine the time when the sample data is generated, minimizing the impact of the network transmission delay on sample data.

3. Key Telemetry technologies

As shown in Figure 7.12, Telemetry collects raw data based on the YANG model, encodes the collected raw data using Google Protocol Buffers (GPB), and then sends the encoded data to the SDN controller using the Google Remote Procedure Call (gRPC) protocol through encrypted channels. Thus, the data collection, modeling, encoding, and transmission processes are integrated.

a. Hardware-based data collection

Telemetry's effect is determined by hardware capabilities of devices. On Huawei intent-driven campus networks, chips of network devices such as switches, WACs, and wireless APs come with Telemetry-based data collection, which collects large amounts of data in real time and reports the collected data to the SDN controller. When a device chip detects an environment change, it immediately reports the change to the SDN controller according to the interrupt mechanism. In this manner, the SDN controller can obtain data within several microseconds as well as obtaining the accurate time of the change, and can display the accurate occurrence exception time when

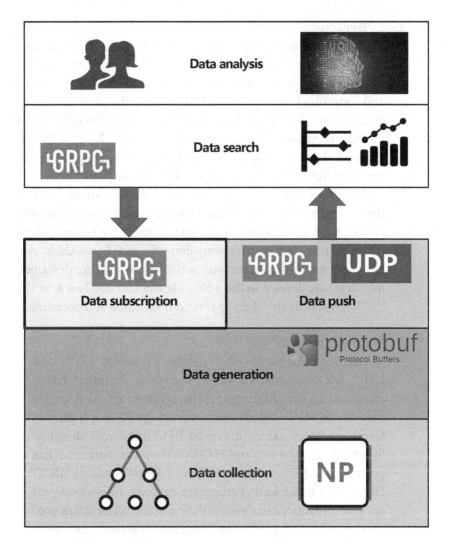

FIGURE 7.12   Telemetry process.

analyzing exceptions. In traditional mode, an exception is often detected tens to hundreds of milliseconds after it occurred. Such a delay may prevent the fault from being located in time.

b. YANG model

The SDN controller uses Telemetry to collect sample data based on Huawei's YANG model, which is compatible with the openconfig-telemetry.yang model defined by OpenConfig.

c. GPB encoding

GPB is a language-neutral, platform-neutral, and extensible mechanism for serializing structured data for communication protocols and data storage. It features high parsing efficiency and consumes less traffic when transmitting the same amount of information (2–5 times more efficient encoding/decoding than JSON). The size of GPB-encoded data is 1/3 to 1/2 that of JSON-encoded data, thereby ensuring the data throughput performance of Telemetry and saving CPU and bandwidth resources.

Compared with the common output formats (XML and JSON), the GPB format has lower readability due to being binary. Because of this, data in GPB format are only suitable for being read by machines to deliver high transmission efficiency. In the O&M system of an intent-driven campus network, raw data are described in a structure defined in the .proto file and encoded based on the YANG model. Table 7.4 compares GPB encoding and decoding.

d. Transport protocol

Telemetry supports two transport protocols: gRPC and UDP. Telemetry-enabled network devices in intent-driven campuses use the gRPC protocol to report the encoded sample data to the SDN controller for storage. gRPC is a high-performance, open-source, universal RPC framework designed for mobile applications and HTTP/2. It also supports multiple programming languages and SSL encrypted channels. gRPC essentially provides an open programming framework, based on which vendors can develop their own server or client processing logics using different languages to shorten the development cycle for product interconnection. Figure 7.13 shows the gRPC protocol stack layers, with Table 7.5 describing each of the layers.

The Telemetry-based data collection system provides the SDN controller with accurate, real-time, abundant data sources, laying a foundation for intelligent O&M. The Telemetry-based data reporting system, on the other hand, enables wired and wireless devices on the entire campus network to efficiently collect and display data, contributing to intelligence and automation of the O&M system.

TABLE 7.4   Comparison between GPB Encoding and Decoding

GPB Encoding	GPB Decoding
`{`	`{`
`1:"HUAWEI"`	`"node_id_str":"HUAWEI",`
`2:"s4"`	`"subscription_id_str":"s4",`
`3:"huawei-ifm:ifm/interfaces/`	`"sensor_path":"huawei-ifm:ifm/`
`interface" 4:46`	`interfaces/interface",`
`5:1515727243419`	`"collection_id":46, "collec-`
`6:1515727243514`	`tion_start_time":"2018/1/12`
`7{`	`11:20:43.419",`
`1[{`	`"msg_timestamp":"2018/1/12`
`1: 1515727243419`	`11:20:43.514",`
`2{`	`"data_gpb":{`
`5{`	`"row":[{ "timestamp":"2018/1/12`
`1[{`	`11:20:`
`5:1`	`43.419",`
`16:2`	`"content":{ "interfaces":{`
`25:"Eth-Trunk1"`	`"interface":[{`
`}]`	`"ifAdminStatus":1, "ifIndex":2,`
`}`	`"ifName":"Eth-Trunk1"`
`}`	`}]`
`}]`	`}`
`}`	`}`
`8:1515727243419`	`}]`
`9:10000`	`},`
`10:"OK"`	`"collection_end_time":"2018/`
`11:"CE6850HI"`	`1/12 11:20:43.419",`
`12:0`	`"current_period":10000,`
`}`	`"except_desc":"OK", "prod-`
	`uct_name":"CE6850HI",`
	`"encoding":Encoding_GPB`
	`}`

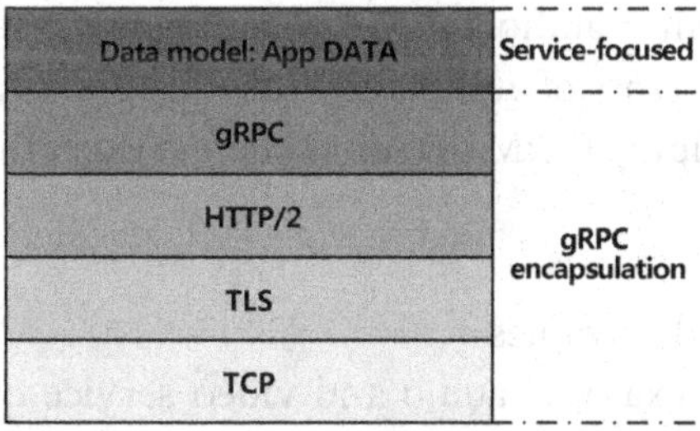

FIGURE 7.13   gRPC protocol stack layers.

TABLE 7.5   gRPC Protocol Stack Layers

Layer	Description
TCP layer	This is an underlying communications protocol, which is based on TCP connections
TLS layer	This layer is optional. It is based on the TLS 1.2-encrypted channel and bidirectional certificate authentication
HTTP/2 layer	gRPC is carried over the HTTP/2 protocol and uses HTTP/2 features such as bidirectional communication, flow control, header compression, and request multiplexing over a single connection
gRPC layer	This layer defines the protocol interaction format for remote procedure calls
Data model layer	Communication parties need to know each other's data models before they can correctly interact with each other

7.2.3 Audio and Video Quality Awareness Using eMDI

Let us assume that 20 employees are in a video conference, and one of them encounters severe frame freezing. Afterward, the employee complains that the poor network quality and frame freezing impact communication with customers; however, the network O&M personnel are unable to replicate this fault. Through scenario reviews and data queries, they determine that most participants do not encounter network issues during video conferences. In this case, the network O&M personnel need to analyze historical data to check for packet loss. This process is time-consuming and is not guaranteed to help locate the fault. If video conferences could be monitored in real time and historical records were saved, the network O&M personnel could quickly locate the fault, determine the time of fault occurrence, query relevant device data, and efficiently rectify the problem.

To address such video issues, Huawei SDN controller has been designed with an audio and video monitoring system. Using certain algorithms to monitor audio and video traffic, the system calculates relevant quality indicators of audio and video programs and displays their service quality, helping O&M personnel detect exceptions and repair the network.

1. Audio and video frames

    Before we examine audio and video service quality monitoring technologies, let us look at the following concepts:

a. I-frame, P-frame, and B-frame

Currently, audio and video frames are encoded in compliance with the Moving Picture Experts Group (MPEG) and H.26x standards, with each video segment consisting of a series of consecutive image frames. According to these standards, three types of frames are used in video compression: intracoded frames (I-frames), predicted frames (P-frames), and bidirectional predicted frames (B-frames). An I-frame contains all information related to its image, and intraframe coding technology is used to recover the image. A P-frame recovers its image based on the previous I-frame or P-frame using a related algorithm. A B-frame recovers its image based on the previous I-frame or P-frame as well as the subsequent I-frame or P-frame.

b. Media stream packet format

The MPEG-2 standard is generally used to encapsulate media stream packets. Figure 7.14 shows the MPEG-2 packet format. Typically, each video frame is encapsulated with a Packetized Elementary Stream (PES) header, which records the presentation timestamp of the video frame. Each PES payload is fragmented and encapsulated into Transport Stream (TS) frames. A TS frame consists of 188 bytes (including a 4-byte header). The TS header contains a TS sequence number, which can be used to calculate the frame loss rate. MPEG-2 packets are generally carried over UDP and encapsulated with

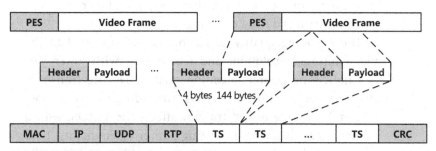

Note: CRC indicates Cyclic Redundancy Check.

FIGURE 7.14   MPEG-2 packet format.

a Real-time Transport Protocol (RTP) header. The RTP header contains an RTP sequence number that can be used to calculate the IP packet loss rate.

RTP, defined in RFC 1889, is used to provide end-to-end real-time transmission for various multimedia data that need to be transmitted in real time on IP networks, such as voice, image, and fax data. Typically, RTP is carried over UDP and the RTP packet header contains two key fields: sequence number and timestamp.

TS is a data stream format defined in the H.222.0/ISO/IEC 13818-2 and ISO/IEC 13818-3 protocols of the International Telecommunication Union-Telecommunication Standardization Sector (ITU-T). It is used for transmission and storage of audio and video data, and contains timestamp and system control information.

2. Audio and video quality monitoring technologies

Within the industry, audio and video services are primarily monitored using Video Mean Opinion Score (VMOS), Media Delivery Index (MDI), and eMDI technologies.

a. VMOS

VMOS is a subjective method of evaluating video quality from the perspective of user experience. The VMOS is measured on a continuous scale of 1–5, representing unsatisfactory, poor, fair, good, and excellent video quality, respectively. The VMOS of a video is related to the encoding quality of video streams received by the monitoring point, as well as the packet loss rate and jitter of video streams during transmission.

The encoding quality of video streams is related to the video bitrate, resolution, and frame rate, with lower values indicating poorer encoding quality. During transmission of video streams, packet loss (including the packet loss rate and the type of lost frames) affects the VMOS, with a higher packet loss rate indicating poorer video quality. Loss of I-frames has the greatest impact on video quality, while loss of B-frames or P-frames will result in less severe issues. The packet loss rate includes the RTP packet loss rate and TS packet loss rate. However, the VMOS algorithm mainly takes the RTP packet loss rate into account. The TS packet loss rate is only calculated when video streams do not contain RTP-encapsulated frames. Figure 7.15 shows how the packet loss rate affects video quality.

In Figure 7.15, the first and second circles from left to right both represent the loss rate of I-frames, with a higher packet loss rate indicating a lower VMOS. The second and third circles from left to right both represent a packet loss rate of 40%. The second circle represents the loss rate of I-frames, and the third circle represents the loss rate of B-frames. The VMOS at the second circle is lower.

The jitter of video streams during transmission also affects the VMOS. If packets are sent to a device at high speeds, a buffer overflow may occur, resulting in packet loss and artifacts. If

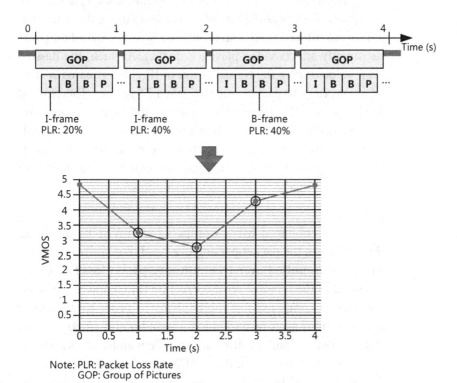

FIGURE 7.15 Impact of the packet loss rate on the VMOS.

packets are sent at low speeds, no video stream in the buffer can be played, leading to frame freezing. The jitter reflects the frame freezing and a larger amount of jitter leads to a lower VMOS.

b. MDI

The MDI is a set of measures that can be used to quantify the transmission quality of streaming media on the network. The MDI is typically displayed as two numbers separated by a colon: Delay Factor (DF) and Media Loss Rate (MLR).

The DF represents the latency and jitter of monitored video streams, in milliseconds. It converts video stream jitter changes into buffer requirements for video transmission and decoding devices. A greater DF value indicates a larger amount of video stream jitter. According to the WT126 standard, the DF value of video streams during transmission should not exceed 50 ms.

The MLR, expressed in the number of media data packets lost per second, indicates the rate of packet loss during the transmission of monitored media streams. The loss of video data packets directly affects video playback quality. As such, the desired MLR value during the transmission of IP video streams is zero. Typically, each IP packet contains seven TS frames. Therefore, if one IP packet is lost, seven TS frames (excluding empty ones) are lost. According to the WT126 standard, the maximum acceptable MLR for standard definition/Video on Demand videos is five media packets per 30 minutes, and that for high definition videos is five media packets per 240 minutes.

c. eMDI

eMDI is enhanced MDI. Compared with VMOS and MDI, eMDI reduces the packet parsing overhead. For audio and video services transmitted over UDP, effective packet loss factors are proposed in the forward error correction (FEC) and retransmission (RET) compensation mechanisms in order to accurately describe the impact of packet loss on audio and video services and improve the fault locating accuracy. For audio and video services transmitted over TCP, the SDN controller analyzes information such as the TCP sequence number and calculates the packet loss rate and latency of upstream and downstream TCP flows, which help with fault locating.

Statistics about monitoring indicators of target flows are collected by an eMDI instance, which is a basic unit of eMDI. Each eMDI instance is composed of the target flow, monitoring interval, lifetime, and alarm threshold. An eMDI-enabled device obtains monitoring data at a specified monitoring interval and periodically sends the obtained data to the SDN controller. eMDI supports real-time quality monitoring and fault locating for services transmitted over TCP and UDP.

The monitoring indicators of services transmitted over UDP are obtained by analyzing fields of the RTP packet header. We can determine the service type-based specific fields in the header. Also, we can obtain the packet loss rate and out-of-order packet rate from the sequence number field, and jitter from the timestamp field. Table 7.6 describes the monitoring indicators for services transmitted over UDP.

For services transmitted over TCP, the monitoring indicators are calculated by analyzing TCP packets. TCP is a connection-oriented, reliable, and byte stream-based transport-layer communications protocol. The average bitrate within the monitoring interval is calculated based on the length of TCP packets transmitted during that interval, while the upstream

TABLE 7.6   Monitoring Indicators for Services Transmitted over UDP

Monitoring Indicator	Calculation Method
Packet loss rate within the monitoring interval	Packet loss rate within the monitoring interval = number of lost packets/(number of received packets + number of lost packets − number of out-of-order packets)
	If the difference between sequence numbers of consecutive two RTP packets is greater than 1, packet loss has occurred. If the sequence number of an RTP packet is less than the largest sequence number of all previously received packets, the packet is an out-of-order packet
Out-of-order packet rate within the monitoring interval	Out-of-order packet rate within the monitoring interval = number of out-of-order packets/(number of received packets + number of lost packets − number of out-of-order packets)
Jitter within the monitoring interval	Jitter = $T_2 - T_1$
	In the above formula, T_1 indicates the time difference between sending the first packet and the last packet, and T_2 indicates the time difference between receiving the first packet and the last packet

and downstream packet loss rates are calculated based on the sequence number in the TCP packet header. The average two-way latency in the upstream and downstream directions is calculated based on the timestamp and sequence number in the TCP packet header. Table 7.7 describes the monitoring indicators for services transmitted over TCP.

TABLE 7.7   Monitoring Indicators for Services Transmitted over TCP

Monitoring Indicator	Calculation Method
Average bit-rate within the monitoring interval, in kbit/s	Average bitrate within the monitoring interval = total length of packets received within the monitoring interval/ monitoring interval
Upstream packet loss rate within the monitoring interval	If no packets are lost, the sequence number of a packet plus the packet length is the expected sequence number of the next packet If the sequence number of a packet is greater than the expected sequence number, packet loss occurs on the upstream device, and the number of lost packets can be calculated based on the average packet size Upstream packet loss rate within the monitoring interval = number of lost upstream packets/(total number of received packets + total number of lost packets)
Downstream packet loss rate within the monitoring interval	If no packets are lost, the sequence number of a packet plus the packet length is the expected sequence number of the next packet. If the sequence number of a packet is less than the expected sequence number, the packet is considered to be a retransmitted packet, and the number of retransmitted packets is considered to be the total number of lost packets Number of lost downstream packets = total number of lost packets – number of lost upstream packets Downstream packet loss rate within the monitoring interval = number of lost downstream packets/(total number of received packets + total number of lost packets)
Average downstream two-way latency within the monitoring interval	Average downstream two-way latency within the monitoring interval = $T_2 - T_1$ A received nonretransmitted packet is randomly selected, and its timestamp is recorded as T_1. The expected sequence number of the next packet is calculated based on the sequence number and length of the current packet. When the sequence number of an upstream packet sent from a downstream device is greater than or equal to the expected sequence number, the timestamp of this packet is recorded as T_2

3. Application of eMDI in audio and video service monitoring

As shown in Figure 7.16, the eMDI-based audio and video quality awareness function is deployed on a campus network. This function enables the campus network to detect Session Initiation Protocol (SIP)- and RTP-based audio and video streams in real time, display the full-process quality of these streams, detect the setup and termination of audio and video sessions in real time, automatically start audio and video quality analysis, and display the analysis results. When poor-quality audio and video sessions are detected, O&M personnel can analyze the root cause and locate faults.

On a campus network, switches and APs can function as detection nodes in the eMDI-based audio and video awareness system. To efficiently locate and demarcate audio and video issues, eMDI must be deployed on the start and end nodes of the eMDI interaction process, as well as on intermediate nodes based on device performance. It is recommended that eMDI probes be deployed on all eMDI-capable devices. As more detection nodes are deployed, the paths become more specific and fault demarcation becomes more accurate.

Huawei SDN controller supports hardware-based eMDI, which ensures the accuracy and real-time availability of sample data. Data collected by network devices are reported to the SDN controller through Telemetry. The SDN controller then summarizes the eMDI statistics reported by each device and displays the full-process quality of audio and video services. The SDN controller provides the

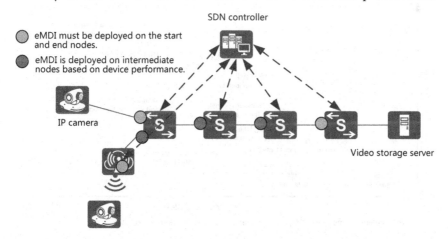

FIGURE 7.16   Application of eMDI in audio and video service monitoring.

fault analysis function for poor-quality audio and video streams, comparing the RTP sequence numbers of UDP streams in order to determine whether packet loss or out-of-order occurs in Inter-Process Communication (IPC) or on network devices. For TCP streams, the SDN controller compares the upstream and downstream packet loss rates to determine whether faults occur on IP cameras or network devices. In this manner, failure points can be accurately identified. The fault analysis function, in conjunction with the KPI correlation analysis function, enables accurate fault locating, identification of fault root causes, as well as visualization of the quality of audio and video streams.

7.2.4 Big Data and AI Processing Models

In the traditional network O&M model, collected data are simply displayed, but in-depth analysis of data relationships cannot be performed, let alone automatic locating of fault root causes. Manually locating fault points and root causes necessitates the need of highly skilled O&M personnel. On the other hand, the intelligent O&M model introduces AI algorithms shown in Figure 7.17, which, by interworking with big data analytics technology, enable machines to learn large amounts of O&M data, and automatically perform correlation analysis, identify anomalies and network faults, predict potential faults, and locate root causes of faults, achieving network

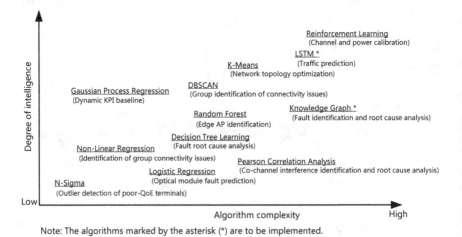

Note: The algorithms marked by the asterisk (*) are to be implemented.

FIGURE 7.17   Application of AI algorithms in intelligent O&M.

self-healing upon faults. Based on such intelligent O&M model, machines can perform automatic data analysis, self-learning, and auto-adjustment.

1. Fault pattern library based on expert experience and machine learning

Through years of campus network O&M, Huawei's network experts have accumulated extensive troubleshooting experiences and built a comprehensive fault library covering more than 90% of campus network fault scenarios, involving networks, devices, terminals, and applications. The intelligent O&M system leverages this fault library to automatically identify network anomalies, greatly improving O&M efficiency and effectively transferring these inherited O&M and troubleshooting experiences to machines. As a result, a network fault pattern library based on expert experience is created and gradually imported into the intelligent network O&M system, enabling the system to intelligently and automatically identify faults as effectively as network O&M experts. Figure 7.18 shows the basic implementation of the fault pattern library.

First, O&M experiences of network experts are digitalized into patterns that can be identified by machines to form a basic network fault pattern library.

These patterns are then loaded into the pattern engine. When the engine receives real-time network O&M data, it determines whether a network fault has occurred based on the determination criteria and outputs the determination result.

In addition, the pattern engine continuously adds to the network fault pattern library through machine learning, optimizes the fault

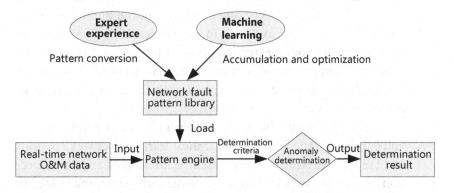

FIGURE 7.18   Intelligent fault pattern library building based on expert experience.

determination criteria, and continuously expands the fault identification scope to improve the fault identification accuracy.

As faults can often be the result of multiple factors, they should also be analyzed from multiple aspects based on expert experience. As with most industries, a large gap exists between the fault identification success rate of a novice and that of an experienced expert. Huawei has summarized and modeled fault locating and identification methods accumulated throughout years of R&D experience and embeds them into the SDN controller to form a fault library. Consequently, the SDN controller delivers expert capabilities and allows each network O&M engineer to share and utilize expert experience.

For example, if users are unexpectedly disconnected from a network, Huawei SDN controller will add this data to the fault library. The causes that may trigger such disconnection are identified based on experience and organized as fault points, and include configuration errors, forcible disconnection of users upon timeout, and air interface exceptions. In this manner, when a user is disconnected from the network unexpectedly, the SDN controller collects network data and can quickly identify the cause based on the fault library.

2. Dynamic baseline trend prediction based on supervised learning

Certain symptoms on the network may represent different issues in different scenarios and might actually be normal in some cases while posing a major threat to user experience in others. For example, in scenario A, 10–20 stations (STAs) are connected to each AP in normal circumstances, with no more than 15 STAs typically connected. In high-density office scenario B, the number of STAs connected to each AP is 30–40 in most cases. If 45 STAs are connected to an AP for an extended period, this is viewed as normal in scenario B but abnormal in scenario A. The network capacity and bandwidth requirements will change with the sudden increase of connected STAs.

As such, to enable dynamic adjustment of the network capacity and bandwidth, the SDN controller must learn the network environment in order to build dynamic baselines for network anomaly detection. A dynamic indicator baseline changes with time, and is used to define the normal range of an indicator and predict its future trends.

Figure 7.19 compares the data trend predicted based on the dynamic baseline and the data generated on the network (within the dashed lines). If the data exceed the dynamic baseline, the system preliminarily determines that an anomaly has occurred and identifies potential faults.

As shown in Figure 7.20, the dynamic baseline algorithm process consists of three phases: data set preprocessing, dynamic baseline generation, and anomaly detection.

a. Data set preprocessing

The SDN controller preprocesses collected data to ensure accuracy and completeness, while also focusing on the elimination of noise data. The data set preprocessing phase includes initial data construction based on expert experience, data cleaning and multidimensional combination, supplementing for lost data, removing unnecessary and abnormal data, correcting tail data, and rationalizing data with large fluctuations.

b. Dynamic baseline generation

Dynamic baselines are generated using algorithms. Threshold models are trained and predicted using the periodic sequence algorithm, the Gaussian process regression algorithm, and a

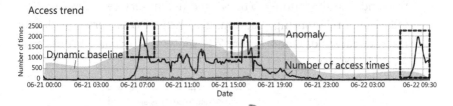

FIGURE 7.19   Anomaly detection based on dynamic baselines.

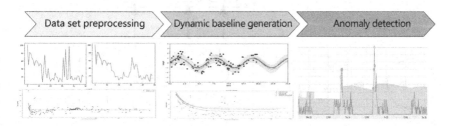

FIGURE 7.20   Dynamic baseline algorithm process.

multivariate power-exponential function algorithm. The threshold edge is calculated and adjusted by determining the mean and variance values. Finally, the sensitivity is corrected based on experience.

c. Anomaly detection

The SDN controller collects the current indicator data and determines whether it exceeds the dynamic baseline. If so, the SDN controller considers an anomaly to have occurred and performs multidimensional identification.

3. Root cause locating for associated KPIs

A campus network has multiple types of service nodes and network nodes, and each node has many indicators. These nodes and indicators are associated one another, resulting in complex and changeable network fault patterns. In some cases, the same fault symptom may have different root causes. For example, a user authentication failure on a Wi-Fi network may be caused by a weak Wi-Fi signal or a certificate error. In contrast, the same root cause may contribute to different fault symptoms. For example, co-channel interference may cause poor bandwidth experience and/or high channel usage on a Wi-Fi network. As a result, a data association system has been designed based on large amounts of network O&M data to identify the association and causal relationship between different service nodes from the O&M data relationship diagram. This helps O&M personnel locate the root causes of faults. Figure 7.21 shows the correlated root cause analysis process.

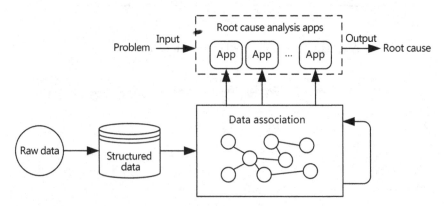

FIGURE 7.21   Correlated root cause analysis.

Raw network O&M data are preprocessed and then converted into structured data. The data association model is then designed based on the network service relationships to associate anomalies from multiple dimensions, and usable structured data are loaded to the model. Based on the data association model, root cause analysis applications can be developed at the service layer. When a network anomaly occurs, the possible root cause can be analyzed based on the service relationship after a specific symptom or relevant data have been input.

7.3 INTELLIGENT O&M PRACTICES

O&M starts after a network is deployed and starts operating. O&M personnel need to ensure network availability and troubleshoot faults immediately after they occur. Network O&M tasks include routine monitoring and O&M, as well as fast fault diagnosis and troubleshooting. In routine monitoring and O&M, O&M personnel learn the running status of the entire network by monitoring the status and running indicators of network devices. Through evaluation of the findings, they can learn the network health status in real time and perform network adjustment as needed. Routine monitoring and O&M are performed on a regular basis to discover potential faults in advance through network diagnosis and continuously optimize the network.

Routine monitoring and O&M, however, only help O&M personnel learn the overall running status of a network and discover common network issues. For systematic or process-related issues, professional personnel are required to perform specific fault analysis, which is time-consuming. To cope with this, intelligent O&M is introduced to free O&M personnel from manual fault analysis. As shown in Figure 7.22, the SDN controller adopts big data analytics and AI algorithms to automatically identify network faults and analyze root causes based on expert experience and continuously optimized fault patterns. These merits make devices smart and intelligent, thereby simplifying network O&M.

7.3.1 Real-Time Health Monitoring Based on KPIs

Faults are rarely seen on a network that is running properly; therefore, the main concern of O&M personnel is the overall running status of the entire network but not troubleshooting. A large campus network typically has hundreds or even more than a thousand branches and therefore may

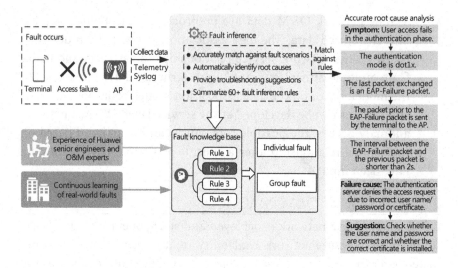

FIGURE 7.22 Fault analysis model in intelligent O&M.

have tens of thousands of devices. During routine network monitoring and O&M, O&M personnel need to view the status of all network devices, so they can get insights into the overall status of the network.

The SDN controller visualizes the health status of the entire network from multiple dimensions such as sites, devices, and users, as shown in Figures 7.23 and 7.24. The SDN controller displays the evaluation scores of the device running quality, device signal quality, user experience, and overall network quality in a method that is easy to understand, helping customers identify the network health status. It also displays monitoring indicators in trend charts, so that the historical network and device running status can be traced during fault locating.

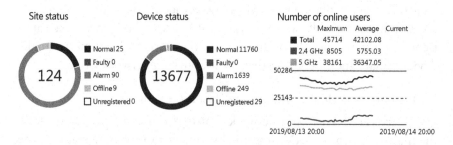

FIGURE 7.23 Tenant-level network status.

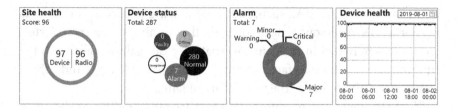

FIGURE 7.24   Site health status.

After logging in to the SDN controller, O&M personnel can view the network health evaluation of each listed site and view site details such as the health and device registration status. The SDN controller displays the following network health information:

- Network health status of tenant sites, which is displayed in different colors representing different health status (excellent/good/medium/poor). The health status of a tenant site is evaluated based on the comprehensive evaluation scores of the device health and radio health.

- Quantities of normal devices, offline devices, unregistered devices, faulty devices, and devices reporting alarms. This means administrators can quickly view the running status of devices at each site.

- Alarms with different severities at each site, including critical, major, minor, and warning alarms, so that administrators can quickly view and handle alarms.

- Overall health curve and historical health status of devices.

The SDN controller also displays the running data of each single device in real time, including the CPU, memory, traffic, and user quantity, and provides historical data curves for tracing. The running data of a single device gives insights into the device's health status, based on which O&M personnel can easily identify abnormal devices and adjust the network efficiently.

As more campus networks are going wireless, radio monitoring is increasingly important during network maintenance. Wireless networks differ greatly from wired networks because wireless networks transmit data using electromagnetic waves. Therefore, wireless networks are vulnerable to interference from surroundings, especially when there are a large number of interfering devices, such as Bluetooth devices, microwave ovens, or private APs. As such, radio monitoring is crucial for campus networks. O&M personnel need to obtain the running status and interference status of radios in real time, so that they can gain insights into the running status of air interfaces and make network predictions and adjustments promptly to reduce the probability of faults.

The SDN controller monitors KPIs and trends of radios, as shown in Figures 7.25 and 7.26. Radio KPIs include the channel usage, interference rate, noise, packet loss rate, and RET rate, which signify the overall air interface performance of all devices at a site. For example, O&M personnel can intuitively view the radio with a high interference rate and then check the radio details to determine whether there are interference sources around the AP based on the AP's deployment location. If so, they can eliminate the interference sources promptly to improve the air interface quality. The radio monitoring curves display the KPI trends of radios, such as the proportion of busy radios, as well as the proportions of radios with high RET rate, high packet loss rate, high noise, and severe interference. According to these trends, administrators can learn the historical network status and predict the future network status.

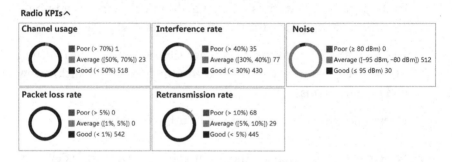

FIGURE 7.25   Radio KPIs.

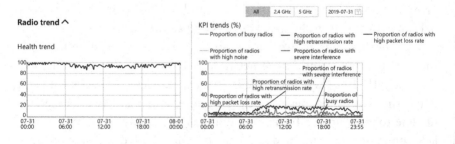

FIGURE 7.26   KPI trends of radios.

7.3.2 Individual Fault Analysis

An individual fault typically affects only a single user. For example, when a user enters an incorrect password and fails to access a network, this fault affects only the user. To assist in troubleshooting individual faults, the SDN controller provides journey analysis, access analysis, experience analysis, and application analysis, as shown in Figure 7.27. These functions help O&M personnel accurately identify individual faults and ensure services of key users.

Protocol tracing and poor-QoE user analysis have been outlined in Section 7.1.1. This section focuses on user journey analysis as well as audio and video service quality analysis.

1. User journey analysis

    When encountering service problems such as poor Internet access experience, failure to open web pages, or Internet access failures, users may do not report the problems immediately, but instead wait until they are free. In traditional O&M, after receiving complaints from a user, O&M personnel need to search through massive numbers of logs to find those relevant to the user's fault and view information about faulty devices. If the problem occurs when the user uses a mobile terminal, O&M personnel need to determine the APs to which the user has connected and check logs of these APs. This O&M mode is time- and labor-consuming. Additionally, key details may be missed, causing failures to determine root causes of faults.

    In intelligent O&M, the SDN controller collects large amounts of user and device data using Telemetry, automatically analyzes correlations within the data and creates visualized full journeys for each

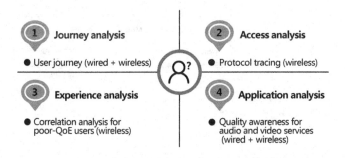

FIGURE 7.27   Individual fault analysis.

user. In this manner, the entire journey of a user's access to a Wi-Fi network can be precisely displayed, as shown in Figure 7.28. If a user encounters a fault during network access, O&M personnel can easily obtain detailed O&M data generated in the process, including the time when the user accesses the network, the AP to which the user connects, user experience, and specific fault. They can then quickly locate the exact time when the fault occurred and root cause, and optimize the network accordingly.

The SDN controller displays network data related to user experience, such as the signal strength, negotiated rate, latency, and packet loss rate, as shown in Figure 7.29, from which O&M personnel can easily check whether and when the access signal becomes weak.

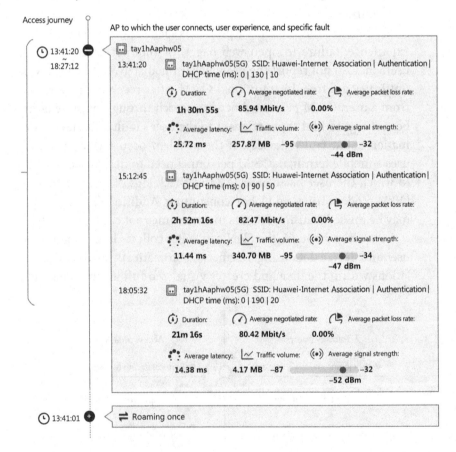

FIGURE 7.28 Sample user access journey.

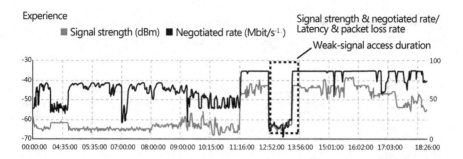

FIGURE 7.29 Example of user experience.

Typically, if the access signal becomes weak, the negotiated rate decreases. As a result, video freezing, slow web page loading, and other problems may occur, degrading user experience. In this case, O&M personnel need to check whether the problem occurs because the user leaves the network coverage area or the network has coverage holes. For example, they can carry out site surveys and tests to determine there are coverage holes to be filled.

2. Audio and video service quality analysis

By leveraging Telemetry and eMDI technologies, the SDN controller can analyze the quality of audio and video services, and displays the session paths of audio and video services on the network and the quality analysis result. In this manner, O&M personnel can efficiently detect the real service experience of users. The audio and video KPIs change accordingly when the quality of audio and video services deteriorates. In addition, the time range when the service quality deteriorates and network devices that cause the quality deterioration are also displayed. O&M personnel can locate root causes of faults based on the correlative analysis result and fix faults according to troubleshooting suggestions.

The SDN controller provides three functions to visualize quality of and analyze faults in audio and video services. Figure 7.30 illustrates quality analysis for audio and video services.

a. Proactive perception of the audio and video service quality

With the SIP snooping technology, the SDN controller can proactively perceive SIP+RTP audio and video streams and detect the setup and teardown of audio and video sessions in real

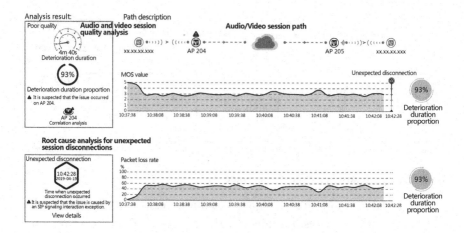

FIGURE 7.30   Quality analysis for audio and video services.

time. It also automatically uses eMDI technology to monitor and analyze the quality of audio and video streams during the session and identify poor-quality audio and video streams.

b. Root cause analysis for poor-quality audio and video services

The SDN controller analyzes the correlation between the Mean Opinion Score (MOS) of audio and video streams and a series of indicators to identify the root cause of poor quality. The indicators include:

i. Air interface indicators, such as the signal strength, interference rate, channel usage, negotiated rate, and backpressure queue count

ii. Wired interface indicators, such as the packet loss rate and cache usage of interfaces

iii. Device indicators, such as the CPU usage and memory usage

c. Root cause analysis for unexpected session disconnections

The SDN controller can automatically analyze root causes of unexpected audio and video session disconnections. Unexpected audio and video session disconnections may be caused by roaming faults, Wi-Fi disassociation, network-side indicator deterioration, or an exception in signaling interaction.

For example, O&M personnel check the audio and video session list in an office area and find that an employee has a poor session quality. O&M personnel further check the details about the poor-quality session, and find that the fault occurs on the AP to which the employee connects. Through correlation analysis, O&M personnel find that neighbor APs interfere with this AP due to improper channel configuration. Therefore, O&M personnel can adjust the channel configuration to eliminate air interface interference and restore the quality of audio and video sessions.

7.3.3 Group Fault Analysis

Group faults occur on a large scale on a network and may affect all users. For example, an authentication server fault will cause authentication failures of a large number of users, while insufficient Wi-Fi coverage of an AP will lead to poor signal quality of all terminals connected to the AP.

When it comes to intelligent O&M, the SDN controller can automatically and accurately identify group faults, which are classified into connection, air interface performance, roaming, and device issues, as shown in Figure 7.31.

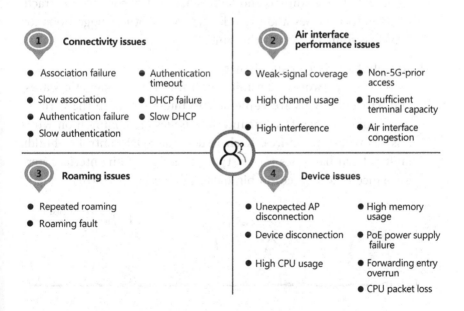

FIGURE 7.31   Group fault classification.

1. Connectivity issues

When a large number of users fail to access the network or access the network at low speeds, a group fault may have occurred. Group faults, especially authentication and access failures, have significant impacts on the network. Such faults must be rectified as soon as possible; otherwise, a large number of users will be affected.

Users fail to access the network due to various reasons, and not all the access failures are caused by network faults. As shown in Figure 7.32, the SDN controller uses machine learning algorithms to generate baselines by training a significant amount of historical data, implementing intelligent detection of abnormal network access behaviors and accurate identification of network faults.

The SDN controller generates an access baseline by training historical big data. Access failures and anomalies within the baseline are considered individual issues. When the number of access failures exceeds the baseline, the SDN controller can automatically determine that an anomaly has occurred and then identify the fault pattern and analyze the root cause. In addition, the SDN controller abstracts characteristics of terminals that fail to access the network and performs group fault analysis using clustering algorithms. Based on login logs of terminals and KPI data, the SDN controller extracts possible root causes and provides troubleshooting suggestions for O&M personnel to rectify the faults.

2. Air interface performance issues

Wireless networks transmit data using electromagnetic waves, and wireless signals are vulnerable to interferences in the wireless environment. Therefore, air interface issues are common on wireless networks. To address such issues, the SDN controller builds fault pattern baselines for six typical models of air interface performance issues based on big data and collects performance data

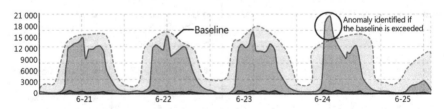

FIGURE 7.32 Dynamic baseline and anomaly detection.

reported by network devices. The SDN controller compares the collected data with the baselines in real time. If an anomaly is identified, the SDN controller automatically diagnoses, analyzes, and records the anomaly.

Weak-signal coverage is a common issue which negatively affects air interface performance in wireless network scenarios. To identify weak-signal coverage issues, the SDN controller needs to analyze the RSSIs of terminals and denoise data sets, for example, excluding data sets about terminals in sleep state, connected only for a short period of time, and with small service volumes, as well as APs with small service volumes (with small numbers of terminals connected). Based on the fault signature database, the SDN controller can automatically identify data with abnormal RSSI patterns among user performance data collected in seconds.

Figure 7.33 shows an example of weak-signal coverage analysis. In this example, employees suffer from low Internet access speed, signal weakness, and long delay and freezing in voice conferences. These negative effects occur for several consecutive days. To locate the fault, O&M personnel access the list of weak-signal coverage issues to check the RSSI distribution of terminals connected to the current AP. By analyzing the RSSIs of terminals, they find that the RSSIs of many terminals in the area are lower than −70 dBm. This is obviously lower than the average value. To enhance the signal strength, O&M personnel can adjust the AP's transmit power or add more APs to improve users' network experience.

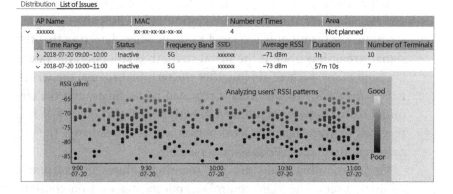

FIGURE 7.33   Weak-signal coverage issue analysis.

3. Roaming issues

As access locations of wireless terminals change constantly, the terminals frequently roam between devices. In comparison, the majority of terminals on wired networks access the network at fixed locations. Even if the access location of a user changes, the user needs to go offline first and go online again after a period of time. It is normal that wireless terminals change their access locations because they may move anytime. Therefore, roaming of wireless terminals is a basic service feature for wireless networks. User roaming involves interactions between the user and multiple devices and may require re-authentication. Therefore, the roaming process is complex. User roaming experience is also related to the performance of user terminals, with terminals having different roaming sensitivity. In the same area, terminals with higher roaming sensitivity and aggressiveness roam preferentially. Terminals with poor aggressiveness may stick to the originally connected AP for a prolonged time, resulting in poor experience. Therefore, the roaming process of wireless terminals is complex and prone to exceptions. The SDN controller can analyze roaming issues and identify common issues, so that O&M personnel can optimize the network accordingly to improve user access and roaming experiences. Roaming issues are classified into repeated roaming issues and roaming faults.

In repeated roaming, a terminal roams multiple times between APs within a short time period and the terminal's KPIs before and after roaming are poor. Roaming faults include roaming failures and long roaming duration.

Repeated roaming is typically caused by poor signal coverage. In an area, when a terminal is covered by multiple APs with poor signal strength, the terminal is triggered to roam when detecting that the signal strength is poor. If the signal strength is still poor, the rate is low, and many packets are lost after the terminal roams, the terminal roams again. This process is repeated many times, and the terminal roams between multiple APs with poor signal quality. When repeated roaming occurs, O&M personnel need to enhance the signal strength or add APs as soon as possible.

Group roaming faults are not roaming failures or exceptions of single users. The SDN controller collects statistics on continuous roaming data of an AP. This is done to determine whether a large number of users encounter roaming faults on this AP. If they encounter roaming failures and roaming faults on this AP, the SDN controller identifies that a group roaming fault has occurred. The SDN controller then can analyze the AP's CPU, memory, and signal coverage to identify the root cause, helping troubleshooting.

The following uses an example to describe the repeated roaming issue. According to the fault statistics shown in Figure 7.34, O&M personnel find that terminal roaming occurs between AP 213 and AP 218 for a total of 172 times within 4 hours. This is abnormal and indicates that a repeated roaming issue has occurred. After checking the AP deployment diagram in this area, O&M personnel find that AP 213 and AP 218 are not adjacent and there are two APs between them, as shown in Figure 7.35.

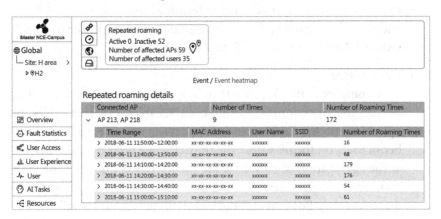

FIGURE 7.34 Repeated roaming analysis.

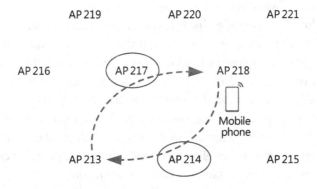

FIGURE 7.35 Locations of APs during repeated roaming.

When a terminal moves from AP 218 to AP 213, there is a high probability that the terminal roams to AP 214 or AP 217 first, instead of repeatedly roaming between AP 218 and AP 213. The reason for repeated roaming occurring is predominantly poor signal coverage. After checking the AP power configuration, O&M personnel find that the transmit power of AP 213 is 21 dBm and the transmit power of other neighbor APs is 16 dBm. A wireless terminal near AP 218 detects that AP 213 has the highest signal strength. As such, the terminal will preferentially connect to AP 213. However, the terminal has a fixed transmit power and is located far away from AP 213. Therefore, the RSSI in packets sent from the terminal to AP 213 is weak. After receiving the packets, AP 213 determines that the terminal has poor signal quality. Then, the terminal is triggered to roam to AP 218 again. As a result, the terminal roams between AP 213 and AP 218 repeatedly. To prevent repeated roaming of the terminal, reduce the transmit power of AP 213.

4. Device issues

A complex network consists of a significant number of physical devices that are connected using coaxial cables, Ethernet cables, or optical fibers. Any fault on any of the devices may affect a large number of users on the network. The impact scope is larger if the devices at the aggregation or core layer are incorrectly connected. The SDN controller classifies device issues into seven types, including device disconnection, Power over Ethernet (PoE) power supply failure, forwarding entry overrun, and high CPU usage. The SDN controller displays device issues in graphics, helping O&M personnel quickly rectify network faults to recover the network.

The SDN controller displays the network running status in a wired and wireless integrated topology. It then identifies faults on the GUI. In traditional network O&M, when a fault occurs, the fault cannot be directly located on the specific device. Instead, the fault needs to be gradually located step by step based on personal experience until it is located on the specific device. This process is time- and labor-consuming. Figure 7.36 shows the topology of a network where a large number of wireless terminals, including portable computers and mobile phones, fail in access authentication.

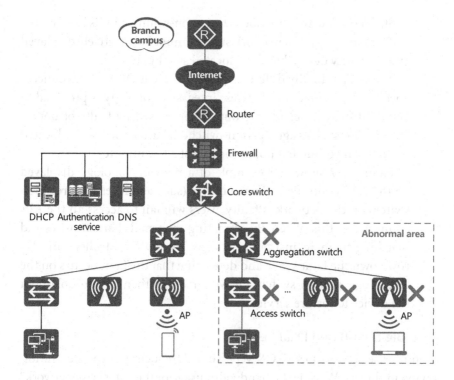

FIGURE 7.36   Network with user authentication failures.

The traditional fault locating roadmap is as follows:

- Check the device configuration. No problem is found.

- Check the authentication server. No problem is found because user authentication in other areas is normal.

- Check the networking based on the user feedback, locate the area where the fault occurs, and find the aggregation switch connected to the APs in the area.

- Check the aggregation switch. Due to a network cable fault, some ports on the aggregation switch have a large number of bit errors, including cyclic redundancy check errors and port queue stacking errors. As a result, data transmission fails. Fault locating requires approximately 90 minutes in total.

- Replace network cables and clear bit errors to restore services.

Such fault locating method is time-consuming. This is because the network topology and network status cannot be intuitively displayed and the faulty device has to be located step by step.

The story is totally different when intelligent O&M is introduced. Specifically, a wired and wireless integrated topology is provided to vividly display lower-layer group issues caused by faults of upper-layer nodes (such as aggregation switches). Such issues can be located by analyzing group fault patterns in just a few minutes.

Figure 7.37 shows an example of a network topology displayed on the SDN controller. When a device (such as a level-1 aggregation switch) on the network is faulty, users will fail to access all the APs connected to this switch. By extracting abnormal characteristics and matching them against fault patterns, the SDN controller can narrow down the fault scope and determine that the fault occurs on the level-1 aggregation switch. O&M personnel then can troubleshoot the device to restore services.

7.3.4 Big Data-Based Data Reports

In enterprise network O&M, O&M personnel frequently encounter the following problems: What is the bandwidth usage of the enterprise network?

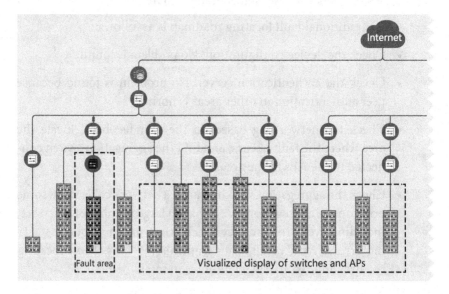

FIGURE 7.37 Group device fault detection on the wired and wireless integrated topology.

Is capacity expansion needed? Which devices in the network system are overloaded for a prolonged time and do they need to be replaced? Which devices have the highest failure rate? Do they need to be replaced? Which security events occur frequently and what policies can be applied to avoid security events?

Any of these problems is a difficulty for O&M personnel. It is almost impossible to manually collect and analyze data of a large network with thousands or even tens of thousands of devices. In addition, evaluation errors may occur. However, with the network data report function, these problems can be solved easily. Network data reports are used for statistical analysis and network evaluation. This is of great significance to the Information and Communications Technology (ICT) construction of enterprises. This is because the running status of the ICT system can be learned through statistical analysis, and problems with the network architecture can be found through network bottleneck analysis. So that, network optimization can be performed accordingly to improve the network running efficiency and user experience.

The SDN controller can display network data statistics from multiple dimensions based on tenants or sites, flexibly demonstrating the network running status. Figure 7.38 shows an example of data report. Currently, the SDN controller can collect and display statistics on devices, terminals,

A	B	C	D
Site	Time	Uplink Rate	Downlink Rate
allSites	2019-08-13 21:00	689.07 Mbit/s	10071.18 Mbit/s
allSites	2019-08-13 21:05	694.85 Mbit/s	9758.43 Mbit/s
allSites	2019-08-13 21:10	717.67 Mbit/s	9650.51 Mbit/s
allSites	2019-08-13 21:15	716.77 Mbit/s	9689.35 Mbit/s
allSites	2019-08-13 21:20	698.81 Mbit/s	9414.10 Mbit/s
allSites	2019-08-13 21:25	668.76 Mbit/s	9298.15 Mbit/s
allSites	2019-08-13 21:30	619.10 Mbit/s	8900.34 Mbit/s
allSites	2019-08-13 21:35	635.85 Mbit/s	8744.93 Mbit/s
allSites	2019-08-13 21:40	597.87 Mbit/s	8642.83 Mbit/s
allSites	2019-08-13 21:45	576.06 Mbit/s	7898.16 Mbit/s
allSites	2019-08-13 21:50	594.35 Mbit/s	6964.58 Mbit/s

◀ ▶ |Traffic Statistics|Attack Detection|Number of Online Users (Peak)| Network Rate ➕

FIGURE 7.38 Data report.

traffic, network rate, applications, and attacks. By querying the traffic of top applications or the traffic and real-time rate of network egresses, O&M personnel can learn the overall network traffic and determine whether the network bandwidth is sufficient and whether capacity expansion is needed. In addition, the traffic statistics are helpful in selecting the egress bandwidth for a new network.

All the collected statistics can be displayed on the GUI and exported into data reports for comparison and demonstration. Data reports can also be customized and exported as needed by data type, time range, and site. Administrators can perform secondary data development based on the generated data reports, read desired data, and build data O&M models for flexible network evaluation and optimization.

7.4 INTELLIGENT O&M OF WIRELESS NETWORKS

The Wi-Fi network system, similar to other wireless communications systems, is a self-interference system, where different APs can interfere with one another, affecting overall system capacity. For example, a high-power AP may interfere with adjacent APs if they work on overlapping channels. In this case, radio calibration is required to reduce inter-AP interference and improve overall system capacity and performance. In addition, Wi-Fi networks use spectrum resources on the free Industrial, Scientific and Medical (ISM) band, where short-range wireless communications devices (such as Bluetooth or ZigBee devices) and electromagnetic devices (such as microwave ovens) may also generate non-Wi-Fi interference. The radio calibration function can dynamically adjust the channels and power of APs managed by the same WAC and avoid non-Wi-Fi interference, ensuring that APs consistently work at the optimal performance.

7.4.1 Basic Radio Calibration

1. Basic radio calibration concepts

   In wireless network deployment, if radio parameters are not adjusted and all wireless APs work on the same channel and use the maximum transmit power, the APs will interfere with each other, and STAs may be associated with remote APs with high transmit power. In this deployment policy, due to the carrier sense multiple access with collision avoidance (CSMA/CA) mechanism and the hidden node problem of Wi-Fi networks, STA uplink and downlink communications become extremely unreliable. In addition, STAs may encounter frequent instances of extended latency and access issues, which severely affects user experience.

To address these problems, the following key radio parameters must be adjusted after Wi-Fi network deployment:

a. Transmit power

The transmit power of an AP determines the radio coverage and isolation between cells, with a higher transmit power indicating a higher downlink signal-to-noise ratio (SNR) and easier STA access. However, it should be noted that the transmit power of STAs is limited and is significantly lower than that of APs. If the transmit power of an AP is excessively high, STAs can receive data sent by the AP, but the AP cannot receive the data sent by the STAs.

b. Frequency band

The working frequency band of an AP determines the radio capacity and coverage. Currently, wireless local area networks (WLANs) use the 2.4 and 5 GHz frequency bands. As the 2.4 GHz frequency band has fewer channel resources and lower path loss than the 5 GHz frequency band, when APs are densely deployed, intrachannel interference on the 2.4 GHz frequency band is much stronger than that on the 5 GHz frequency band. In addition, a large amount of non-Wi-Fi interference exists on the 2.4 GHz frequency band from devices (such as cordless phones, Bluetooth devices, and microwave ovens). As such, the amount of information transmitted on the 2.4 GHz frequency band is significantly less than that on the 5 GHz frequency band.

c. Channel

The working channel of an AP indicates the frequency at which the AP works. If two adjacent APs work on the same channel, they will compete for channel resources, decreasing throughput and wasting the resources of other channels.

d. Bandwidth

The bandwidth of an AP determines its maximum rate and the channel's capacity. Obviously, the channel capacity for the 20 MHz bandwidth differs greatly from that for the 80 MHz bandwidth, and so the bandwidth must be properly allocated.

When planning radio parameters, we should be clear about the AP's location, interference, and running services. Only by doing so can we configure the most appropriate transmit power, frequency band, channel, and bandwidth. This approach, however, is time- and labor-consuming, and is unable to respond immediately if severe interference occurs abruptly.

To address these problems, the automatic radio calibration function is utilized. When this function is enabled, a Wi-Fi network automatically detects neighbor relationships between APs, interference on each channel, and load information within a given period. Based on the detected information, the Wi-Fi network automatically calculates radio parameters and delivers them to APs.

2. Key technologies used in radio calibration

a. Obtaining network status information

i. Radio topology and interference identification

After radio calibration begins, all radios are scanned for a given duration. This involves each radio switching to other channels to send Probe Request frames, receive Probe Response frames, listen to Beacon frames and other frames, and obtain the transmit power by exchanging packets between APs over the air interface. The transmit power is carried in the vendor-defined field, and the difference between the obtained transmit power and receive power is the path loss between the APs. The physical distance between two radios can then be calculated using the path loss calculation formula provided by the IEEE TGac.

During channel scanning, an AP obtains a large number of packets by listening to 802.11 frames, including air interface packets from other APs managed by the same WAC, and packets from APs or wireless routers managed by other WACs. The AP considers the APs or wireless routers managed by other WACs as external Wi-Fi interference sources, and stores their MAC addresses, receive power, and frequencies.

Also during channel scanning, the AP enables the spectrum scanning function to identify non-Wi-Fi interferences through time domain signal collection, fast Fourier transformation, and spectrum profile matching, and stores the non-Wi-Fi interference information together with the receive power and frequency information.

As shown in Figure 7.39, each AP periodically uploads collected neighbor information as well as Wi-Fi and non-Wi-Fi interference information to the WAC, which then generates a topology matrix, a Wi-Fi interference matrix, and a non-Wi-Fi interference matrix after signal filtering.

ii. Radio load statistics

Load information is easy to collect. This is because APs record wireless and wired traffic statistics, and periodically

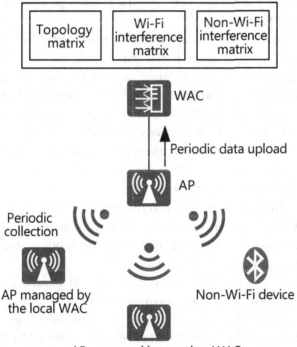

FIGURE 7.39   Radio topology and interference identification.

upload these statistics and user quantity information to the WAC. The WAC then can calculate radio load information.

By scanning spectrum resources and scanning and parsing 802.11 frames, the WAC can obtain topology information, channel interference information, and radio load statistics, enabling it to allocate channels and bandwidths to radios accordingly. The WAC preferentially allocates a channel with the minimum interference and the maximum bandwidth to a heavily loaded radio.

b. Automatic transmit power adjustment

In multi-AP networking scenarios, adjusting the transmit power of each radio can prevent coverage holes and overlapping, and can enable neighboring APs to fill any coverage holes upon an AP radio failure. As STAs cannot detect Wi-Fi signals in a coverage hole, avoiding coverage holes is extremely important.

Overlapping coverage generates interference and affects users' roaming experience. If the overlapping coverage area of two intrafrequency APs is large, co-channel interference increases and the APs' throughput is affected. For example, if a STA associated with AP 1 is also located in the core coverage area of AP 2, the STA does not proactively roam when the signal strength of AP 1 is strong. However, the user experience is poor because AP 1 cannot receive uplink packets from the STA due to unbalanced uplink and downlink power.

Utilizing the automatic transmit power adjustment algorithm, an AP can determine the coverage boundary by a certain SNR and automatically adjust its own transmit power to ensure both the coverage area and signal quality.

An AP's transmit power decreases as the number of neighbor APs increases, as shown in Figure 7.40, where each circle represents the coverage area of the corresponding AP. After AP 4 is added, the transmit power of the other APs decreases due to automatic power adjustment.

However, if an AP goes offline or fails, the transmit power of its neighbor APs increases, as shown in Figure 7.41.

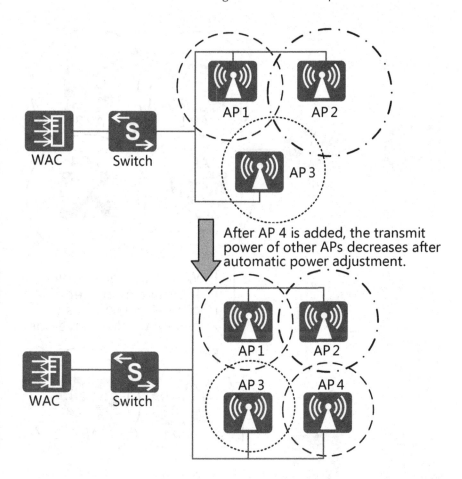

FIGURE 7.40 An AP's transmit power decreases when the number of neighbor APs increases.

c. Automatic channel and frequency bandwidth adjustment

After obtaining the neighbor relationship topology, external interference information, and long-term load statistics, a WAC can use the Dynamic Channel Assignment (DCA) algorithm to allocate channels and bandwidths to APs. Due to the limited number of channels on the 2.4 GHz frequency band, the bandwidth of each 2.4 GHz channel is typically fixed at 20 MHz. However, as there are more channels available on the 5 GHz frequency band and IEEE 802.11 standards support wider bandwidth, 5 GHz channel resources should be fully utilized to maximize system bandwidth, improve system throughput, and

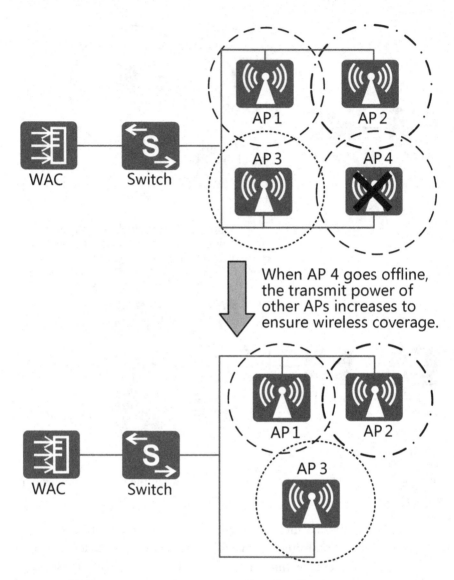

FIGURE 7.41   When an AP goes offline or fails, the transmit power of neighbor APs increases.

satisfy customer demands. Utilizing the DCA algorithm, a WAC can dynamically allocate 5 GHz channels and bandwidths to APs based on the topology, interference, and load information.

On a Wi-Fi network, adjacent APs must work on non-overlapping channels to avoid radio interference. Figure 7.42

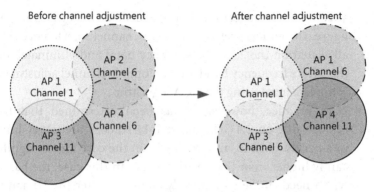

Note: Each circle indicates the AP's signal coverage range.

FIGURE 7.42   Before and after channel adjustment.

shows an example of channel distribution before and after channel adjustment. Before channel adjustment, both AP 2 and AP 4 use channel 6. After channel adjustment, each AP is allocated an optimal channel to minimize or avoid adjacent-channel and co-channel interferences, ensuring reliable data transmission on the network. As such, AP 4 now uses channel 11 so that it does not interfere with AP 2.

Figure 7.43 shows a distribution example of 2.4 and 5 GHz channels for seven APs that provide continuous signal coverage

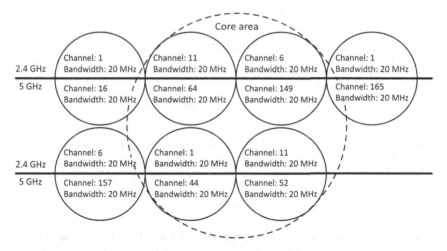

FIGURE 7.43   Channel allocation results on the 2.4 and 5 GHz frequency bands.

after the DCA algorithm is implemented. According to the figure, the co-channel and adjacent-channel interferences are alleviated on the 2.4 GHz frequency band and eliminated on the 5 GHz frequency band after automatic channel adjustment is implemented using the DCA algorithm.

In practice, however, it has been determined that the radio traffic on 2.4 GHz channel 44 reaches the upper limit in a period of time, and the traffic in the core area is significantly higher than that in edge areas. In this case, the bandwidth needs to be dynamically adjusted. After the Dynamic Bandwidth Selection (DBS) algorithm is enabled, frequency bandwidths are dynamically allocated to APs.

After the frequency bandwidth of the radio on 5 GHz channel 44 is adjusted to 80 MHz, as shown in Figure 7.44, service capability is greatly improved and the radio traffic no longer approaches the limit. In addition, the user access capacity in the core area can be improved when APs are not densely deployed. The frequency bandwidths of other radios in the core area can also be increased from 20 to 40 MHz based on service demands, enabling all the 13 channels on the 5 GHz frequency band to be fully utilized.

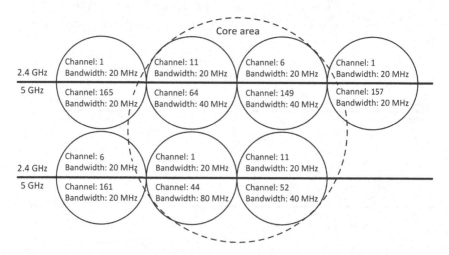

FIGURE 7.44   Bandwidth allocation results on the 2.4 and 5 GHz frequency bands.

d. Automatic frequency band adjustment

In high-density scenarios, dual-band APs capable of supporting both 2.4 and 5 GHz frequency bands are usually deployed to meet the higher capacity requirements of customers. In most cases, these APs are deployed within a closer proximity to one another. As only four nonoverlapping channel combinations are available on the 2.4 GHz frequency band, a large number of AP radios utilize these channels, restricting system capacity gains and leading to severe co-channel interference between APs. Such radios are known as redundant radios.

Huawei's dual-band APs support 2.4 GHz-to-5 GHz radio switchover, which prevents overlapping of 2.4 GHz radios and improves overall system capacity. For AP models that do not support 2.4 GHz-to-5 GHz radio switchover, redundant 2.4 GHz radios can be disabled or switched to monitoring mode to reduce co-channel interference through Dynamic Frequency Selection (DFS) or Dynamic Frequency Assignment (DFA).

3. Radio calibration framework

Figure 7.45 shows the basic radio calibration framework.

The configurations that must be performed for basic radio calibration on the customer interface include enabling radio calibration and delivering related parameter settings. When radio calibration is triggered on the device side — such as global calibration triggered by commands or at a scheduled time, partial calibration triggered when APs experience severe interference, and coverage hole compensation upon AP faults or recovery — the APs enable the air scan function, send the detected air interface data to the WAC for processing, and obtain the network topology and radio load statistics. The WAC then delivers the obtained information to the core calibration algorithm module for processing, obtains the radio calibration result, and delivers the result to the APs as a configuration item. By this point, the entire radio calibration process is complete.

4. Customer benefits of radio calibration

Radio calibration is more and more widely applied on Wi-Fi networks, and offers the following customer benefits:

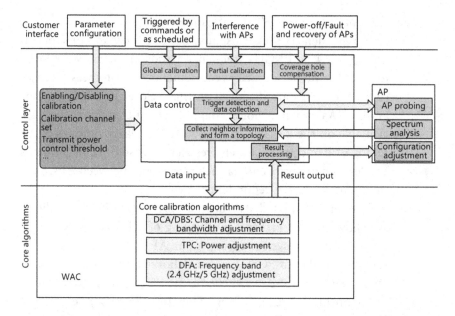

FIGURE 7.45   Basic radio calibration framework.

a. Retains optimal network performance. Radio calibration enables real-time intelligent radio resource management, enabling Wi-Fi networks to quickly adapt to environment changes and maintain optimal network performance.

b. Reduces network deployment and O&M costs. Radio calibration automatically manages radio resources, lowering the skill requirements for O&M personnel and reducing labor costs.

c. Improves Wi-Fi network reliability. Radio calibration can quickly eliminate the impact of network performance deterioration, improving system reliability and user experience through automatic radio monitoring, analysis, and adjustment.

7.4.2 Big Data-Based Intelligent Calibration

1. Issues addressed by big data-based intelligent calibration

Traditional network calibration is performed periodically and passively based on radio detection, which identifies changes in the surrounding environment. The collected data are primarily radio interference signals, and network calibration is not based on actual service quality. As such, traditional network calibration is rigid and inflexible. Using a transportation network as an example, traffic lights have different signal durations on trunk and branch roads, and traffic volumes at the same intersection vary depending on the time of day. If the same signal duration is used regardless of these factors, vehicles will likely need to wait at intersections for an unnecessarily long time, leading to low traffic efficiency. Even if the signal durations have been optimized, traffic efficiency remains low if the traffic lights have not been adjusted based on the actual traffic volume on the roads. However, if intelligent vehicle sensing technology is incorporated into the traffic lights system, the traffic volumes for each intersection can be predicted in real time. As a result, the signal duration of traffic lights can be dynamically adjusted based on traffic volumes, which greatly improves traffic flow efficiency.

This approach is also true for wireless network optimization. During intelligent optimization, the wireless network can intelligently predict the number of users and volume of traffic on the network based on the network load, and then determine the optimal radio parameters based on historical data to ensure the optimal air interface efficiency. The big data-based intelligent calibration system optimizes data from the following aspects:

a. Network topology completeness

    Network topology information is collected through radio scanning after radio calibration is enabled. This involves all APs randomly leaving the current working channel and scanning other channels. When an AP scans the current working channel of another AP that is also scanning channels, the APs cannot discover each other's air interface neighbors. As a result, the collected network topology is incomplete or inaccurate. To address this issue, big data-based topology collection, a long-term and gradually stable topology collection method, ensures the authenticity and reliability of topology data and allows for network adjustment based on the actual topology.

b. Network load prediction

    Network calibration is service-oriented, with traditional methods aiming to eliminate interference. However, even if the interference is minimized, the subsequent service experience is not always improved. The intelligent calibration system evaluates and predicts historical experience data such as traffic and user quantity, and optimizes the network based on real services to meet service requirements.

c. Collection of historical interference data

Frequent calibration affects user services, as some STAs have poor compatibility and protocol compliance and may go offline after the connected APs switch channels. For periodic calibration, the interval must be extended, or set to perform network calibration at a specified time at night when service traffic volumes are low and interference sources do not exist. As a result, the actual daytime network load cannot be reflected, and the calibration effect is inaccurate. Calibration based on historical data can reflect the actual calibration effect.

d. Identification and processing of edge APs

On Wi-Fi networks, APs at the edge of the logical or physical topology usually can often become nomadic STAs. Such STAs connect to the network but do not generate traffic and will soon leave the network. However, as large numbers of nomadic STAs will affect the communications quality of the network, the Wi-Fi network must accurately identify and process edge APs to prevent such an impact from occurring.

2. Big data-based intelligent calibration solution

In the big data-based intelligent calibration framework shown in Figure 7.46, the SDN controller accurately identifies the network topology and edge APs by analyzing a large amount of data reported by devices, and predicts the network load in the next calibration period by analyzing historical data. When the network calibration begins, the latest prediction data are used by the calibration algorithm to perform optimization calculation based on the real-time network quality in order to obtain the optimal calibration effect.

The following should be noted regarding the big data-based intelligent calibration solution:

a. The network administrator must enable big data-based calibration and scheduled calibration, and enable APs to periodically report KPI data.

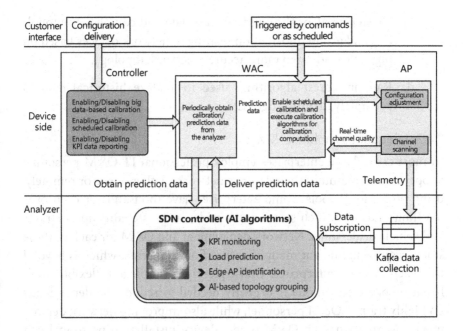

FIGURE 7.46   Big data-based intelligent calibration framework.

b. The SDN controller receives and stores network KPI information and optimization-related parameter information reported by APs, calculates and predicts data by analyzing historical data, and delivers the obtained calibration prediction data to devices.

c. APs report network KPI information and calibration-related parameters to the SDN controller through Telemetry. In addition, APs periodically obtain calibration prediction data in the next optimization period from the SDN controller, and perform calibration calculation based on the real-time channel quality and calibration prediction data.

3. AI algorithms used in big data-based intelligent calibration

a. Gaussian regression algorithm: used to calculate and predict AP loads in the next calibration period based on historical data.

b. Neural network algorithm: used to calculate and predict AP loads in the next calibration period based on historical data.

    c. Clustering algorithm: used to calculate and complete the topology and group information relating to network devices based on big data, leading to more accurate network topology.

    d. Random forest algorithm: used to analyze historical data to accurately identify edge APs.

7.4.3 Ubiquitous Mobile O&M

In most cases, large enterprises employ professional IT O&M personnel to operate and maintain their campus networks, either onsite or remotely, using professional tools. Some enterprises have thousands or even over ten thousands of branch stores nationwide, such as chain supermarkets and vehicle dealerships. Network deployment and O&M for each of these stores require significant manpower and financial costs, which is beyond the reach of most enterprises. Consequently, a simple and flexible intelligent mobile O&M mode is urgently needed to reduce the dependency on highly trained O&M personnel, while also improving network deployment efficiency. In such O&M mode, device installation personnel can complete network deployment and acceptance.

To help you better understand when and how to perform mobile O&M, the following introduces the roles of campus network O&M personnel, as shown in Figure 7.47.

- Authorized Service Partner (ASP)/Certified Service Partner (CSP): As engineering contractors, ASPs or CSPs assist in device installation or directly install devices at the initial stage of a project and are also responsible for subsequent project O&M.

- Network installation/maintenance engineer: Large ASPs/CSPs usually employ construction teams for network installation, while smaller ASPs/CSPs arrange their own engineers for device installation. If customers choose to self-install devices, ASP/CSP engineers only participate in device deployment and replacement and primarily deploy and connect physical network devices.

- Technical service engineer: A device or system service provider's technical support personnel. Large-scale projects require on-site technical support from technical service engineers, while the

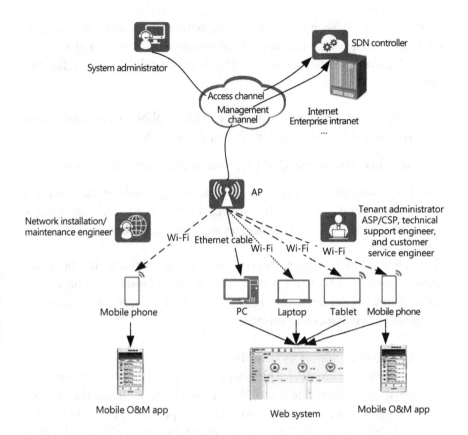

FIGURE 7.47 Mobile O&M scenario.

on-site technical support of small-scale projects is provided by ASPs/CSPs.

- Network administrator of the customer: The customer's network administrator can only perform simple O&M on local devices and troubleshooting of common issues. When more complex network anomalies or faults are encountered, the help from ASPs/CSPs or technical service engineers is required.

Professional technical O&M personnel, such as technical service engineers and some O&M engineers from ASPs/CSPs, possess high O&M expertise and comprehensive O&M tools. In contrast, the network O&M personnel of tenants often lack professional skills as well as network O&M and detection tools. As such, a simple, easy-to-obtain, and easy-to-operate network O&M tool is vital for these O&M personnel. To meet their requirements, Huawei launches the mobile O&M app that delivers the following advantages:

- High convenience and simplicity: Mobile phone-based operations are more intuitive than their PC-based equivalent.

- Mobility: The mobile O&M app can be used anytime, anywhere.

- Easy near-end connection to devices: In wireless network scenarios, mobile phones can easily connect to devices for local operations regardless of whether the devices are online or offline.

The mobile O&M app currently serves wireless networks, with support for wired network O&M also planned. The app offers four main functions: information import by scanning barcodes, device deployment, network diagnosis, and network monitoring.

1. Information import by scanning barcodes

To implement device plug-and-play, network device information must be imported to the SDN controller. In addition, the tenants and sites using the SDN controller must possess registration information about devices in order to ensure their validity. After we plan a Wi-Fi network, only an AP position bitmap is generated, while the mapping between AP positions and their actual installation positions is unavailable. In this case, we need to consider how to mark APs efficiently at the correct positions on the bitmap. If only a small number of APs exist, we can drag these APs can be directly to corresponding deployment positions on the bitmap after they go online. If there are a large number of APs, we are unable to mark their deployment positions on the bitmap using drag-and-drop mode because we cannot obtain the mapping between AP positions and their actual installation positions. In traditional mode, a Wi-Fi network is deployed as follows:

a. Network planning personnel carry out site survey and network planning, and provide a network planning report. The report includes floor plan details, the position and model of each AP, as well as the type, direction, power, and gain of antennas used by each AP which is numbered in the report.

b. Configuration personnel, such as ASPs, CSPs, authorized channel partners, or Huawei engineers, deploy services on APs.

c. Project construction personnel install APs and route cables according to the drawings provided in the network planning report. They need to record the APs' MAC addresses and mark AP IDs on the bitmap.

d. Deployment personnel import the information recorded by the project construction personnel to the SDN controller, mark the AP positions on the SDN controller, and configure services for the APs.

Currently, this procedure involves the following problems:

i. Construction personnel must manually record APs' MAC addresses and IDs, prone to errors.

ii. Deployment personnel must import the mapping between AP MAC addresses and AP IDs into the SDN controller, which is time-consuming and error-prone.

iii. The deployment process is divided into hardware installation and deployment configuration, each of which is completed by a separate team, thereby introducing additional complexity and communication.

To address these issues, Huawei provides app-based deployment by scanning barcodes. Figure 7.48 shows how to import device information by scanning barcodes using the mobile O&M app.

When the mobile O&M app is used for deployment, it automatically identifies an AP's equipment serial number (ESN) and MAC address by scanning its barcode, before accurately marking the AP's location. In addition, the app directly uploads AP information and its location to the SDN controller, greatly simplifying AP deployment and configuration while reducing personnel skills requirements and minimizing the risk of configuration errors.

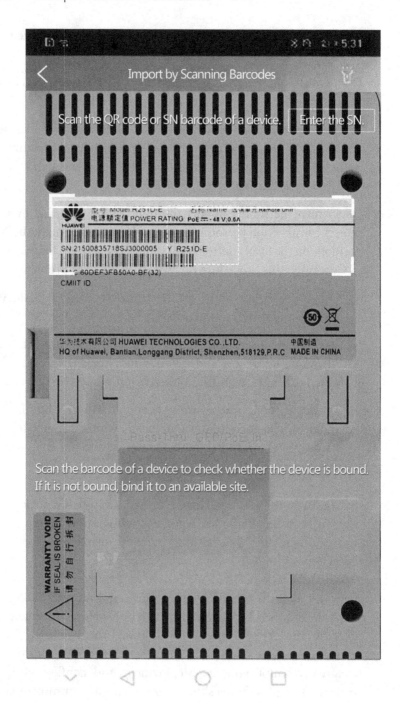

FIGURE 7.48   Information import by scanning barcodes.

2. App-based deployment

In small and micro branches, a single AP is often capable of meeting network requirements. Network uplinks are usually Internet lines provided by carriers and a Point-to-Point Protocol over Ethernet (PPPoE) account needs to be configured for Internet access in dial-up mode. Traditionally, O&M personnel can log in to an AP from a laptop or PC and perform account and service configurations on the web system. However, if APs need to be deployed in a large number of stores and the construction and deployment personnel do not possess necessary network O&M skills or tools, professional deployment personnel are required. This increases the overall deployment time and costs. As shown in Figure 7.49, after the mobile O&M app connects to an AP and simple configurations are performed, the AP is ready for deployment. This simplified process dramatically improves deployment efficiency and reduces associated costs, while also requiring minimal deployment skills. Thanks to the mobile O&M app, even installation personnel can successfully complete deployment and bring APs online.

FIGURE 7.49   Networking for app-based deployment.

Before performing app-based deployment, install the mobile O&M app on a mobile phone. Then, connect the mobile phone to the SDN controller through the 3G/4G network or the onsite Wi-Fi network and obtain an account for logging in to the SDN controller. Currently, app-based deployment is applicable only for deploying APs. Via this app, AP information import and deployment configuration can be completed in a one-off process, as shown in Figure 7.50.

3. Network diagnosis

Network services can be configured in advance by network administrators or configured after network devices are deployed and registered. When the network is operating properly, deployment personnel can perform simple network acceptance or diagnosis tasks

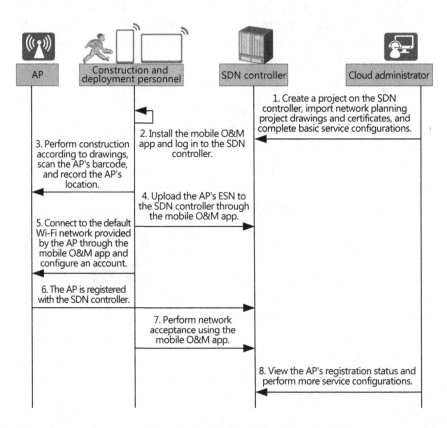

FIGURE 7.50 App-based deployment process.

to ensure that services are running correctly. For example, they can check whether wireless signals can be connected, access the external network to check network connectivity, or carry out roaming tests to check the network coverage status. If a network fault occurs, local IT O&M personnel can use the mobile O&M app for troubleshooting. They can also test network connectivity and verify the quality of video and gaming user experiences to proactively determine network faults and quickly remedy simple issues.

The mobile O&M app leverages the diagnosis function of network devices to remotely display network diagnosis information. The network diagnosis functions include:

a. Ping: APs can ping STAs or other devices to check connectivity between the local and external networks. Also, the SDN controller can ping APs to check connectivity between the APs and SDN controller.

b. Throughput testing: The AP or network throughput can be tested to check network performance.

c. Roaming testing: Network roaming performance can be tested, including the roaming time and roaming effect.

d. Game testing: The gaming experience on the current network can be tested by simulating games using the mobile O&M app to evaluate the network performance.

e. Video testing: The video transmission performance of the current network can be tested by playing network videos and examining the video playback effects.

f. Intelligent diagnosis: CPU, memory, and other anomalies of a specific device can be diagnosed.

4. App-based monitoring

In most cases, O&M personnel are required to log in to the management system using a PC or laptop in order to perform network O&M. With the mobile O&M app, O&M personnel can monitor the network using mobile phones anytime and anywhere regardless of being off duty or on business trips. Figure 7.51 shows the device and traffic monitoring screens on the mobile O&M app.

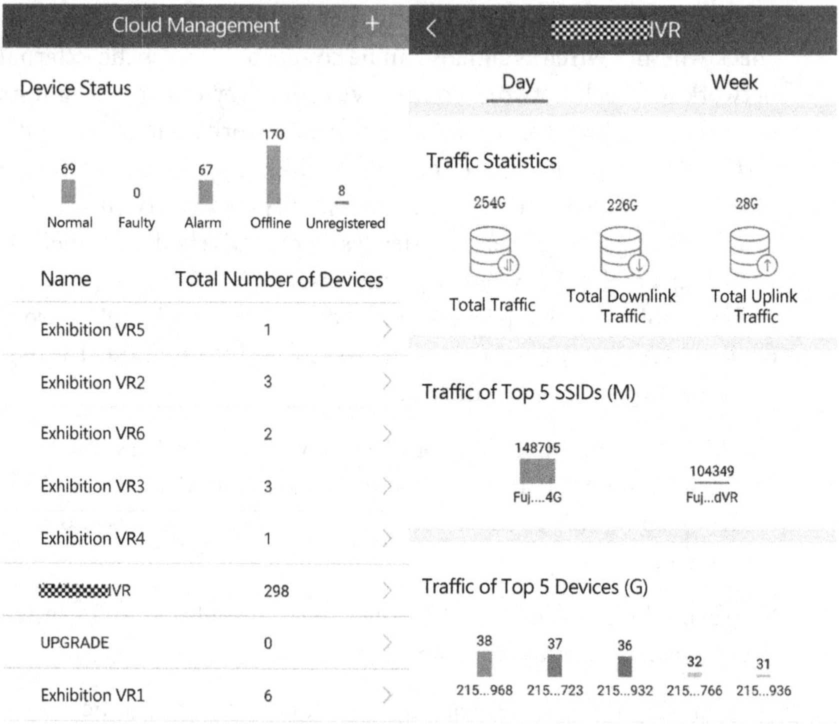

| Cloud Management | + | < | IVR |

Device Status

	170		
69	0	67	8
Normal	Faulty	Alarm	Offline	Unregistered

Name	Total Number of Devices	
Exhibition VR5	1	>
Exhibition VR2	3	>
Exhibition VR6	2	>
Exhibition VR3	3	>
Exhibition VR4	1	>
IVR	298	>
UPGRADE	0	>
Exhibition VR1	6	>

| | Day | Week |

Traffic Statistics

| 254G | 226G | 28G |
| Total Traffic | Total Downlink Traffic | Total Uplink Traffic |

Traffic of Top 5 SSIDs (M)

| 148705 | 104349 |
| Fuj....4G | Fuj...dVR |

Traffic of Top 5 Devices (G)

| 38 | 37 | 36 | 32 | 31 |
| 215...968 | 215...723 | 215...932 | 215...766 | 215...936 |

FIGURE 7.51 Device and traffic monitoring screens on the mobile O&M app.

O&M personnel use the mobile O&M app to check basic monitoring data, enabling them to monitor the network status anytime and anywhere. The mobile O&M app displays the following monitoring data:

a. Site device status, including the registration and running status of all site devices of a tenant

b. AP information, such as the IP address, version number, running status, and connected terminals

c. Traffic data, including traffic statistics for the current day or week, top SSIDs, and top APs

d. Terminal information, including the IP address, access time, access duration, signal strength, accumulated traffic volume, and RET rate of terminals

CHAPTER 8

E2E Network Security on an Intent-Driven Campus Network

N ETWORK SECURITY IS USED to prevent hardware, software, and system data on information networks from being damaged, modified, and disclosed either accidentally or maliciously. It ensures continuous, reliable, and normal system operations as well as uninterrupted information services. In a complex information system, such as a campus network, ensuring network security is a system engineering task. That is, no individual device, node, technology, or configuration can ensure the security of the entire network by itself. Instead, network security is an organic whole that consists of many physical devices, security technologies, security solutions, and others, which are linked based on the appropriate configurations.

8.1 CAMPUS NETWORK SECURITY THREAT TRENDS AND CHALLENGES

Nowadays, mobile terminals, represented by smartphones and tablets, are more often used in mobile offices, besides their common function of accessing the Internet anytime and anywhere. Mobile applications, Web 2.0, and social networking websites have all become indispensable on the Internet. What's more, cloud computing and software-defined networking (SDN) technologies are being deployed to accelerate service deployment and allow networks to change on demand. All of these Information and Communications Technology (ICT) transformations have greatly improved enterprise communication efficiency.

However, from the perspective of enterprise information security, mobile offices blur the boundary of enterprise networks, and hackers can more easily intrude into enterprise IT systems using mobile terminals. In addition, because traditional security gateways can only implement security and protection control through IP addresses and ports, they cannot cope with the increasing number of application threats and web threats. Therefore, it is clear that enterprise campus networks are facing unprecedented challenges in information security.

8.1.1 Network Security Threat Trends

In the 1990s, with the development of the Internet, network attacks extended from labs to the Internet and kept evolving. However, as attack techniques evolved, so did the defense techniques used to respond to threats. As shown in Figure 8.1, attack techniques have evolved from simple scanning and overflow attacks to advanced persistent threats (APTs). Accordingly, defense techniques have evolved from simple malformed packet filtering to network-wide intelligent security defense based on big data and artificial intelligence (AI).

In recent years, the types and intensity of network attacks have increased exponentially. Specifically, the types of attacks against application-layer protocols, such as Hypertext Transfer Protocol (HTTP), Hypertext Transfer Protocol Secure (HTTPS), Session Initiation Protocol (SIP), and Domain Name Service (DNS), have experienced continuous evolution. As the campus network is the major service operations network of an enterprise, its security is of vital importance. Therefore, to safeguard the campus network, we must deploy a security defense solution that defends against mass attacks at the Internet egress and application layer.

8.1.2 Traditional Security Defense Model Based on Security Zones

A security zone (or zone for short) consists of one or more interfaces, and users in the same security zone have the same security attributes. Firewalls use security zones to divide networks and mark the routes of packets. A security check is triggered when packets travel between different security zones, and security policies are enforced on firewalls accordingly.

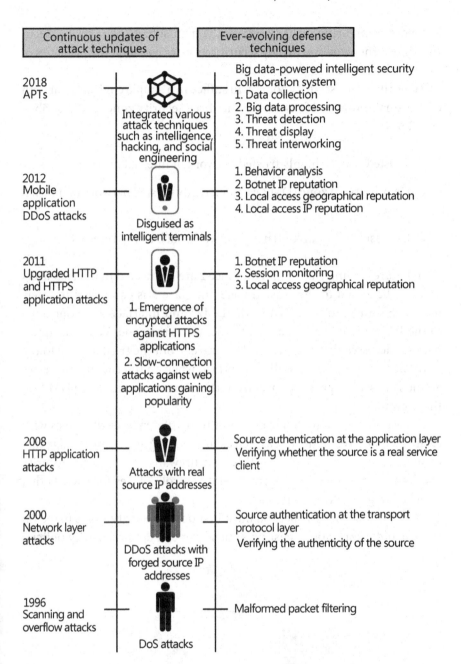

FIGURE 8.1   Evolution of network attack and defense techniques.

Before configuring security services, you must first create related security zones. Then you can deploy security services based on the priorities of the security zones.

Typically, three security zones will suffice in a simple environment with only a small number of networks. The three security zones are described as follows:

- Trusted zone: a highly trusted network of internal users

- Demilitarized zone (DMZ): a moderately trusted network of internal servers

- Untrusted zone: an untrusted network, such as the Internet

In Figure 8.2, interface 1 and interface 2 are connected to internal users and can be added to the trusted zone; interface 3 is connected to internal servers and can be added to the DMZ; and interface 4 is connected to the Internet and can be added to the untrusted zone. When an internal user accesses the Internet, packets travel from the trusted zone to the untrusted zone on the firewall. In contrast, when an Internet user accesses an internal server, packets travel from the untrusted zone to the DMZ on the firewall.

Firewalls are used to divide networks into different security zones with clear levels and relationships, as well as functioning as a secure bridge that connects multiple networks. In this way, firewalls can perform security checks and implement management and control policies for packets that travel through multiple networks.

To define a security zone, you must first define security assets based on service and information sensitivity; then you can define security policies

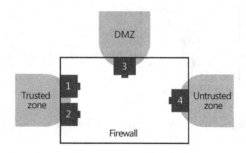

FIGURE 8.2   Security zone division.

and security levels for the security assets. A security zone consists of security assets with the same security policies and security levels. Finally, you can design different security defense capabilities for different security zones based on their potential risks.

It is recommended that security devices, such as firewalls, be deployed between different security zones to ensure border defense and isolation, as shown in Figure 8.3. For example, in a campus network scenario, even though it is considered a secure network, it will inevitably confront security challenges due to being connected to the Internet. Therefore, when creating security zones, we allocate the Internet to the untrusted zone and the campus network to the trusted zone; and deploy security devices at the campus network egress to isolate the campus network from the Internet and defend against external threats. In addition, we usually allocate the data center to the DMZ, and deploy security devices in the DMZ to isolate traffic between the campus intranet and servers in the data center.

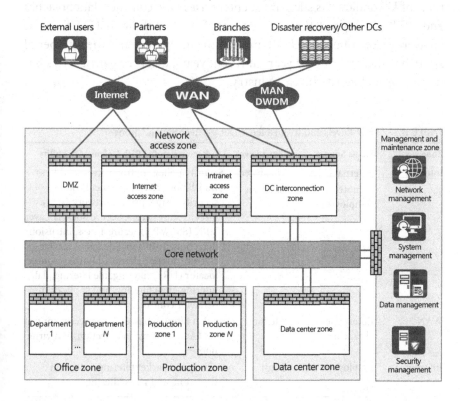

FIGURE 8.3   Security zone-based border defense.

However, if some areas inside a campus network require special security protection, they can be divided even further into different zones and security devices can be deployed at the egress of each zone.

After security zone design is completed, we analyze the security threats that each zone may face and deploy the corresponding security defense approach. Table 8.1 lists the trust levels of common security zones and the recommended risk defense design to be deployed.

8.1.3 APTs Compromise Traditional Security Defense Models

Security threats have changed dramatically in recent years. Hacker attacks, for example, have shifted from conventional pranks and technical flaunting to profit-making and commercialization. One such attack occurred on May 12, 2017, when a large-scale ransomware attack, known as WannaCry, was unleashed, impacting more than 45000 hosts across 74 countries within less than 20 hours, and resulting in the loss of billions of US dollars. As such, most enterprises now consider data breaches and APTs the biggest threats they face. Statistics show that APTs cause an average loss of nearly US$10 million to enterprises, and the number of APTs has increased every year. In recent years, defense against APTs has been a topic of debate in the industry.

TABLE 8.1   Trust Levels and Risk Defense Design Recommendations

Zone	Access Source	Trust Level	Recommended Risk Defense Capability
Internet	External users	Untrusted	Security policy control, network address translation (NAT), Internet Protocol security virtual private network (IPsec VPN) secure access, Secure Sockets Layer VPN (SSL VPN) secure access, intrusion detection, distributed denial of service (DDoS) attack defense, uniform resource locator (URL) filtering, file filtering, data filtering, mail filtering, and application behavior control
	Remote employees	Medium	
WAN	Enterprise branches	Medium	Packet filtering based on the access control list (ACL), intrusion detection, and virus filtering
Intranet	Employees	High	ACL-based packet filtering, intrusion detection, and virus filtering
	Guests	Low	

An APT is a network attack and intrusion behavior aimed at intercepting an enterprise's core data. An APT integrates various attack techniques such as intelligence, hacking, and social engineering to launch persistent attacks on specific targets for an extended period and access an enterprise network, obtain data, and monitor the target computer systems for extended periods without being noticed. Table 8.2 compares traditional security threats and APTs.

The APT's lifecycle consists of four stages: conducting reconnaissance, establishing a foothold, escalating privileges, and causing damage or exfiltrating data, as shown in Figure 8.4.

Step 1 Conducting reconnaissance: Attackers collect all intelligence information related to the specified target. This may include the target's organizational structure, office location, products and services, employee address book, administrative email address details, executive meeting agenda, portal website directory structure, internal network architecture, deployed network security devices, open ports,

TABLE 8.2   Comparison between Traditional Security Threats and APTs

Item	Traditional Security Threats	APTs
Attacker	Opportunists, hackers, or cybercriminals	Professional and organized criminals, illegal companies, hackers, and competitors, with strong organizational capabilities and resource assurance
Targeted user	Unspecified	Targets are clearly specified, and may include national security data and trade secrets
Purpose	Financial benefits, identity theft, fraud, spam, and defamation of reputation	Market manipulation, strategic advantages, damage to critical infrastructure, and political purposes
Frequency	One-off attacks that last only a short period	Continuous attacks that remain undetected for an extended period
Approach	Spreading malicious software that already exists for profit	Exploiting a zero-day vulnerability or using tailored malware with a focus on attacking and disrupting
Characteristics	Known characteristics, easy to capture, high detection rate	Hard to detect due to a long-term lack in samples

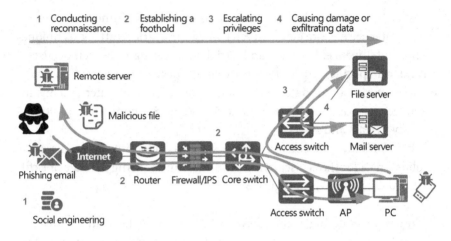

1 Conducting reconnaissance   2 Establishing a foothold   3 Escalating privileges   4 Causing damage or exfiltrating data

FIGURE 8.4   APT's lifecycle.

office and email systems used by enterprise employees, and the systems and versions used by corporate web servers.

Step 2 Establishing a foothold: Attackers employ social engineering to implant malware in the target network using phishing emails, web servers, and USB flash drives, and wait for target users to open email attachments, URL links, and files in USB flash drives, or access "watering hole" websites.

Step 3 Escalating privileges: Once the attackers have established a foothold, malware unleashes a range of viruses. For example, it can install the remote access Trojan (RAT) in the target system or initiate encrypted connections to the server specified by the attacker to make the server download, install, and run even more malicious programs. Then, malicious programs escalate their permission or add administrators, and they set themselves to start upon system startup or start as a system service. Some malicious programs may even modify or disable the firewall settings of their victims in the background to ensure they remain undetected for an extended period. When attackers successfully establish a foothold in the target network and obtain the corresponding permissions, the victim computers become zombies, leaving them with no choice but to wait for the attackers' further exploitation.

Step 4 Causing damage or exfiltrating data: After all preparations are complete, attackers can wait for the right moment to cause damage or exfiltrate data. For example, attackers can use the RAT's keystroke logging and screen recording functions to obtain users' domain, email, and server passwords and remotely log in to various internal servers, such as internal forums, team spaces, file servers, and code servers, through zombies and steal valuable information.

Next-generation threats represented by APTs pose unprecedented challenges to the industry. Therefore, new threat detection and defense technologies are urgently needed to compensate for the following weaknesses of traditional threat defense approaches:

1. Delayed threat detection

   Traditional security detection tools cannot quickly detect threats hidden inside encrypted traffic without decrypting the traffic. In addition, malicious code patterns change rapidly as attackers work to evade existing defense technologies, thereby further delaying the threat detection period.

2. Single-point passive defense system

   A traditional security defense system is usually deployed at the network border. Without the assistance of other systems, such a system can only cope with threats individually. That is, once the security border is breached, viruses can easily flood the enterprise intranet and cannot be controlled.

3. Complex security service management

   Traditional security network element (NE) management standards vary depending on the specific vendor; however, most require manual configuration, and usability is often poor. Security NE management is also often complex and requires information such as IP addresses, ports, and physical locations. In addition, traditional security NE management systems cannot provide customized security protection for tenants.

8.2 CORE CONCEPTS AND OVERALL ARCHITECTURE OF BIG DATA-POWERED SECURITY COLLABORATION

Compared with traditional security defense systems, the intelligent security collaboration solution powered by big data and AI features evolution from discrete sample processing to holographic big data analytics, from manual analysis to automatic analysis, and from static signature analysis to dynamic signature, full-path, behavior, and intention analysis. This solution provides a comprehensive security defense system to ensure the security of campus networks and services.

8.2.1 Core Concepts of Security Collaboration

By comparing the characteristics of traditional and new security threats, we can identify substantial differences in the way that attacks are carried out. Therefore, it is clear that new concepts and policies must be deployed to cope with new security threats.

Traditional security threats can be viewed in a similar way to thieves. That is, they usually damage the network in a relatively violent manner and cause specific anomalies that are easy to capture and identify. Like threat detection systems, the police can easily detect and stop thieves from committing violent acts. On the network, security devices, such as firewalls, are equivalent to the police and they can easily cope with traditional security threats.

However, new security threats work more like swindlers because they often carry out their crimes in a seemingly civilized manner. The evidence of their "crime" is hidden deep within the network and is hard to find without gathering and analyzing a wide range of information. Therefore, when dealing with swindlers, it is not enough to rely solely on the strength of the police. Instead, we need witnesses to provide clues, which are comprehensively analyzed by the police to discover and stop swindlers in a timely manner. Using this example, the NEs (such as routers and switches) that are distributed in every corner of the network are like witnesses, because they provide a large number of valuable clues and intelligence.

The core concept of big data-powered security collaboration is to collect a large amount of security-related source data from each NE, which is then comprehensively analyzed on the big data analytics platform to accurately identify security threat events. Then, the associated network controller is instructed to perform security handling. In this way, the enterprise campus network can be proactive when defending the network and its security defense capabilities are more effective.

Figure 8.5 illustrates the intelligent security collaboration system based on big data and AI. This system is so far the best solution in the industry to cope with APTs. First, in terms of time, the system can monitor each

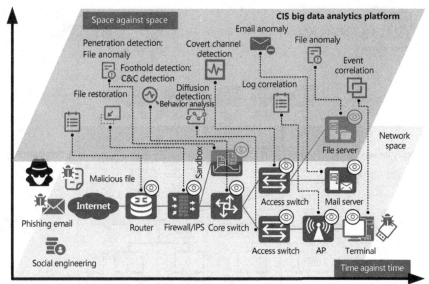

FIGURE 8.5 Big data and AI-based intelligent security collaboration system defending against APTs.

phase of an APT and analyze the obtained information. Second, in terms of space, the system can monitor every corner of the entire network and analyze the information obtained from different parts of the network. Finally, based on big data analytics technologies, the system can analyze the collected information, attack behaviors, and intentions, and perform intelligent judgment and associated processing actions. This allows the system to constantly monitor the entire network, comprehensively analyze network data, and defend against APTs.

8.2.2 Overall Architecture of Security Collaboration

Figure 8.6 shows a campus network's overall security collaboration architecture based on big data and AI. Compared with traditional defense solutions, big data-powered security collaboration implements network-wide monitoring and in-depth mining of network behavior data to detect and handle threats in a timely manner. This helps customers improve the intelligence and automation of security analysis and O&M, and ensures the stability of customers' key infrastructure and service continuity.

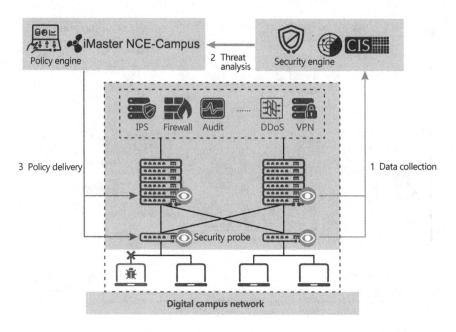

FIGURE 8.6   Overall architecture of security collaboration on a campus network.

1. Accurate identification of network threats

An "armed" enterprise usually deploys all security defense devices from the network border all the way to terminals. However, single-point defense devices are isolated from one another and do not form a defense system for joint operations. As a result, the enterprise cannot accurately detect threat events or effectively monitor unknown threats. As such, single-point detection is considered unsatisfactory. Regarding multipoint exception analysis on the entire network, enterprises still rely on Security Information and Event Management (SIEM) devices to monitor logs on the entire network for network-wide threat control. Mature SIEM products in the market perform excellently in log collection, coverage, adaptation, and parsing, but they are weak in log analysis. In addition, logs on the network only record the alarms of single-point events and cannot reflect the correlation between multi-point exceptions for the intrusion and attack chain globally.

The big data-powered security collaboration solution builds AI- and big data-based security analyzers to convert network infrastructure into security detection sensors. As sensors, network devices such as switches, routers, and firewalls provide information such as traffic, NetFlow data, metadata, logs, and files to the security analyzers. In addition, scripts are prepared based on the network topology and threat scenarios to study the intention and path of hacker intrusions and construct threat detection models and rules. The big data analytics tools are also leveraged to perform multidimensional comprehensive threat analysis on a large amount of information, thereby helping administrators converge large amounts of original event logs, automatically and accurately detect threats, and even predict threats.

Based on the network-wide multiscenario and multidimensional comprehensive analysis model, the big data-powered security collaboration solution combines subscenarios such as spear phishing, web penetration, hacker remote command and control (C&C), account exception, internal traffic exception, and data theft for comprehensive analysis, and calculates the threat type, severity, and trust level of each attack behavior to better understand hackers' attack intentions. The solution then matches the attack behaviors against the attack behavior pattern database provided by the expert system, dynamically adjusts the model based on hackers' intrusion behaviors, identifies and predicts the attack type and trust level, and reduces the occurrence probability of single-point original threat events to less than 0.0001. This effectively reduces the time for log analysis and source tracing, and helps save the required investment in professional security analysis personnel.

As sensors, network devices can also be instructed by security analyzers to help detect threats. For example, when an exception occurs on a network, we can instruct network devices to import suspicious traffic to the analysis platform for further analysis, thereby ensuring accurate judgments are made. With the support of network devices, threat detection and identification are like a submersible with a large number of probes for deep-sea detection, leaving undercurrents no place to hide.

2. Network collaboration for quick threat handling

Forty-two percent of new vulnerabilities will be exploited by hackers within 30 days after being disclosed. However, because an enterprise's threat response time is usually slow, combating threats can be viewed as a race against time. Therefore, reducing the time from threat intrusion to damage repair is key to reducing economic and data loss. As mentioned above, the WannaCry ransomware caused losses to more than 240000 victims; however, compared with highly complex attacks such as Stuxnet, WannaCry was not technically powerful; rather, it was able to spread quickly and had a wide range of infections.

Even though vendors claim that they can detect the WannaCry virus, customers are more concerned about whether infected computers can be quickly located and isolated to prevent the virus from spreading to the internal network. They also require that infected computers can be quickly repaired. Therefore, the automatic response and repair capability need to be advanced to satisfy the needs of customers.

The big data-powered security collaboration solution uses controllers to collaborate with network devices and quickly handle threats. This solution uses the AI-based threat analysis capability to quickly detect and respond to unknown worms, for example, blocking port 445 of the egress firewall and router, and updating the IPS signature database. The solution's key mechanism uses network devices as executors which collaborate with access switches through controllers to isolate infected computers in a timely manner. It also collects traffic through network nerve endings, locates the infection path, instructs associated terminal software to automatically clear worms, pushes patches in batches to assist O&M personnel in fixing vulnerabilities, and automatically releases tools to restore encrypted files.

3. Automated policy O&M

Larger enterprises usually have more complex networks. Security devices guarding networks against threats accumulate massive amounts of security policies as time goes by, making security policy O&M the top issue for large- and medium-sized organizations and enterprises. For example, in the network of a financial services customer, the data center may include more than 500 firewalls, with each firewall including tens of thousands of security policies. The data center's firewall policies need to be adjusted each time services are updated; therefore, thousands of policies may need to be updated on a daily basis, thereby making policy O&M extremely difficult. In addition, notifications such as service offline and IP address reclamation cannot be delivered to the network security department in a timely manner. As a result, a large number of expired security policies can be accumulated. During network migration, security policy migration is also challenging. If policies need to be reconfigured, it would require several weeks if manual operations are performed.

Big data-powered security collaboration implements service-driven security policies, upgrades the IP-based machine language to application-based advanced language, and establishes automatic mappings between applications and IP addresses. In addition, controllers closely bind security policies with service lifecycles. When services go online, change, or go offline, big data-powered security collaboration detects service changes in real time and automatically translates service policies into IP policies that can be executed by terminals, without requiring manual intervention.

More importantly, the big data-powered security collaboration uses controllers to visually analyze application mutual access relationships and automatically generate a policy whitelist during equipment room migration, eliminating the need for manual resetting. By observing and analyzing application mutual access flows and policy matching rates, the solution dynamically adjusts and optimizes network-wide policies, deletes duplicate policies, and ensures expired policies are brought offline in a timely manner. It also uses dynamic traffic analysis to verify the validity of pre-online policies and ensure that the policies executed are the same as those defined.

8.3 INTELLIGENT SECURITY THREAT DETECTION BASED ON BIG DATA

8.3.1 Big Data Analytics Process

The big data analytics platform is the core of the security collaboration solution. Using Huawei cybersecurity intelligence system (CIS) as an example, Figure 8.7 shows the big data analytics process during security collaboration.

1. Data collection

    Data collection includes both log collection and traffic collection, which are implemented by the log collector and the flow probe, respectively.

    The log collection process includes log receiving, categorization, formatting, and forwarding, while the traffic collection process

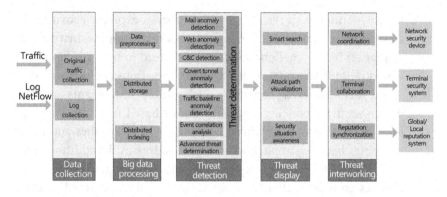

FIGURE 8.7   Big data analytics process during security collaboration.

includes traffic collection, protocol resolution, file restoration, and traffic metadata reporting.

2. Big data processing

Big data processing includes data preprocessing, distributed storage, and distributed indexing. Data preprocessing formats the normalized logs reported by the collector and the traffic metadata reported by the flow probe, supplements related context information (including users, geographical locations, and areas) and releases this formatted data to the distributed bus. Distributed storage stores the formatted data and classifies heterogeneous data of different types, such as normalized logs, traffic metadata, and Process Characterization Analysis Package (PCAP) files for storage. These stored data are mainly used for threat detection and visualization. Distributed indexing creates indexes for key formatted data, providing keyword-based quick search services for visualized investigation and analysis.

3. Threat detection

The analyzer performs multidimensional threat analysis on the data that have been collected and processed based on big data, leading to the identification of threats.

4. Threat display

The analyzer displays the results of threat identification on the Graphical User Interface (GUI), enabling users to intuitively understand the entire network's security situation. However, some security threats still require manual analysis and identification.

5. Threat interworking

The analyzer generates an interworking policy based on suspicious analysis results and delivers it to all NEs on the network. This policy contains precise control instructions, enabling the NEs to block any suspicious threats.

8.3.2 Principles of Big Data Analytics

1. Mail anomaly detection

Mail anomaly detection extracts mail traffic metadata from historical data. It analyzes Simple Mail Transfer Protocol (SMTP), Post Office Protocol Version 3 (POP3), and Interactive Mail Access Protocol (IMAP) information such as the recipient, sender, mail server, mail body, and mail attachment. It then detects mail anomalies in offline mode, such as sender and recipient anomalies, malicious mail downloads, mail server access anomalies, and mail body URL anomalies, based on sandbox file inspection results.

2. Web anomaly detection

Web anomaly detection recognizes web penetration and abnormal communication. It extracts HTTP traffic metadata from historical data and analyzes HTTP fields, including the URL, User-Agent, Refer, and message-digest algorithm 5 (MD5) values of uploaded and downloaded files, in order to detect anomalies in offline mode, such as malicious files, access to unusual websites, and non-browser traffic, based on sandbox file inspection results.

3. C&C anomaly detection

C&C anomaly detection analyzes DNS, HTTP, Layer 3 protocol, and Layer 4 protocol traffic to detect C&C communication anomalies. DNS traffic-based C&C anomaly detection adopts a machine learning method; performs training based on sample data to generate a classifier model; and uses the classifier model to identify communication anomalies that access Domain Generation Algorithm (DGA) domain names in the customer network to discover zombie hosts or abnormal APT behavior in the C&C phase. C&C anomaly detection based on Layer 3/Layer 4 protocol traffic analyzes the characteristics of information flows between C&C Trojan horses and external devices, differentiates between C&C communication information flows and normal information flows, and performs traffic detection to discover C&C communication information flows existing in the network. HTTP traffic-based C&C anomaly detection uses statistical analysis, recording each time an intranet host accesses the same destination IP address and domain name, calculating the length of time between each connection, and periodically checking for any changes that might reveal abnormal external connections from the intranet host.

4. Covert tunnel anomaly detection

Covert tunnel anomaly detection identifies the transmission of unauthorized data by compromised hosts using normal protocols and tunnels. The detection methods include Ping Tunnel, DNS Tunnel, and file anti-evasion detection. Ping Tunnel detection analyzes and compares Internet Control Message Protocol (ICMP) payloads transmitted between a pair of source and destination IP addresses within a certain time window to detect abnormal Ping Tunnel communications. DNS Tunnel detection checks the validity of domain names in DNS packets between a pair of source and destination IP addresses within a certain time window, and analyzes the DNS request and response frequency to detect abnormal DNS Tunnel communications. File antievasion detection analyzes and compares file types in traffic metadata to detect inconsistencies between file types and file name extensions.

5. Traffic baseline anomaly detection

Traffic baseline anomaly detection identifies abnormal access between intranet hosts or regions (between intranet and extranet regions, between an intranet region and the Internet, between intranet hosts, between an intranet host and the Internet, and between an intranet host and a region). The traffic baseline is a rule for access between intranet hosts, between regions, or between the intranet and external network. It specifies whether access is allowed within a given time range and, if allowed, the access frequency range and the traffic volume range.

The traffic baseline can be obtained through system autolearning or defined by users. System autolearning refers to the system automatically collecting the access and traffic information between intranet hosts, between regions, and between the intranet and external network within a time period (for example, one month) and generating a traffic baseline from the information gathered (an appropriate floating range is automatically set for the traffic data). A user-defined traffic baseline refers to a user manually configuring the access and traffic rules between intranet hosts, between regions, and between the intranet and external network. Traffic baseline anomaly detection loads the auto-learned and user-defined traffic baselines to memory, collects statistics on and analyzes traffic data in online mode, and exports anomaly events once inconsistencies have been detected between network behaviors and the traffic baselines.

6. Event correlation analysis

Event correlation analysis determines the correlation and time sequence relationship between events, in order to detect effective attacks. Event correlation analysis uses a high-performance traffic computing engine, which obtains normalized logs directly from the distributed messaging bus, stores them in memory, and analyzes the logs based on correlation analysis rules. Some correlation analysis rules are preset in the system, but users can also customize their own specific correlation analysis rules. If multiple logs match the same correlation analysis rule, the system considers these logs to be correlated, exports an anomaly event, and records the original logs in the event.

7. Advanced threat determination

Advanced threat determination correlates, evaluates, and determines anomalies in order to generate advanced threat characteristics, providing data for threat monitoring and attack chain visualization. Specifically, it identifies and classifies anomalies based on attack chain stages, and establishes the time sequence and correlation relationships of anomalies through host IP addresses, file MD5 values, and URLs based on the time an anomaly occurred. It then determines whether advanced threats exist based on the predefined behavior determination mode, provides scores and evaluation results based on the severity, impact scope, and credibility of the associated anomalies, before finally generating threat events.

8.3.3 Efficiently Detecting Security Threats with Deep Machine Learning

Traditional technologies used in the field of malicious file analysis are based on the matching of file signatures, also known as static signatures.

However, as the range of malicious files types continues to expand, and evasion techniques become even more sophisticated, it is widely acknowledged within the industry that traditional one-to-one static signature technologies are no longer fit for purpose. Consequently, a variety of sandbox products have been developed to effectively identify malicious files and are gradually evolving to provide dynamic detection and defense against unknown threats. The core of their effectiveness lies in their ability to identify threats by checking the actions of malicious files during normal operation. Sandbox vendors write malicious behavior pattern libraries (also called behavior sequences) to assist in identifying malicious files. The following is an example behavior sequence: invokes the encryption Application Programming Interface (API) to generate keys, opens multiple document files for read operations, and then rewrites these files. This sequence has been designed to operate similar to ransomware.

While defining such behavior sequences is useful, it is just one step toward the level of threat defense actually required. To more accurately determine malicious files, data analysis researchers have proposed the implementation of machine learning techniques and have attempted to apply these advanced techniques in the malicious file analysis field. However, although researchers possess extensive data analysis foundations, they lack an understanding of the binary reverse engineering applied in the traditional security field. As a result, after isolating malicious files, researchers extract only static characteristics of file metadata, such as the field size and data of the PE header, and send them to statistical algorithms for malicious classification. This k-fold cross-validation method may be effective for samples collected within a fixed period. However, it proves less effective in current network environments due to file encryption and packing.

As shown in Figure 8.8, the aforementioned machine learning techniques actually use static behavior scoring to determine malicious files. For example, assuming that registry modification adds five points, writing files onto a disk adds two points, adding a start-up entry in the Windows registry adds five points, and Internet connection adds five points, software is considered high-risk when a score of more than 70 points is achieved. This method has a critical defect: judging future threats through a global straight line that is manually defined in advance. If a total score is above

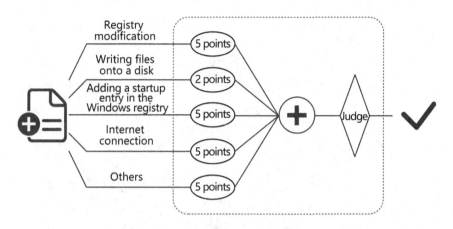

FIGURE 8.8   Using static behavior scoring to identify malicious files.

the straight line, a threat exists. Otherwise, no threat exists. As such, this method proves unreliable in terms of detection and false positive rates.

To address this, Huawei's security team have meticulously analyzed the advantages and disadvantages of dynamic behavior and static data, and have taken the lead in adding dynamic behavior to the technical route of machine learning, which they call machine learning of dynamic behavior. Simply put, based on the experience of security experts, dynamic behavior (including function names, parameters, and return values) is digitized through probability statistics to form eigenvectors. The supervised learning method is used to add the random forest machine learning algorithm and establish a detection model for each family. This new approach leads to significantly improved detection rates.

However, during the process of continuously improving detection rates, the random forest algorithm also encountered a roadblock: how can nonlinear problems be addressed? The team of experts then employed the deep learning route, utilizing the convolutional neural network (CNN) algorithm based on backpropagation (BP), as shown in Figure 8.9. Through meticulous design and exploration of convolutional layer parameters, and selection of filters for each layer, the team finally achieved effective results.

As more general-purpose graphics processing units (GPUs) are invested in computing, the parameters of each CNN layer are no longer restricted, contributing to scalability and automatic improvement of the BP-based CNN algorithm. For example, as shown in Figure 8.10, through machine learning of dynamic behavior, the BP-based CNN algorithm transforms the

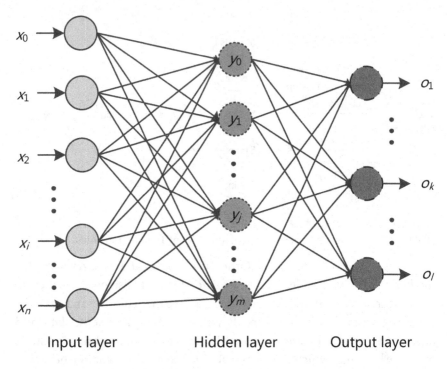

FIGURE 8.9 BP-based CNN algorithm.

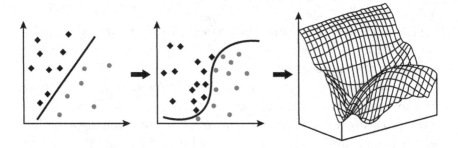

FIGURE 8.10 Threat determination based on curves, or curved surfaces within multidimensional spaces.

threat determination standard from a straight line in the industry to a curve, or even a curved surface within multidimensional space. More samples lead to stronger general-purpose GPU computing capabilities, more inflection points on the curved surface, and a higher dimension (second-order derivative, third-order derivative, and so on) of inflection points. Files above the curved surface are considered malicious, while those below are not.

On a high-dimensional, multiinflection-point curved surface of N-dimensional space, the black-and-white calibration within the sample space may be more refined and possess more accurate classification. To outline the process, the first step is to collect and calibrate eigenvectors. This is achieved by running the behavior sequence of millions of malicious samples, digitizing the behavior sequence to form millions of eigenvectors, and then marking these eigenvectors as black. The same method is then used to run the behavior sequence of nonmalicious samples in order to obtain millions of white eigenvectors. The second step involves designing an algorithm model. In the TensorFlow framework, the number of CNN layers is defined. The neurons between adjacent layers are in a linear relationship, similar to a polynomial that requires each parameter to be determined. In this way, a neural network model is designed. The third step involves training, in order to determine the optimal solution for these polynomial parameters. The eigenvectors of millions of samples are sent to the algorithm model framework from the second step. As the neurons between adjacent layers are in a polynomial relationship, a multilayer CNN results in a tensor flowing all the way to the last layer, where the actual result is produced. As such, millions of eigenvectors are iterated repeatedly, and a tensor travels through the CNN algorithm framework like a flow. For black (malicious) eigenvectors, the final output must be 1, while white (nonmalicious) eigenvectors must have an output value of 0. To fulfill this requirement, millions of black and white eigenvectors are sent to the algorithm framework, with the output value limited to either 1 or 0. Based on the iteration of each parameter in the polynomial, the optimal solution for each parameter is obtained. This allows us to create an analogy with a curve fitting the traditional data analysis.

Efficient calculation of optimal parameters is achieved by adopting deep learning BP technology. This technology introduces the feedback mechanism, enabling tensor flowing and iteration of polynomial parameters to be quickly converged, and optimized parameters to be quickly obtained. As a result of the above, we can conclude that sandbox technology is indeed very suitable for deep learning. At the core of this approach is collaboration between security and AI experts, resulting in the design of eigenvectors and the deep learning algorithm model framework based on shared experience, as well as obtaining high-quality samples and feeding them into deep learning models. When the results fail to meet the expected standards, security and AI experts can jointly check the samples, adjust algorithm models and parameters, and perform iterative tests to achieve the optimal model for product release.

It must be emphasized that machine learning is not a single algorithm or software package. Instead, it resembles a process. It requires sample collection, eigenvector design, algorithm model design, and repeated iteration and optimization. It must deal with data constantly aging as time elapses. This technology currently focuses on supervised learning, where a model is trained based on historical calibrated data in order to calibrate and classify future data.

8.4 IDENTIFYING MALICIOUS ENCRYPTED TRAFFIC WITH ECA TECHNOLOGY

Thanks to digital transformation, many enterprises have begun to utilize encryption technologies to safeguard their data and application services. For example, the HTTPS service is often used instead of the traditional HTTP service. Encryption has become an important method for ensuring our privacy, protecting sensitive data against possible snooping, and effectively preventing criminals from illegally obtaining credit card information, details relating to application usage habits, or passwords. For specific types of traffic, encryption has even become a mandatory legal requirement. According to statistics, more than 70% of modern traffic is encrypted.

However, traffic encryption introduces entirely new risks to network security. From a sender's perspective, traffic encryption can effectively prevent communication data from being stolen or maliciously tampered with by hackers. The recipient, however, can never be completely sure that the received encrypted traffic is secure. Hackers can use encryption technologies to hide malicious files and commands in encrypted traffic, enabling them to evade detection and engage in malicious activities. According to the Gartner forecast, more than 70% malicious traffic will use some type of encryption in 2020.

8.4.1 Detecting Threats in Encrypted Traffic

Currently, solutions for detecting encrypted traffic primarily use man-in-the-middle (MITM) technology to decrypt the traffic, analyze its behavior and content, and then re-encrypt it once again. However, such solutions have the following limitations:

- Encryption aims to protect the privacy of data. Using the MITM technology to decrypt traffic goes against the original intention and introduces further risk to the data channel integrity.

- MITM-based decryption usually leverages devices such as higher-performance firewalls to view traffic. However, this method is time-consuming and requires additional devices on the network.

- If the encrypted traffic continues to increase, decryption will consume a large amount of resources and deteriorate existing network performance.

Although the contents of encrypted traffic cannot be viewed without first decrypting the data, detailed analysis and research of such traffic has found that normal encrypted traffic and malicious encrypted traffic, regardless of whether they are on the client or server, exhibit many differences in terms of data packet time sequence. For example, normal encrypted traffic usually makes use of modern encryption algorithms and parameters, while malicious encrypted traffic often uses outdated and weaker encryption techniques. Such characteristics can be extracted and used by machine learning algorithms to design and train a model capable of distinguishing between normal and malicious encrypted traffic.

8.4.2 Logical Architecture of ECA Technology

Encrypted communications analytics (ECA) is a traffic identification and detection technology. Without compromising data integrity and privacy, it can identify both encrypted and non-encrypted traffic on a network, extract the characteristics of encrypted traffic, and send them to the security analyzer, or CIS, for malicious traffic detection. This technology enables users to quickly detect and handle threats hidden in encrypted traffic. Figure 8.11 shows the logical architecture of an ECA system, which consists of the ECA flow probe and ECA system.

The ECA flow probe extracts characteristics of encrypted traffic and sends them to the ECA system for threat determination. The ECA flow probe can be deployed independently on a general-purpose server or embedded in a firewall or switch.

The ECA system is integrated into CIS, and detects malicious encrypted traffic based on the ECA detection model.

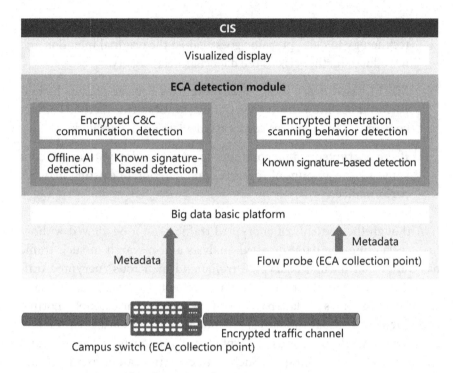

FIGURE 8.11   Logical architecture of an ECA system.

8.4.3 Working Principles of ECA Technology

As shown in Figure 8.12, a complete ECA process includes the following steps:

First, millions of black and white samples are collected. Based on open-source threat intelligence, the characteristics of these samples are extracted, including Transport Layer Security (TLS) handshake information, Transmission Control Protocol (TCP) flow statistics, and associated DNS/HTTP information. Machine learning then utilizes such characteristic information to form an ECA detection model.

The ECA flow probe deployed on each key network node then extracts the characteristics of encrypted traffic on the existing network and sends them to CIS.

Finally, CIS performs big data processing and analytics on the received encrypted traffic characteristics, based on the built-in ECA detection model, to determine whether malicious encrypted traffic exists. If so, CIS collaborates with other devices to handle the threats.

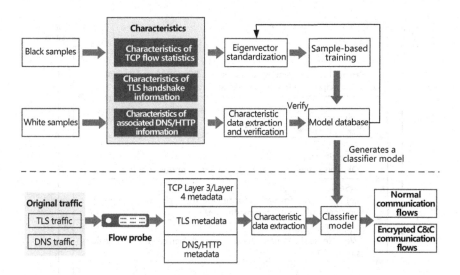

FIGURE 8.12   Working principles of an ECA system.

8.4.4 Data Characteristics Extracted by ECA Technology

During the process of forming an ECA detection model, it is of vital importance to extract and analyze the data characteristics of black and white samples, and obtain a typical characteristic model of black and white samples through comparative research. The ECA flow probe sends the extracted characteristics of data traffic on the live network to CIS, which then analyzes the characteristics based on the ECA detection model to identify malicious encrypted traffic. The ECA flow probe extracts TLS handshake information, TCP flow statistics, and associated DNS/HTTP information from packets. Here, TLS handshake information is used as an example.

Figure 8.13 shows the TLS negotiation process. In the TLS protocol, all information, excluding TLS handshake information, is encrypted. As a result, characteristics can only be extracted from the TLS handshake information and its context information.

In the TLS handshake phase, packets are exchanged in clear text, so the certificate information and the encryption method selected by both parties can be extracted. Unencrypted metadata in a TLS handshake flow includes a data fingerprint that a hacker cannot hide, which can be used to train detection algorithms.

Handshake information consists of two parts: client fingerprint and server certificate.

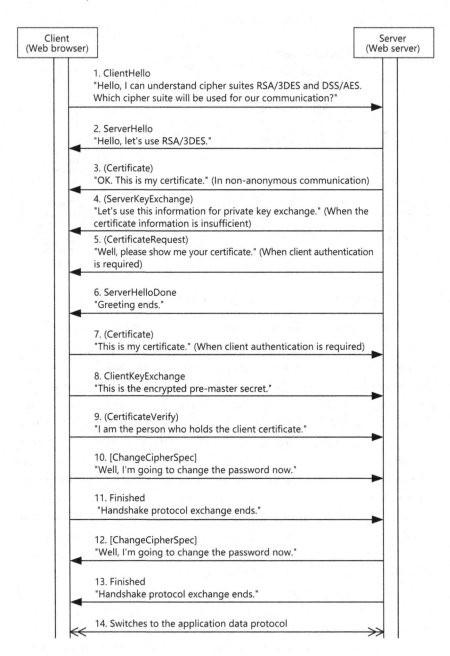

FIGURE 8.13   TLS negotiation process.

For client fingerprint information, the cipher suite information used by the client during the TLS handshake is analyzed. Figure 8.14 shows the proportions of some cipher suites in black and white samples. According to the distribution of the cipher suites used by the client, it is evident that their usage varies significantly. Under normal circumstances, each client has several cipher suites. According to Table 8.3, the cipher suites 000a, 0005, 0004, and 0013, which are not recommended in the current Secure Sockets Layer (SSL) version, account for a large proportion in the black samples.

Regarding server certificate information, the black samples also reveal some special characteristics.

As shown in Figure 8.15, Huawei's statistical analysis of a large number of black- and white-sample certificates has confirmed that three empty fields are the threshold between white-sample and black-sample certificates. Most certificates above this threshold are black-sample certificates.

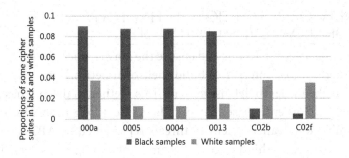

FIGURE 8.14 Proportions of some cipher suites in black and white samples.

TABLE 8.3 Description of Cipher Suite Algorithms

Cipher Suite	Algorithm
000a	TLS_RSA_WITH_3DES_EDE_CBC_SHA
0005	TLS_RSA_WITH_RC4_128_SHA
0004	TLS_RSA_WITH_RC4_128_MD5
0013	TLS_DHE_DSS_WITH_3DES_EDE_CBC_SHA
C02b	(Recommended) TLS_ECDHE_ECDSA_WITH_AES_128_GCM_SHA256
C02f	(Recommended) TLS_ECDHE_RSA_WITH_AES_128_GCM_SHA256

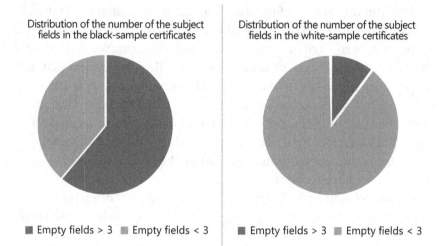

Distribution of the number of the subject fields in the black-sample certificates
Distribution of the number of the subject fields in the white-sample certificates

■ Empty fields > 3 ■ Empty fields < 3 | ■ Empty fields > 3 ■ Empty fields < 3

FIGURE 8.15 Statistical analysis of black-sample and white-sample certificates.

8.5 NETWORK DECEPTION TECHNOLOGY FOR PROACTIVE SECURITY DEFENSE

To prevent hacker intrusions and ensure the normal operation of information systems, most enterprises build their own network security system, known as the in-depth defense system. This method of defense has played a pivotal role on enterprise networks for quite a long time. However, as the number of advanced security threats represented by APTs continues to rapidly multiply, this approach becomes increasingly obsolete. The in-depth defense system offers protection based primarily on the identification of security events and only becomes active after such an event has occurred, regardless of whether it leverages signature-based threat defense or big data-based threat detection.

To compensate for the weakness of in-depth defense, a new type of security defense technology, known as deception, has emerged. Deception is defined by the use of deceits, decoys, and/or tricks designed to thwart or throw off an attacker's cognitive processes, disrupt an attacker's automation tools, delay an attacker's activities, or detect an attack.

8.5.1 Architecture and Principles of Network Deception Technology

Based on the concept of deception, Huawei has launched its own network deception technology capable of proactively interacting with attack sources. Through network deception and service simulation, attack

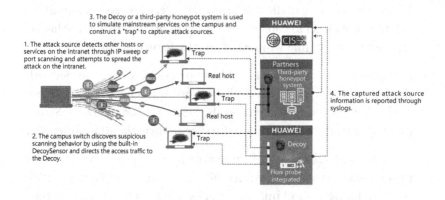

FIGURE 8.16   Architecture of the network deception technology.

sources can be identified as early as in the intranet scanning phase, and quickly isolated through collaborative handling, preventing services from being affected. Figure 8.16 shows the architecture of network deception technology. Key components include the DecoySensor, Decoy, and CIS, whose functions will be described in Table 8.4.

Network deception is a proactive security defense technology that operates using the following defense process:

- Deployed network-wide as a built-in component of a switch or firewall, the DecoySensor can capture pre-attack scanning and sniffing behaviors from any network location.

TABLE 8.4   Key Components of the Network Deception Technology

Key Component	Function
DecoySensor	The DecoySensor is built into a campus switch or firewall in order to identify IP sweeps or port scans against unused IP addresses or disabled ports, and to respond to such attacks, deceiving attackers into attacking the Decoy
Decoy	The Decoy simulates mainstream services on the campus network to construct deception traps. For example, it simulates services such as HTTP, Server Message Block (SMB), Remote Desktop Protocol (RDP), and SSH, allowing it to interact with attack behaviors, record the interaction process, and capture payload scripts. The Decoy integrated into CIS, or a third-party honeypot system, can be used
CIS	CIS performs correlation analysis, displays threats based on alarms, protocol interaction, and file behavior reported by the Decoy, and collaborates with network controllers to handle threats

- When detecting a malicious scan, the DecoySensor responds on behalf of the target to construct a fake network, confusing the hacker and delaying the time window of attacks against actual services.

- If the hacker accesses an IP address or port on the fake network, the DecoySensor views this behavior as an intended attack. When the hacker's access traffic reaches the DecoySensor, the DecoySensor diverts the traffic to the Decoy to further confirm the attack intention.

- The Decoy simulates actual mainstream enterprise services, which expose obvious security vulnerabilities. When the hacker's traffic reaches the Decoy and initiates attacks, the Decoy captures all associated behaviors and creates a profile of the attacker.

- The Decoy submits the attacker profile to CIS, which implements security isolation through collaborative processing.

8.5.2 Network Obfuscation and Simulation Interaction

1. Network obfuscation

Network obfuscation technology presents a large number of false resources to attackers in order to prevent them from obtaining actual resources and vulnerability information. In this solution, the DecoySensor is embedded in a switch, which is closer to the protected network and can be widely deployed. Compared with traditional honeypot deployment, DecoySensor deployment offers the following advantages: lower cost, higher density, wider coverage, and improved defense.

Once the deception function is enabled, the DecoySensor on a switch displays a large number of false resources on the network, responds to attackers' network scanning behaviors, and displays a false topology that merges fake and real network elements to the attackers. This technique effectively slows down the attack speed of automatic attack programs, such as scanners and worms, and prevents attackers from collecting system information and vulnerabilities.

Using port resource utilization as an example, when an attacker scans ports that are not open on a device, the DecoySensor responds on behalf of these ports to deceive the attacker and entice them into further accessing the ports. Once the attacker accesses a port that is not opened, the DecoySensor diverts the attacker's traffic to the Decoy to further confirm the attacker's intention.

2. Simulation interaction

Simulation interaction uses fake resources to implement attack-defense interaction, accurately identifying attack intentions and exposing attackers. The Decoy simulates comparable services based on the peripheral environment, exposing obvious vulnerabilities and deceiving attackers who may then launch attacks on these services. The attacker's intention can be identified through attack interaction. For example, a scanner or crawler usually responds to the scanned resources but does not attack their vulnerabilities. In addition, a simulated service can often deceive attackers into attacking other simulated services, trapping them in a loop and buying more time for attack defense. By simulating the interaction process, the network deception system can capture additional attack information, which the CIS can utilize to take more accurate defensive behavior and further reduce the probability of attacks on the real system, minimizing losses.

The Decoy supports the interactive simulation of HTTP, Server Message Block (SMB), Remote Desktop Protocol (RDP), and Secure Shell (SSH) applications. For example, for HTTP applications, the Decoy can simulate a full range of web services similar to the real thing. However, more services are opened in the simulated version, adding an extra incentive to attackers who may try to exploit these vulnerabilities. Once a vulnerability on the Decoy is triggered, the initiator is considered to be an attacker.

CHAPTER 9

Open Ecosystem for an Intent-Driven Campus Network

IN THE DIGITAL ERA, how open an enterprise's campus network determines its office and production efficiency, and even the success of service decision-making and execution. As digital transformation deepens, networks and upper-layer service systems become inseparable, resulting in an urgent need for applications that cater to specific industry scenarios. Against this backdrop, openness is especially important for network platforms. It is a concern for customers and partners whether more effective value-added applications can be customized based on an open campus network.

9.1 DRIVING FORCE FOR AN OPEN CAMPUS NETWORK

For higher education institutions, campus networks not only need to support teaching, office work, and students' Internet access, they must also pioneer in network technology research and innovation. The campus networks of higher education institutions are the perfect testing ground for all aggressive network innovations, which are seen as the origin of most existing network technology innovations and their subsequent spread. For example, the concepts of Software-Defined Networking (SDN) and OpenFlow were first proposed and implemented by professors and students of Stanford University. Due to the importance of universities in scientific research and innovation, they need an open campus network architecture more than others do.

For service providers that offer Software as a Service (SaaS) to small- and medium-sized enterprises, in order to stand out from the competition, they must continuously improve the competitiveness of their services. Service providers can opt to introduce more partners and provide more software and services, and to establish themselves as best of the rest. Alternatively, they can develop differentiated functions and services for segmented markets to set a leading position in these market segments. All these require a sufficiently open cloud service software architecture to support service providers as they strive toward offering better software and services.

Constantly emerging new services also pose challenges to enterprises' network Operations and Maintenance (O&M) teams, who must react quickly and integrate new services into their networks without adjusting the network architecture. In most scenarios, traditional network solution providers are capable of quickly providing integration solutions. However, from time to time, the solutions offered by traditional network solution providers fail to meet service requirements. In this case, an O&M team needs to perform integration independently or seek help from a third party. Therefore, to accelerate the integration of new services, enterprises need an open campus network architecture.

For network solution providers, it is impossible for them to continuously improve the competitiveness of their solutions and meet the varying requirements of segmented markets by merely relying on their own software and hardware products. Instead, they must build a vibrant ecosystem that utilizes open solutions to attract more valuable third-party partners for the joint development of end-to-end solutions.

Similarly, in order to maximize benefits, third-party developers prefer solutions that have a thriving ecosystem to build platforms for cooperation. New functions and features will be preferentially or exclusively released on these platforms.

To sum up, a campus network is not only a service platform that provides network infrastructure, but is also, more importantly, a network operating system. Competition among operating systems is largely centered on open ecosystems. For campus networks, an open ecosystem should provide the following capabilities:

- Fully open network architecture: The openness of a campus network architecture is represented by application programming interfaces (APIs). A larger number of APIs usually indicate a network has stronger open capabilities, which can be further indicated by the width and depth of Application Programming Interface (API) openness. In terms of width, being open means that a campus network architecture should provide APIs that are different and comprehensive, such as network, value-added service (VAS), and authentication APIs. The depth in openness is determined by the degree to which the resource abstraction models of different layers in the network architecture model are open. It is especially important to determine whether the network layer allows upper-layer applications to invoke network layer resources through APIs to implement end-to-end service management.

- Ecosystem construction capability of a campus network: This capability is measured based on the comprehensiveness of the solution's existing ecosystem and whether a complete array of functions and applications are provided for third parties. Solution providers should first try to build a favorable ecosystem by fully leveraging their open capabilities. If the solution ecosystem is favorable and has a large number of partners, the solution is more likely to integrate a wider range of third-party functions and applications. From the solution users' point of view, solutions need to be as complete and easy to use as possible.

- Joint innovation capability based on the developer community: This capability can be illustrated by a vendor having a good platform development and support environment as well as good business monetization potential.

A network operating system has a more complex development environment compared to a stand-alone operating system. For a stand-alone operating system, we can easily set up a development and test environment, whereas a network operating system treats the entire network as the programmable object. For individual developers and even small- and medium-sized development teams, building a development and test environment is expensive. Therefore, the support service for a network solution will involve providing a development environment, for example, an online remote lab, in addition to traditional expert support.

A network operating system's commercial monetization capability also differs from a stand-alone operating system because profit generation through third-party applications relies heavily on network solution providers. Therefore, network solution providers need to offer comprehensive operations and marketing platforms that can help developers promote and sell third-party applications, in order to attract continuous investment from developers in platform application development.

9.2 FULLY OPEN CAMPUS NETWORK ARCHITECTURE

Huawei's intent-driven campus network solution provides open capabilities at all layers, for all services, and in all scenarios, and by leveraging these open capabilities, industry partners can quickly develop industry applications. Third-party applications can run as components at the application layer of the intent-driven campus network solution to supplement existing industry-specific standard applications at the application layer.

9.2.1 Intent-Driven Campus Network Openness

The platform layer of the intent-driven campus network provides the application layer with more than 150 APIs that can be classified into four categories: basic network APIs, VAS APIs, third-party authentication APIs, and location based service (LBS) APIs, as shown in Figure 9.1.

1. Basic network APIs

   Basic network APIs are used to manage, control, and maintain campus network services. Through basic network APIs, upper-layer applications can invoke encapsulated network resources to implement interaction between networks and services. For example, video services can invoke network resources to ensure optimal end-to-end service quality. Basic network APIs can also be used to configure, manage, and maintain networks. For example, higher education institutions or other scientific research institutes can use basic network APIs to verify network architecture and application

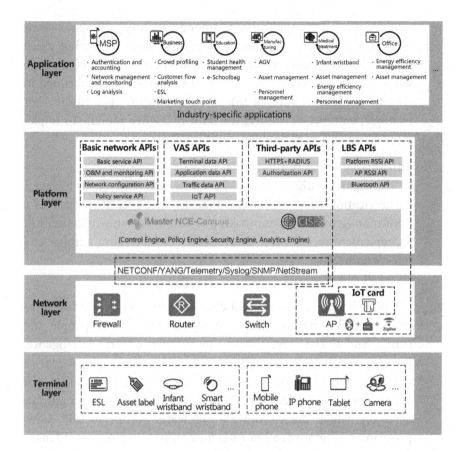

FIGURE 9.1   Intent-driven campus network openness.

innovations, and basic network APIs can directly invoke south-bound Network Configuration Protocol (NETCONF) interfaces to control the network layer.

2. VAS APIs

VAS APIs are used to access big data on a campus. The platform layer has a built-in big data engine, which uses southbound Telemetry and NetStream interfaces and integrates traditional Syslog and Simple Network Management Protocol (SNMP) interfaces to collect and store the big data of terminals, applications, and traffic on the entire network. Third-party applications can access and analyze the big data through VAS APIs, to provide VASs accordingly. For example, business organizations can use big data mining to implement intelligent business applications, among which typical applications are crowd profiling, customer flow analysis, and marketing touch point. VAS APIs also include Internet of Things (IoT) APIs, through which users can access and control the IoT modules embedded in APs to implement various IoT-based intelligent applications. For example, enterprises can use the built-in ZigBee module to implement energy efficiency management in offices.

3. Third-party authentication APIs

Third-party authentication APIs allow external systems to access authentication information and manage user identities, such as a hotel management system invoking a third-party authentication API to implement room access control based on the room number and guest's name.

4. LBS APIs

LBS is a type of service that emerged with Wi-Fi and IoT technologies. LBS enables campus networks to perceive people and things and is therefore the key to automation and intelligence. For example, various navigation services in industrial scenarios can be provided based on location information. In addition, the key performance index of positioning technology is its geographic accuracy, which determines the application scope of LBS. Low-precision positioning can be used for remote campus security management, whereas high-precision positioning helps in automated guided vehicle (AGV) navigation and geo-fence implementation in industrial automation.

9.2.2 Typical Openness Scenarios for an Intent-Driven Campus Network

Authentication and authorization, LBS, crowd profiling, network O&M, and smart IoT are typical service functions of campus networks. If a shopping mall or supermarket wants to know which aisles or areas are the most popular, it can use LBS to identify the location of its users that are using its campus network. Similarly, smart IoT can be used by a higher education institution to manage IoT terminals on its campus as well as count and locate certain assets. This section takes five typical functions of campus networks as examples to illustrate how applications can be implemented based on the open capabilities of an intent-driven campus network.

1. Authentication and authorization

In retail scenarios such as shopping malls or supermarkets, retailers may want to push advertisements to potential customers when they try to access retailer's networks. This can be achieved through applications that are developed based on open authentication and authorization APIs. Figure 9.2 illustrates how the SDN controller, and third-party authentication and authorization platform are connected.

The SDN controller can connect to a third-party authentication and authorization platform to implement the following functions:

a. Verifies the users' identity, authorizes user access, and pushes advertisements to these users.

b. Charges online users based on their duration and traffic.

The SDN controller can connect to a third-party authentication and authorization platform in either of the following modes.

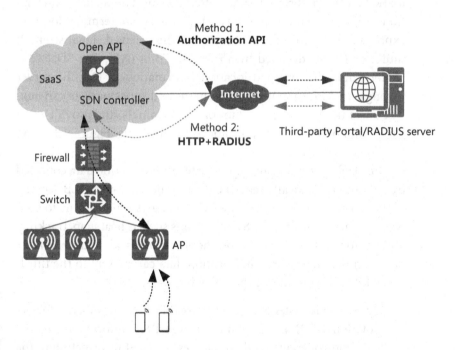

FIGURE 9.2 Interconnection with a third-party authentication and authorization platform.

Shopping malls and supermarkets can select the most suitable mode for themselves.

a. Authorization API: After authenticating a user terminal, the third-party authentication and authorization platform invokes the authorization API to instruct the SDN controller to authorize the user terminal.

b. HTTPS+RADIUS: The SDN controller functions as a RADIUS client to interwork with the third-party authentication and authorization platform using RADIUS.

2. LBS

For shopping malls and supermarkets, the key to Wi-Fi VASs is to obtain the location of user terminals. By locating terminals, shopping malls and supermarkets can provide LBS, such as navigation service, for customers. Terminal location data also gives insights into customers' consumption habits, allowing shopping malls and supermarkets to target promotional content. Terminal location information can be broken down into two types: accurate terminal location expressed using coordinates (x, y), and the received signal strength indicator (RSSI) detected by an Access Point (AP). The SDN controller sends the terminal location information to the LBS server through an API, so that shopping malls and supermarkets can analyze customer flow, push promotions to terminals, and provide navigation and other related applications to terminals. Figure 9.3 shows the LBS architecture.

The SDN controller can aggregate terminal location data collected by APs and periodically send the data to the third-party LBS server. The third-party LBS server applies data analysis algorithms to analyze the data, providing VAS applications such as heat map, tracking, and customer flow analysis for shopping malls and supermarkets based on the analysis result. Location data can be sent to the third-party LBS server in any of the following ways:

a. Hypertext Transfer Protocol Secure (HTTPS) + JavaScript Object Notation (JSON): APs send detected RSSI information to the SDN controller, which then releases the RSSI information to the third-party LBS server.

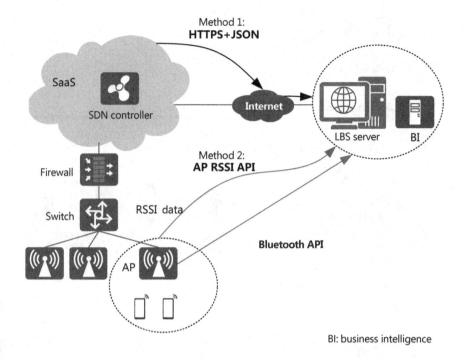

FIGURE 9.3   LBS architecture.

   b. AP RSSI API: APs periodically send detected RSSI information to the third-party LBS server.

   c. Bluetooth API: APs obtain the location of user terminals using the Bluetooth technology and periodically send the location information to the third-party LBS server.

3. Crowd profiling

Precision marketing is an important marketing technique in the retail industry that requires retailers to obtain crowd profiles and identify target customers. Once this information has been obtained, retailers can deliver personalized services or push specific information to these target customers. Crowd profiling requires two types of data: one is online data, which can be collected using an online system; the other type of data is offline data, which can be collected using Wi-Fi probes or the network, if the data are carried over the network. As shown in Figure 9.4, the SDN controller provides terminal-related offline data to the big data analytics platform through the LBS API and VAS API.

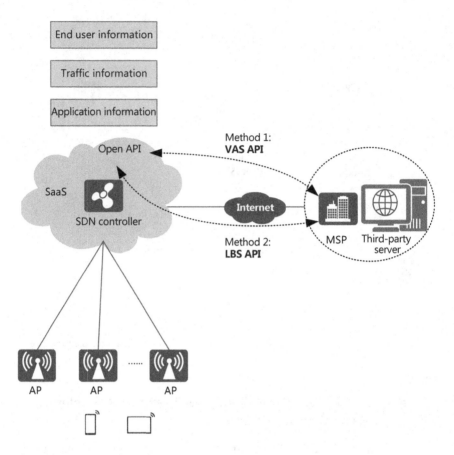

FIGURE 9.4 Crowd-profiling architecture.

4. Network O&M

In some scenarios, customers or Managed Service Providers (MSPs) may need to use a third-party network management system (NMS) to manage or monitor devices managed by the SDN controller. Typical management and monitoring operations on a third-party NMS are creating tenant administrator accounts, managing devices, configuring networks for specified devices, and monitoring device status and alarms. In such scenarios, network service O&M can be implemented in two modes, as shown in Figure 9.5.

a. Network service API: The third-party NMS interconnects with the SDN controller through a RESTful API to manage and monitor devices.

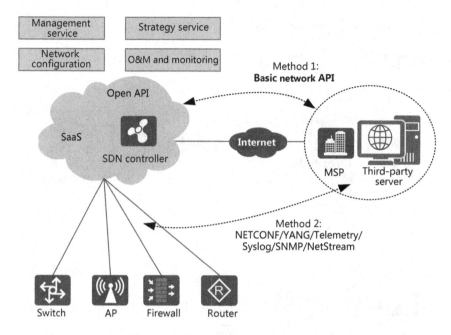

FIGURE 9.5   Interconnection with a third-party NMS.

    b. Traditional NETCONF/YANG/Telemetry/SNMP interfaces: The third-party NMS can directly configure, manage, and maintain devices through these device interfaces.

5. Smart IoT

When partners want to deploy IoT applications (such as electronic shelf label, IoT positioning, energy efficiency management, and asset management), they can use the network infrastructure to provide IoT signal coverage (ZigBee, Bluetooth, Radio Frequency Identification (RFID), etc.), without the need to deploy a second network. This reduces the capital expenditure for customers. Figure 9.6 shows the smart IoT architecture.

In this scenario, Huawei provides the network infrastructure, open AP hardware, and basic management and monitoring functions for IoT cards, while partners develop IoT card applications, card management software, and IoT service software. APs communicate with IoT cards through console ports or network ports, and IoT services can be provided based on the network infrastructure through cooperation between Huawei and its partners.

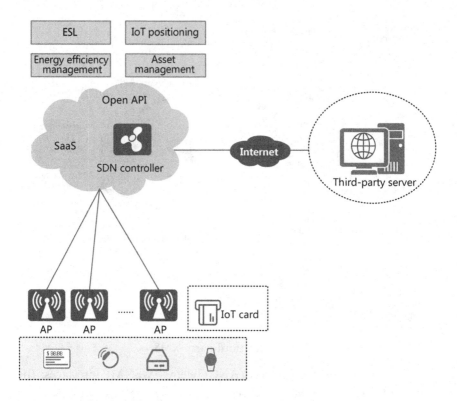

FIGURE 9.6   Smart IoT architecture.

9.3 CONTINUOUS CONSTRUCTION OF THE CAMPUS NETWORK INDUSTRY ECOSYSTEM

An open innovation ecosystem is required to drive the digital transformation of enterprises. As such, enterprises should aim to build an open campus ecosystem platform by using their campus networks as reliable digital infrastructure. By doing so, they can promote close cooperation between industries, stimulate cross-industry innovation, and ultimately improve the speed at which industry-specific end-to-end solutions are built.

9.3.1 Intent-Driven Campus Ecosystem Solution Overview

Based on the intent-driven campus network's fully open architecture, Huawei has developed an intent-driven campus ecosystem that leverages Huawei's extensive experience in the industry and joint innovation with Huawei's customers and partners. At its foundation, this ecosystem involved more than 30 industry partners and integrated more than 50 industry applications. These partners come from a wide range of industries such as retail, education, enterprise office, and manufacturing. In the future, Huawei will work with more partners to build more industry-specific, intent-driven campus applications and solutions.

In the retail industry, the joint electronic shelf label (ESL) solution integrates IoT cards on APs, and enables retail stores to change their commodities' price tags in real time without the need for human intervention, thereby eliminating the need to deploy a dedicated ESL network and reducing the O&M costs for retail stores.

In the enterprise office industry, the joint asset management solution is used to help enterprise operations personnel remotely manage enterprise assets. For example, operations personnel can use the solution to track the precise location of assets and count assets in real time, greatly improving enterprises' operational efficiency.

In the education industry, the joint smart campus solution converges Wi-Fi and IoT networks, and provides typical campus applications such as smart classroom, energy management, asset management, and apartment management to greatly accelerate the IT transformation of campus networks.

Finally, in the manufacturing industry, the joint ultra-wideband (UWB) positioning solution is used to deliver submeter-level positioning capabilities to meet the precise positioning requirements of the warehousing industry. It is widely used in industrial AGV and smart manufacturing scenarios.

9.3.2 ESL Solution for Smart Retail

The development of new retail rejuvenates offline stores; however, it poses new challenges to the operations of shopping malls and retail stores as well. For example, in traditional retail scenarios, paper price tags are replaced manually in ways that are both inefficient and error-prone. In addition, traditional ESL and Wi-Fi networks are deployed separately, resulting in high investment costs and low O&M efficiency. To address these challenges, Huawei and its partners jointly develop the ESL solution. This solution reduces investment and improves O&M efficiency by converging Wi-Fi and ESL networks, and can be used in commercial scenarios such as large shopping malls and supermarkets, retail stores, and shopping malls at airports.

In the ESL solution architecture shown in Figure 9.7, the ESL management system is connected to the customer's Enterprise Resource

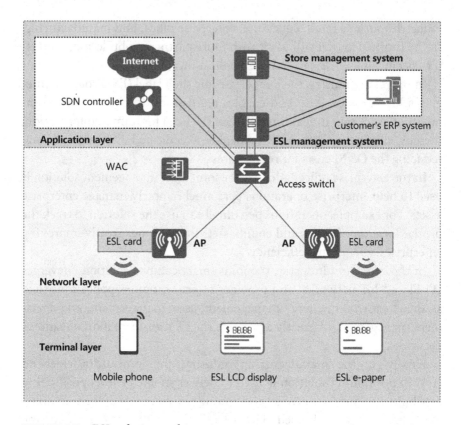

FIGURE 9.7   ESL solution architecture.

Planning (ERP) system. IoT APs provide built-in peripheral component interconnect slots to connect to ESL cards. The ESL management system obtains commodities' price information and sends the information in wireless packets to the ESL through the ESL cards built into the APs. Then, after receiving the information, the ESL displays the commodities' prices as dot matrices. In actual application, retailers typically deploy the ESL management system at their headquarters and deploy ESLs in their stores. After retail management adjusts a commodity's price in the ERP system, the adjustment result is synchronized to the ESL management system, which then adjusts the price accordingly based on the adjustment result and the preset price adjustment plan. ESLs then display the new price and other commodity information such as the validity period and detailed product parameters. Table 9.1 describes the ESL solution's benefits to customers.

TABLE 9.1 Benefits of the ESL Solution

Advantages	Implementation Description
High-quality Wi-Fi coverage	Seamless Wi-Fi coverage ensures a smooth user experience APs can work in either traditional management mode or cloud management mode
Multinetwork convergence and centralized O&M	The wired, wireless, and ESL networks are converged, and the office and application networks are integrated The intelligent interference avoidance technology reduces Wi-Fi and RFID transmission interference, improves the ESL update success rate, and prolongs the ESL's lifespan Unified site survey and network planning as well as centralized O&M and management reduce investment and improve efficiency
ESL and digital marketing	ESLs are connected to the ESL management system to ensure prices are updated in real time Manual update is not required, reducing labor costs and improving update accuracy

9.3.3 Commercial Wi-Fi Solution for Smart Shopping Malls/Supermarkets

The rapid development of mobile Internet and wide use of smart terminals have resulted in a sharp increase in the number of terminals that access networks in shopping malls and supermarkets, and boosted the demand for wireless network services in such scenarios as well. Against this backdrop, large shopping malls and supermarkets deploy Wi-Fi networks to provide convenient and free Wi-Fi services to attract more customers, improve customer satisfaction, and deliver more commercial VAS offerings.

In the commercial Wi-Fi solution architecture shown in Figure 9.8, Huawei is responsible for building a network infrastructure that provides a high-quality Wi-Fi network for consumers and delivers improved user experience. Additionally, ecosystem partners provide a visualized Wi-Fi management platform that delivers functions such as social media authentication, Wi-Fi positioning, and customer segment analysis, helping business customers efficiently interact with consumers and improve their efficiency in sales and operations.

By focusing on improving user experience and adding value to operations, the commercial Wi-Fi solution deeply explores valuable data on

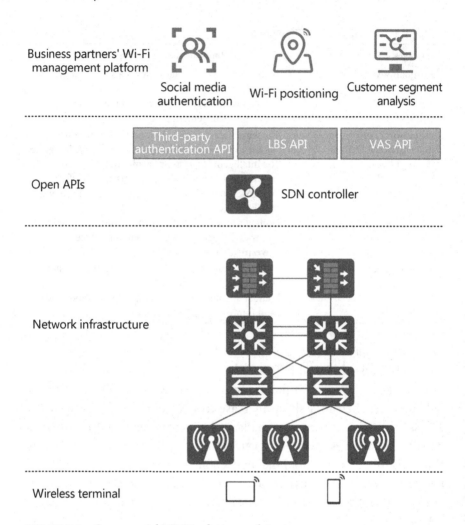

Business partners' Wi-Fi management platform

Social media authentication

Wi-Fi positioning

Customer segment analysis

Third-party authentication API

LBS API

VAS API

Open APIs

SDN controller

Network infrastructure

Wireless terminal

FIGURE 9.8   Commercial Wi-Fi solution architecture.

the network and extracts, compares, and analyzes large amounts of data. The data analysis results provide insight into users' consumption behaviors and the stores or commodities that are more attractive to consumers. Then, enterprise customers from the large shopping mall/supermarket, retail, airport, and hospitality industry can use the insight to deliver a personalized consumption experience to consumers. In this way, enterprise customers can accurately grasp business dynamics, gain more business value, improve sales efficiency, and achieve business success. Table 9.2 describes the benefits of the commercial Wi-Fi solution.

TABLE 9.2 Benefits of the Commercial Wi-Fi Solution

Advantages	Implementation Description
High-quality Wi-Fi coverage	The Wi-Fi network provides seamless Wi-Fi coverage to ensure a smooth Wi-Fi experience
	The cloud management solution provides centralized O&M and management for networks with multiple branches, improving O&M efficiency
User-centered Wi-Fi management	The solution provides free Wi-Fi coverage and integrates the Wi-Fi management platform
	The solution provides authentication and accounting functions and allows the customization of portal pages
Customer flow analysis and precision marketing	Wi-Fi data analysis gives insights into users' consumption habits
	The data analysis results help identify target customers and carry out precision marketing, thereby improving business value

9.3.4 UWB Positioning Solution for Industrial Intelligence

In the mobile Internet era, the Global Positioning System (GPS) positioning function has become a standard capability of smartphones and is now widely used in applications such as map-based navigation and online car-hailing. In addition, with the development of IoT technology, many industries (especially those in the industrial intelligence field) now require higher positioning precision for digitalization and automation. Table 9.3 provides the requirements for positioning accuracy in typical positioning scenarios. From this table, we can infer that

TABLE 9.3 Requirements for Positioning Accuracy in Typical Positioning Scenarios

Marketplace	Typical Scenarios	Required Positioning Accuracy
Mobile broadband/ business applications Low-speed Movement/ indoor and outdoor	People (proactive positioning): high-precision navigation, LBS applications, and AR/VR gaming	Meter/submeter level
	People (passive positioning): fixed-point positioning service (ride-hailing drivers locating their passengers), targeted selling in shopping malls, and information pushing	Submeter level
	Things (low cost and low power consumption): logistics tracking, pet tracking, and bicycle sharing	20 m

(Continued)

TABLE 9.3 (*Continued*)   Requirements for Positioning Accuracy in Typical Positioning
Scenarios

Marketplace	Typical Scenarios	Required Positioning Accuracy
Internet of Vehicles (IoV)/Unmanned aerial vehicles	Automobile: high-precision navigation, IoV, vehicle platooning, and autonomous driving	Submeter level
High-speed movement/outdoor	Unmanned aerial vehicle (UAV): power line inspection and express delivery	Submeter level
Industrial application Low-speed movement/ indoor and outdoor	Factory: industrial visualization products, AGVs, robots, and workers	Submeter level
	Port: intelligent scheduling, cranes, forklifts, and operators	Submeter level

submeter-level high-precision positioning will become a required service in the IoT era.

Figure 9.9 shows the architecture of the UWB positioning solution. In this solution, Huawei builds a network infrastructure to provide high-quality wireless networks for consumers, and ecosystem partners provide UWB positioning services through either SaaS or local systems. IoT cards, provided by ecosystem partners, are integrated into Huawei APs as IoT base stations and send detected location tag information to the APs. These APs then send the data to the third-party UWB positioning platform, which performs positioning based on the positioning algorithm.

As shown in Figure 9.10, the UWB positioning solution can be used in a wide variety of scenarios, such as warehousing and logistics positioning management, automated control of port machinery, smart factory and visualization, intelligent navigation of industrial vehicles, security management in production and O&M, and positioning of special personnel. For example, enterprises in the chemical and manufacturing industries need to take precise and efficient measurements to mark out dangerous areas and restrict, thereby reducing the risk of security accidents. In this case, enterprises can deploy geo-fences using the UWB positioning solution; then, if a worker enters a dangerous area either intentionally or unintentionally, an alarm will be triggered immediately, thereby reducing potential security accidents.

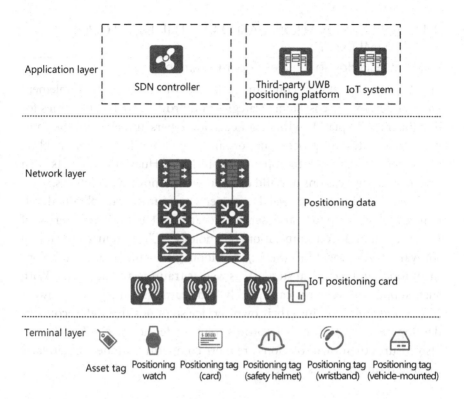

FIGURE 9.9   Architecture of the UWB positioning solution.

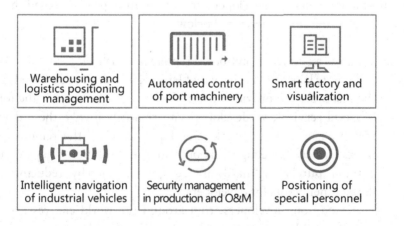

FIGURE 9.10   Typical application scenarios of the UWB positioning solution.

9.4 JOINT INNOVATION BASED ON THE DEVELOPER COMMUNITY

9.4.1 Introduction to the Developer Community

The Huawei Developer community is a platform used to implement Huawei developer ecosystem strategies and provide open capabilities for developers. As a platform that connects developers and Huawei, the community provides full-process support and service for developers, enabling developers to integrate the open capabilities of Huawei's products with upper-layer applications to build differentiated innovative solutions.

Developers in the Huawei Developer community can obtain development-related support and service regarding Huawei's full series of Information and Communication Technology (ICT) products, including software development kits (SDKs), development documentation, development tools, and technical support, as well as training and activities. With such support, developers can quickly and seamlessly integrate Huawei's ICT product features into their own service solutions through secondary development. This not only improves solution development efficiency but also ensures their own solutions are more competitive in the marketplace, driving future business success.

9.4.2 Support and Service Provided by the Developer Community

The Huawei Developer community is committed to building an industry-leading open enablement platform. In an effort to provide comprehensive support and service for developers, the community proposes the innovative LEADS concept, as described below:

1. L: Lab as a Service (cloud-based remote lab service)

   Remote labs have a full suite of Huawei ICT devices deployed, and the resources are provided to developers free of charge in the form of cloud services, so developers can remotely invoke these device resources over the network. Developers can use the remote labs' cloud services to debug the applications and solutions they develop without purchasing hardware equipment, thereby reducing the development costs for developers.

   In addition, developers from around the world can reserve lab resources and connect to the lab closest to their location once they have registered with the Huawei Developer community. The Huawei

Developer community provides an API for reserving remote labs, which developers can use to reserve and access remote lab resources on the Integrated Development Environment (IDE) Graphical User Interface (GUI).

Remote labs provide support for various types of terminals that run different operating systems, such as Windows, Linux, Mac, and Android, allowing these terminals to quickly connect to the remote labs.

2. E: End to End (end-to-end process and platform support service)

The Huawei Developer community is dedicated to building an industry-leading open enablement platform that provides developers with end-to-end services from understanding, learning, development, and verification, to commercialization, helping developers achieve business success.

To achieve this goal, the Huawei Developer community has developed an end-to-end platform to provide full technological and service support. The platform leverages cloud-based remote labs and developer tools to comprehensively support developers at each stage of the development process. This platform can quickly respond to developers' requirements and acts as a fully connected social interaction system to facilitate interaction and mutual assistance, communication and feedback, as well as cooperation and innovation between both developers and the Huawei Developer community and between developers themselves. The platform also serves as a marketplace where developers can roll out and promote their own applications. In every phase of development, the Huawei Developer community adheres to the developer-centric philosophy and provides end-to-end services that address the developers' core requirements.

3. A: Agile (processes and tools that support agile development)

The Huawei Developer community provides a comprehensive set of tools for developers, as shown in Table 9.4.

4. D: Dedicated (24/7 online dedicated service from experts)

The Huawei Developer community uses a variety of methods to provide 24/7 online support for developers from different fields and at different proficiency levels, as described in Table 9.5.

TABLE 9.4   Tools Provided by the Huawei Developer Community

Tool Name	Tool Description
Multilingual SDKs	SDKs are available in multiple languages, such as Java, C++, and C#
IDE plug-in	The eSDK IDE plug-in simplifies eSDK secondary development. Developers use this plug-in to download SDKs or demos, create projects, access remote lab resources, and refer to online API help manuals within a click. These functions enable developers to quickly complete project creation, configuration, and commissioning, improving efficiency in secondary development. The IDE plug-in can run in two mainstream IDEs: Eclipse and Visual Studio
API Explorer	The API Explorer enables developers to quickly experience Huawei's open APIs in the simulated environment of the Huawei Developer community through simple parameter settings
CodeLab	CodeLab provides instructions for hands-on practice of API functions. In addition, a wide array of sample codes is available in CodeLab, and developers can invoke an API by simply copying and pasting the code
Analytics	In Analytics, developers can monitor and analyze API calls from newly developed applications, and leverage data analysis to optimize applications
Open source of Software Development Kit (SDK) core code (Github)	Github provides access to open source SDK core code. Developers can view and download the SDK core code after logging in to Github

TABLE 9.5   Methods to Provide Support in Huawei Developer Community

Method	Description
Offline training	Offline training sessions are organized at irregular intervals to improve developers' and partners' secondary development capabilities
Online videos	Online videos are available for developers from different fields and at different proficiency levels. Developers can select courses to improve service system development capabilities
Developer Center	The Developer Center provides online service portals for cooperation and development support, as well as complaint and suggestion submission. Developers can access the Developer Center through the Huawei official website. Huawei ensures a 100% response rate to feedback within 48 hours and provides point-to-point expert support when required. On the personal home page in the Developer Center, developers can view their feedback's processing progress, as well as score and comment on the processing result. The Huawei Developer community conducts satisfaction assessment by means of customer review and online surveys, thereby continuously improving its service quality and providing developers with faster, more accurate, and more efficient professional services

5. S: Social (various forms of social interactions)

The Huawei Developer community provides rich online platforms and offline activities to promote social interaction between
developers. Online interaction can be conducted on the portal website, forums, and WeChat, while offline activities include Huawei
Developers Gathering (HDG), Developer Challenge, and HUAWEI
CONNECT. All these platforms and activities comprise channels for
feedback, communication, sharing, mutual help, cooperation, and
innovation (Table 9.6).

TABLE 9.6   Means of Social Interaction Provided by Huawei Developer Community

Social Interaction Means	Description
Huawei official website	Developers can directly access the Developer Center from the Huawei official website and obtain the following open capability information of Huawei ICT products: APIs, SDKs, IDEs, remote lab resources, documentation, development tools, technical support, and training services. In the Developer Center, developers can communicate with support experts from the Huawei Developer community and manage their resources in the Huawei Developer community
Forum	To better serve developers, the Huawei Developer community has created a technical forum together with China Software Development Network, the largest IT technical community and service platform in China. Covering 14 ICT ecosystems, this forum works as an online platform for technical discussions, service innovation, cooperation, and communication
WeChat	The Huawei Developer community provides an official WeChat account (Huawei_eSDK) which serves as an information and resource integration platform for developers to stay updated with the latest Huawei news, gain access to technical support and services, and get acquainted with like-minded technicians
HDG	HDG is a technical gathering held offline by the Huawei Developer community every month since April 2016. The gathering provides a relaxed environment for face-to-face communication between Huawei experts and developers to share technologies, ideas, and experiences, and discuss developers' core requirements. During HDG, Huawei also collects developers' opinions and suggestions to improve the Huawei Developer community accordingly

(Continued)

TABLE 9.6 (*Continued*)   Means of Social Interaction Provided by Huawei Developer
Community

Social Interaction Means	Description
Developer Challenge	Huawei Developer Challenge is a large-scale software contest held annually for developers and partners in China. It is committed to discovering innovative ideas and developing them into solutions by leveraging the open capabilities of a full portfolio of Huawei products and end-to-end support services. By rolling out innovative solutions to the market, Huawei helps developers realize their dreams. Since its debut in Beijing on August 15, 2016, Huawei Developer Challenge has developed into a platform for developers and partners to demonstrate their innovative products and solutions and obtain technical and financial support
HUAWEI CONNECT	The Huawei Developer community participates in the annual HUAWEI CONNECT every year. At this conference, not only does the Huawei Developer community display the support and services it provides but it also demonstrates and promotes innovative solutions jointly developed with developers and partners. Huawei uses joint marketing to help developers and partners achieve business success

CHAPTER 10

Intent-Driven Campus Network Deployment Practices

THIS CHAPTER STARTS WITH a description of the overall design process for a campus network, followed by the sample deployment of a university's campus network. This example illustrates the analysis of a customer's requirements, network deployment, and service provisioning of an intent-driven campus network. The design process is based on the university's demand for multi-network convergence, advanced architecture, and on-demand expansion.

10.1 CAMPUS NETWORK DESIGN METHODOLOGY

10.1.1 Network Design Process

Network design is a process of designing a proper network architecture and technical solution based on the network environment and service requirements of customers. Different scenarios have varying network requirements for reliability, security, and usability, so a good understanding of network requirements and network status is the foundation for a successful network design. The network design process starts with a survey of the customer's requirements, followed by a thorough analysis, based on which a network solution can be carefully designed. Figure 10.1 shows the network design process.

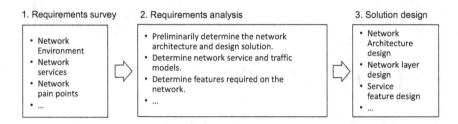

FIGURE 10.1   Network design process.

10.1.2 Customer Requirements Survey and Analysis Guidelines

The requirements' survey and analysis cover these six aspects of network design: network environment, network pain points, network services, network security, network scale, and terminal type. Table 10.1 describes the content of the general survey and its objectives for these six aspects, while Tables 10.2 through 10.7 elaborate on each aspect.

TABLE 10.1   Overview of the Campus Network Requirements Survey

No.	Category	Survey Content	Main Purpose of the Survey
1	Network environment	Network construction, deployment, and usage status, and whether the customer wants to reconstruct a legacy network or build a new one	To preliminarily determine the network architecture and design solution
2	Network pain points	Customers' pain points on the legacy network (for network reconstruction scenarios) or expectations for the new network	To determine network construction requirements and objectives, as well as preliminarily outline the features to be supported on the network
3	Network services	Services to be deployed on the network and their characteristics, network service models, and traffic models	To determine the network bandwidth and service characteristics
4	Network security	Whether services need to be isolated, their corresponding isolation requirements (optional), and network security construction requirements	To determine service isolation and network security protection solutions
5	Network scale	Number of users on the network, and the user growth trend for the next 3–5 years	To finalize the network architecture and design solution
6	Terminal type	Terminal types and access requirements	To determine the network access solution

TABLE 10.2   Overview of the Campus Network Requirements Survey — Network Environment

Requirement	Survey Content	Analysis
1.1	Network construction type: to reconstruct a legacy network or deploy a new one	For network upgrade and reconstruction projects, network design becomes more complex, raising issues such as those related to device compatibility and reusability, smooth service migration, and whether service interruption is allowed
1.2	Network type: wired, wireless, or converged	To determine whether both wired and wireless networks need to be constructed and whether the two networks need to be converged. If wired and wireless networks need to be converged, WAC + Fit AP networking is recommended
1.3	Geographical distribution: centralized or dispersed	Preliminarily determine the basic network architecture. For a campus with closely located buildings, the single-core architecture is recommended. For a campus with geographically dispersed branches or buildings, if each branch or building has a large network scale and traffic between the branches or buildings is heavy, multiple core or aggregation devices may be required
1.4	Customer's organizational structure (with authorization granted)	Preliminarily learn about the usage of the campus network based on the customer's organizational structure, including how the departments of the organization use their networks and how their networks are deployed. This also analyzes whether to isolate networks by department, region, or service, whether multiple core or aggregation devices are needed, whether there are branch access requirements, and, if so, what kind of cables are needed for this
1.5	Distribution of equipment rooms or extra-low voltage (ELV) rooms	If there are many equipment or ELV rooms, distributing core or aggregation devices among them will improve reliability. In most situations, an aggregation device is deployed in each equipment or ELV room. For a multicore interconnection network, the core devices can be distributed in different equipment rooms. In addition to this, you need to consider the device layout and distance between them

TABLE 10.3 Overview of the Campus Network Requirements Survey — Network Pain Points

Requirement	Survey Content	Analysis
2.1	Network speed: whether network congestion occurs	Preliminarily determine the required network bandwidth and switch specifications
2.2	Network quality: whether the network is unstable, for example, service interruptions occur	Use a network management system (NMS) or network analysis software to find out the root causes and eliminate the risks during solution design. For example, use a product with the hardware-based Operation Administration and Maintenance (OAM) function to prevent such risks from occurring. Additionally, analyze the network quality required by services based on the customer's industry characteristics to ensure that the network quality meets service needs
2.3	Network scale: whether the number of access users reaches its limit	Understand the number of users on the customer's network and estimate the user growth trend in the next 3–5 years. Increase the number of access switches or use switches with high-density ports to replace the existing ones while considering network capacity
2.4	Network capabilities: whether wired, wireless, and remote access are required	If a wireless network is required, consider integrating wired and wireless networks to provide unified deployment and authentication. For remote access, determine the VPN solution based on the application scenarios, for example, use Secure Sockets Layer (SSL) VPN for remote access of mobile users, IPsec VPN for fixed access from branches, or both
2.5	Network security: internal network security, terminal security, and protection against threats from external networks	Determine whether to use security devices and the corresponding device models that are required, as well as understand the security level required by customers. If high security is required, use independent security devices. In low-security scenarios, use integrated security devices, such as value-added security service cards

TABLE 10.4 Overview of the Campus Network Requirements Survey — Network Services

Requirement	Survey Content	Analysis
3.1	Common service types: office, email, and internet access	Common office services do not require high network bandwidth, and therefore 200 kbit/s bandwidth will suffice for ordinary network access
3.2	Key service types: data, voice over Internet Protocol (VoIP), video, and desktop cloud	Campus networks are typically LANs, so network latency is not a concern. However, network latency will be a major concern if a campus network carries desktop cloud, video, and VoIP services, and connects to branches, metropolitan area networks (MAN), or wide area networks (WAN). More specifically, if desktop cloud services are required, network reliability or availability must be carefully considered during deployment, and if video services are required, bandwidth requirements are equally important. If VoIP is required, determine whether to deploy VoIP and PC services on the same or different networks, and whether Power over Ethernet (PoE) power supply is needed. These will determine the number and specifications of switches
3.3	VIP services	Design QoS policies to guarantee customers' key services, delivering a good user experience
3.4	Multicast service type	A multicast solution needs to be designed
3.5	New service types for the next 3–5 years	If new services are possible for the next 3–5 years, design a smooth upgrade and capacity expansion solution that avoids unnecessarily wasting resources
3.6	(Optional) Live network environment	Obtain live network information, including running network protocols, network topology, device type and quantity, network quality, and support for running services. The information can be used as a reference during the design stage

TABLE 10.5   Overview of the Campus Network Requirements Survey — Network Security

Requirement	Survey Content	Analysis
4.1	Service security: service isolation and interoperability requirements	Determine whether network services need to be isolated and how to isolate them. Physical isolation means construction of multiple networks and the separate design of each network. VXLAN is recommended for logical isolation. With VXLAN technology, a campus network is virtualized into multiple virtual networks to carry different services. Another thing worth considering is whether interoperability is required between different services. If interoperability is required, develop interoperability policies and solutions in advance
4.2	Security defense against external threats	Deploy security devices such as the firewall, intrusion prevention system (IPS), intrusion detection system (IDS), and network log audit devices to protect network border security. If a customer has high network security requirements, independent security devices are recommended. Otherwise, use integrated security devices, such as Unified Threat Management (UTM) or value-added security service cards
4.3	Internal network security defense	Use online behavior management software or dedicated devices to prevent security incidents caused by internal users
4.4	Terminal network security defense	Terminal network security defense includes terminal access security check and terminal security check, which determines whether a network access control (NAC) solution is required

TABLE 10.6 Overview of the Campus Network Requirements Survey — Network Scale

Requirement	Survey Content	Analysis
5.1	Number of wired users or APs	Determine the port quantity and density of access switches and the approximate network bandwidth requirements based on the service survey findings
5.2	(Optional) Number of wireless users	Determine the WAC specifications and AP quantities, and whether there is a need for high-density access in key areas such as conference rooms
5.3	(Optional) Specifications of the legacy network	Estimate network reconstruction workload and determine the network upgrade solution, covering device reusability, compatibility, and seamless upgrade
5.4	Network scale for the next 3–5 years, or the highest growth rate in recent years	When designing network interfaces, ensure capacity expansion and smooth upgrade are carefully considered so that the design can be future-proof for the next 3–5 years. Network scale includes both the user scale (number of users or terminals) and the service scale (service type, bandwidth, quantity, and scope)
5.5	(Optional) Branch offices	Consider the connection mode (private line or VPN) between the branches and the headquarters and determine whether link backup is required

TABLE 10.7 Overview of the Campus Network Requirements Survey — Terminal Type

Requirement	Survey Content	Analysis
6.1	Wired user terminals, including desktop computers	Consider the network interface card (NIC) rate of terminals
6.2	Wireless user terminals, including portable computers, smart phones, and mobile smart devices such as tablets	Determine the supported wireless protocols/standards, access frequency band, access authentication mode, whether to use unified authentication for wired and wireless networks, whether to allow guest access and which zones are accessible to them, as well as the power supply for terminals
6.3	Dumb terminals, including IP phones, network printers, and IP cameras	Determine the access and authentication solutions for dumb terminals

(Continued)

TABLE 10.7 (*Continued*)   Overview of the Campus Network Requirements
Survey — Terminal Type

Requirement	Survey Content	Analysis
6.4	IoT devices, including radio frequency identification (RFID) terminals used in logistics, wristbands for students, and antitheft wristbands for medical treatment	Determine the protocols and standards supported by terminals and network isolation requirements
6.5	Other terminals, including industrial control computers and field test controllers	These terminals may affect the model selection of access switches. For example, industrial switches may be required for industrial campuses or production networks. Furthermore, outdoor devices may require a special power supply mode
6.6	Special network devices, including dedicated network encryption devices and industrial switches	Consider compatibility and performance of these devices to prevent mismatching specifications

10.2 INTENT-DRIVEN CAMPUS NETWORK DEPLOYMENT CASES

A college or university's campus network is a network platform capable of providing teaching, scientific research, and comprehensive information services for teachers and students. Such a network can account for tens of thousands of users, which is not unusual for large-scale campus networks. This section details typical cases of intent-driven campus networks and how they are deployed in universities and colleges.

10.2.1 Construction Objectives and Requirements Analysis

University A is a technical university that focuses on new technologies and topics in the Information and Communications Technology (ICT) field within the higher education industry. It is a well-known university notable for cultivating technical talent and has even provided ICT consulting services for multiple government ministries over the years.

To accelerate its development, University A plans to construct a new campus and equip it with a high-performance campus network.

1. Construction objectives

    The construction objectives include multinetwork convergence, advanced architecture, and on-demand expansion.

    a. Multi-network convergence: A comprehensive network capable of carrying wired, wireless, and Internet of Things (IoT) services must be constructed to meet the access requirements of various data terminals and sensors at any location.

    b. Advanced architecture: The overall architecture should be industry-leading in terms of performance, capacity, reliability, and technology application, and must be future-proofed to ensure it can meet any and all requirements over the next 5–10 years.

    c. On-demand expansion: The overall network architecture must meet the requirements for coverage area, terminal numbers, and services, and should be expandable on demand without architecture adjustment.

2. Requirements analysis

    University A's computer center is responsible for network construction and Operation & Maintenance (O&M). The center wants to introduce network virtualization and software-defined networking (SDN) technologies in order to build a campus network as a service that can uniformly carry and flexibly deploy multiple services such as teaching and scientific research. Table 10.8 lists the specific network construction requirements.

10.2.2 Overall Network Planning

Figure 10.2 shows the deployment architecture of the intent-driven campus network designed for University A, based on virtualization and SDN technologies. The campus network is logically divided into the terminal layer, access layer, aggregation layer, and core layer. Modules at these layers are clearly defined and the internal adjustment of each module is limited to a small scope, facilitating efficient fault locating.

TABLE 10.8   Analysis of University A's Campus Network Construction Requirements

No.	Key Requirement	Description
1	High reliability and high rate	The new campus network should have carrier-class reliability and provide high-performance basic network services at a high rate
2	Network virtualization	The new campus network should support network virtualization so that it can be divided into multiple VNs, such as office, teaching research, and IoT networks. Communication across VNs must be controlled by firewalls
3	Network automation	Automated network configuration deployment should be enhanced, as University A is accustomed to using self-service network construction
4	Network security	Access security should be strengthened for VNs. University A requires in-depth security checks and on-demand traffic diversion to the security resource pool. All port traffic must be managed and controlled by firewalls in a unified manner
5	Terminal authentication	NAC should be exerted over all terminals. Most terminals use dynamic authentication, such as 802.1X authentication. Some dumb terminals (including printers and cameras) use MAC address authentication for network access
6	Full wireless network coverage	The entire campus should be completely covered by a wireless network to ensure that teachers and students can gain access at any time, regardless of location (dormitories, office buildings, teaching buildings, canteens, libraries, and playgrounds)

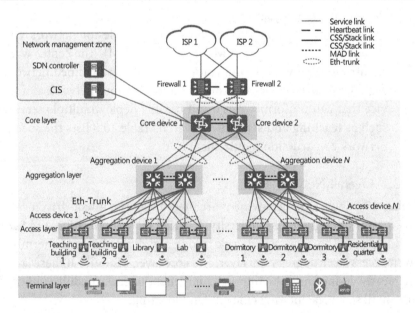

FIGURE 10.2   Intent-driven campus network deployment architecture.

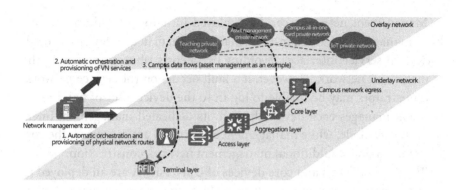

FIGURE 10.3   Abstracted network model for the intent-driven campus network.

Figure 10.3 abstracts a network model for the intent-driven campus network. An ultra-broadband converged transport network is built to connect all service subsystems and terminals on the entire campus, achieving a fully connected campus. From the core layer to the access layer, virtualization technologies are leveraged to construct VNs based on various service plans. Finally, the SDN controller is used for automated network deployment. This design is capable of achieving the required intent-driven campus network goal.

Table 10.9 describes the items to be planned for the intent-driven campus network.

10.2.3 Management Network and Deployment Mode Planning

In the virtualized campus network solution, management network planning involves management network connection and device management by the SDN controller.

TABLE 10.9   Items to Be Planned for the Intent-Driven Campus Network

Category	Items to Be Planned
Management network and deployment mode	Network management and deployment mode
Underlay network	Device- and link-level reliability, as well as OSPF and BGP routing
Overlay network	Device roles, user gateways, network between border nodes and egress nodes, VNs and subnets, and VN communication
Egress network	Firewall security zones, firewall hot standby, and intelligent traffic steering
Service deployment	User access and network policies

Management networks can be deployed in both in-band and out-of-band management modes. In-band management manages devices through their own service interfaces, avoiding extra costs when constructing the management network. However, if a fault occurs on the service network, administrators may be unable to log in to the device. Out-of-band management manages each device through its dedicated management interface. In this mode, while management and control are separated, the cost is increased due to additional management network construction.

The egress devices and core devices of University A are all deployed in the core equipment room, resulting in lower management network construction costs. Consequently, out-of-band management mode is most suitable for our needs. In addition, the services running on the egress and core devices are complex, requiring onsite commissioning by network engineers during deployment. As such, the local command line interface (CLI) or web system is used for deployment. Many devices, including aggregation devices, access devices, and access points (APs), are sparsely deployed below the core layer and feature similar service configurations. To simplify deployment in this scenario, in-band management and plug-and-play deployment are recommended. Table 10.10 provides management network and deployment mode planning.

10.2.4 Underlay Network Planning

1. Device-level reliability planning

The core, aggregation, and access layers use stacking or clustering technology to horizontally virtualize two or more switches into one and provide device redundancy.

Switch stacking or clustering prevents Layer 2 loops on traditional redundant networks, avoiding complex loop prevention protocol

TABLE 10.10 Recommended Management Network and Deployment Modes

Location	Device	Management Network	Deployment Mode
Egress	Firewall	Out-of-band management	Local CLI or web system
Core layer	Core switch	Out-of-band management	Local CLI or web system
Aggregation layer	Aggregation switch	In-band management	Plug-and-play
Access layer	Access switch	In-band management	Plug-and-play
	AP	In-band management	Plug-and-play

configurations. On the Layer 3 network, the stack or cluster system shares the same routing table, reducing route convergence time at the time of a network fault. The stack or cluster system also facilitates network management, maintenance, and expansion. All this positions the stack or cluster solution as the best choice for campus network switches.

2. Link-level reliability planning

Link-level reliability relies primarily on link redundancy. On a campus network, the dual-uplink redundancy design is typically used to improve link reliability between devices. For redundant links, link aggregation technology is leveraged to virtualize multiple physical links into one logical Eth-Trunk link using the Link Aggregation Control Protocol (LACP). The interfaces are then grouped into an Eth-Trunk interface. Link aggregation enhances the reliability of links between devices and increases the link bandwidth without requiring hardware upgrades. Consequently, LACP-based link aggregation is recommended between devices on the campus network.

3. Open Shortest Path First (OSPF) route planning

In the virtualized campus network solution, the underlay network provides a transport network with reachable routes for the overlay network, enabling virtual extensible local area network (VXLAN)-encapsulated service packets to be transmitted between VXLAN nodes. Internet Protocol (IP) unicast routing protocols, such as OSPF and Intermediate System to Intermediate System (IS-IS), can be adopted to implement connectivity on the underlay network. OSPF is recommended due to the following: OSPF routes are primarily used on campus networks to implement IP network communication; OSPF routing technology is mature; and network construction and maintenance personnel possess extensive OSPF experience.

As shown in Figure 10.4, the SDN controller enables automatic orchestration of underlay network routes. After IP network segments used for communication are planned in underlay network resources, the SDN controller automatically orchestrates OSPF routes and delivers them to border and edge nodes, thereby implementing automatic deployment of underlay network routes. During underlay network route orchestration, the network segments of the Border Gateway

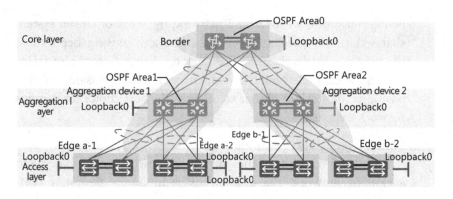

FIGURE 10.4   OSPF route planning on the underlay network.

Protocol (BGP) source interfaces (such as Loopback0) planned on the device are imported to the OSPF area of the underlay network to implement interworking between BGP source interfaces.

4. BGP route planning

The virtualized campus network solution uses VXLAN technology to construct VNs. In this solution, VXLAN uses BGP Ethernet Virtual Private Network to implement data forwarding on the control plane, including dynamic VXLAN tunnel establishment, Address Resolution Protocol (ARP)/neighbor discovery (ND) entry transmission, and routing information transmission. To achieve this, BGP must be deployed on VXLAN tunnel endpoints (VTEPs), such as border and edge devices.

In our solution, BGP is automatically deployed by the SDN controller. When an overlay network is created, if a device is selected as a border node or an edge node, the SDN controller automatically delivers configurations (such as the BGP peer address) to complete BGP routing protocol deployment. In addition, to reduce network and CPU resource consumption, you are advised to select one VTEP as the route reflector (RR) when configuring node roles. Figures 10.5 and 10.6 show BGP routing protocol planning on the underlay network.

10.2.5 Overlay Network Planning

1. Device role and user gateway planning

Figure 10.7 shows how a network is deployed. A VXLAN is deployed between the core devices and access devices, which

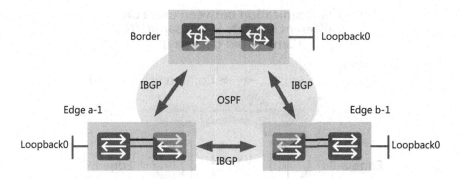

FIGURE 10.5   BGP routing protocol planning on the underlay network (without an RR).

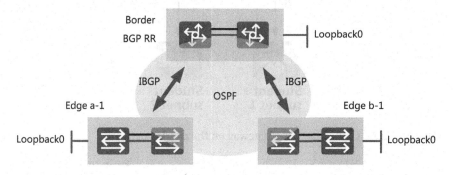

FIGURE 10.6   BGP routing protocol planning on the underlay network (border node as an RR).

function as border nodes and edge nodes, respectively. The entire network uses VNs, simplifying service provisioning and network management. The border node taking on the role of user gateway is deployed in centralized mode.

2. Network planning from the border node to the egress

In the virtualized campus network solution, the border node is responsible for communication between the campus overlay network and external networks. In most cases, the border node in the campus network architecture is connected to the firewall in the northbound direction, in order to implement security policy control on incoming and outgoing traffic. As shown in Figure 10.8, the border node functions as the border gateway to connect to the firewall at Layer 3. This VN-based interconnection can be implemented using either

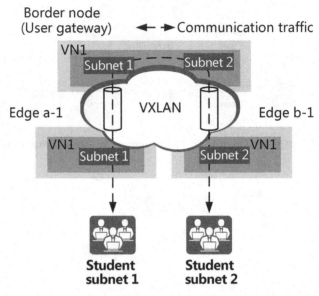

FIGURE 10.7   Centralized user gateway on the overlay network.

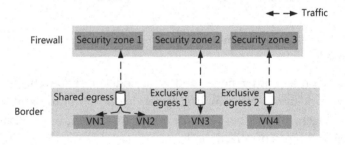

FIGURE 10.8   Traffic model of Layer 3 egresses.

the Layer 3 shared egress or Layer 3 exclusive egress. Table 10.11 describes the application scenarios of the two types of egresses.

3. VN and subnet planning

In the virtualized campus network solution, each VN corresponds to a virtual private network (VPN) and can contain multiple subnets. By default, users on the same VN can communicate with

TABLE 10.11   Application Scenarios of Different Types of Egresses

Egress Type	Application Scenario
Layer 3 shared egress	VNs representing different service networks share the Layer 3 egress and connect to the security zones of the firewall. This mode reduces network configurations but cannot implement service-based refined security policy control. As such, this mode is suitable for networks with simple security policy requirements
Layer 3 exclusive egress	Each VN representing a service network exclusively uses a Layer 3 egress and connects to a firewall security zone through this interface
	VNs, Layer 3 egresses, and security zones can use one-to-one mappings to implement service-based refined security policy control

each other, but routes for users on different VN are isolated. VNs can be planned based on the following principles:

a.   An independent service department is considered a VN.

b.   VNs are not used to isolate users of different levels in the same service department. Instead, intergroup policies of security groups, which are divided based on user roles, achieve this result.

In this deployment case, based on the service characteristics of University A, VNs are divided into teaching private network, campus all-in-one card private network, asset management private network, and IoT private network.

Service data enter VNs from physical networks through edge nodes, according to the VLANs to which users belong. During network design, we must plan the mappings between VLANs of physical networks and subnets of VNs, and configure VLANs for both wired and wireless users. Wired user packets enter a VXLAN network based on VLANs, while wireless user packets are forwarded to a wireless access controller (WAC) (possibly an edge or a border node) through a Control and Provisioning of Wireless Access Points (CAPWAP) tunnel. The WAC decapsulates CAPWAP packets, which then enter the corresponding VXLAN network based on the VLANs to which wireless users belong. Since access to VN subnets depends on user authentication technologies, it is recommended that terminals of the same type be added to the same subnet.

TABLE 10.12   Application Scenarios of VLAN Access Modes for VN Subnets

VLAN Access Mode	Characteristics	Recommended Scenario
Static VLAN	This access mode is secure and meets the requirements of authentication-free access scenarios. However, it lacks flexibility	Access of terminals with fixed positions, for example, access of dumb terminals
Dynamically authorized VLAN	This access mode requires authentication, applies to access of terminals at any location, and delivers good flexibility	Access of terminals with flexible positions, for example, access of mobile terminals such as phones
Voice VLAN	This access mode is dedicated to voice services and supports dynamic authorization and static deployment	Access of IP phones
VLAN pool	VLANs in the VLAN pool are automatically allocated to ports Subnets in the VLAN pool belong to the same VN This access mode supports dynamic authorization and static deployment for wireless users, but only dynamic authorization for wired users	Access in high-density scenarios

There are four VLAN access modes for VN subnets, and Table 10.12 describes the application scenarios of each mode. The dynamically authorized VLAN mode requires users to go online again during Portal authentication and is therefore not recommended for Portal authentication.

4. VN communication planning

In the virtualized campus network solution, VNs are isolated at Layer 3 using VPNs and cannot communicate with each other by default, unless they reside on the same overlay network or communicate through firewalls. Table 10.13 describes the application scenarios of the two VN communication modes.

Figure 10.9 shows how VN communication traffic travels within the overlay network. Here, to achieve VN communication, configure VNs and their subnets and then import each other's routes.

Figure 10.10 shows how VNs communicate through firewalls. Different VNs connect to different security zones on the firewalls through different logical egresses. Interzone policies are deployed on the firewalls to implement VN communication.

TABLE 10.13   VN Communication Modes and Application Scenarios

Communication Mode	Application Scenario
Communication inside an overlay network	Communication between VNs does not require advanced security policy control by firewalls Instead, the edge device is used to implement basic intergroup policies of security groups
Communication through firewalls	Communication between VNs must be controlled by advanced security policies through firewalls

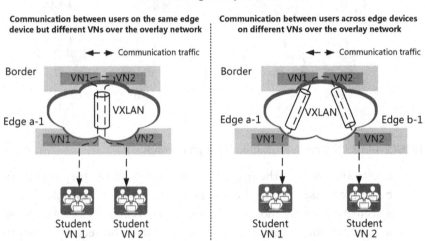

FIGURE 10.9   VN communication on the overlay network.

10.2.6 Egress Network Planning

1. Firewall security zone planning

On a campus network, security zones must be divided on firewalls in order to implement corresponding security policies. In most cases, we add the Internet to the untrusted zone, the campus intranet to the trusted zone, and the data center to the demilitarized zone (DMZ). We deploy firewalls at the campus egress to isolate traffic between the campus intranet and external network. We also deploy firewalls in the DMZ, which are used to isolate traffic between the campus intranet and servers in the data center.

When user gateways are located on the overlay network in the virtualized campus network solution, each overlay network egress connected to external network resources corresponds to a Layer 3

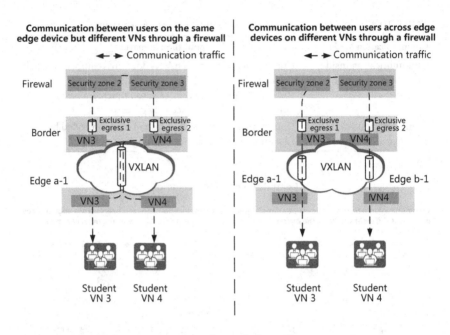

Communication between users on the same edge device but different VNs through a firewall

Communication between users across edge devices on different VNs through a firewall

FIGURE 10.10   VN communication through firewalls.

logical interface on the firewall. In this solution, security zones can be divided based on these egresses, binding each logical interface to a security zone, as shown in Figure 10.11. When user gateways are located outside the overlay network, we must bind security zones to different gateways based on the gateways' security policies.

2. Firewall hot standby and intelligent traffic steering planning

In the virtualized campus network solution, it is recommended that firewalls be used as egress devices to connect to the carrier network and they work in hot standby mode to enhance device-level reliability. Figure 10.12 shows the firewall deployment mode (data used in the figure, such as IP addresses, are for reference only).

The firewalls use Virtual Router Redundancy Protocol (VRRP) to deploy hot standby mode in the downlink direction. Note the following during deployment:

a. The physically connected interfaces on the two firewalls working in hot standby mode must be numbered the same to facilitate management and fault locating.

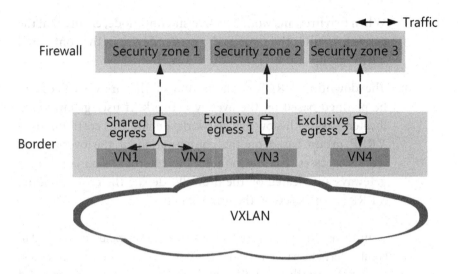

FIGURE 10.11   Firewall security zone division when user gateways are located inside the overlay network.

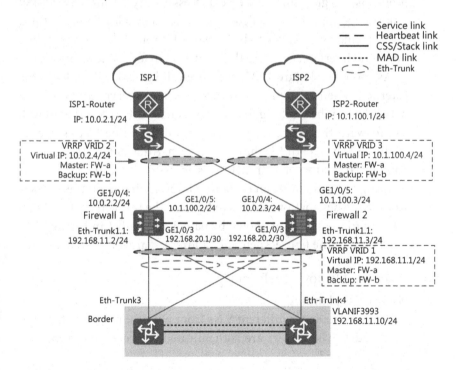

FIGURE 10.12   Firewalls working in hot standby mode.

b. When two firewalls work in active/standby mode, ensure that the active firewall is the same as the master device in the configured VRRP group.

c. The downlink VRRP configuration of the firewalls needs to be planned based on the overlay network. If user gateways are located inside the overlay network, deploy the corresponding VRRP group based on the configuration of the overlay network's egresses that are connected to external resources. If user gateways are located on the firewalls, deploy the corresponding VRRP group based on the user subnet.

Assume that the egresses of University A's network are connected to different Internet service provider (ISP) networks. As users can access network resources through different ISP networks, it is recommended that intelligent traffic steering be configured on the firewalls to ensure proper use of egress links and egress access quality. In this scenario, ISP link selection is recommended. This function enables traffic destined for a specific ISP network to be forwarded through the corresponding outbound interface, ensuring the shortest path for traffic forwarding.

As shown in Figure 10.13, each firewall has two ISP links to the Internet. When a campus network user accesses server 2 on the ISP 2 network, and the working firewall has equal-cost routes, the firewall can forward the access traffic along two different paths to server 2. In this example, path 1 is the optimal path. After ISP link selection is configured, and when intranet users access server 1 or server 2, the working firewall selects an outbound interface based on the ISP network of the destination address to forward the traffic from the shortest path to the server, as shown by path 3 and path 1 in Figure 10.13 (data used in the figure, such as IP addresses, are for reference only).

10.2.7 Service Deployment Planning

1. User access planning

Service data enter different VNs from physical networks through edge nodes, where users are authenticated before access to the network is granted.

802.1X authentication, Media Access Control (MAC) address authentication, and Portal authentication are commonly used on

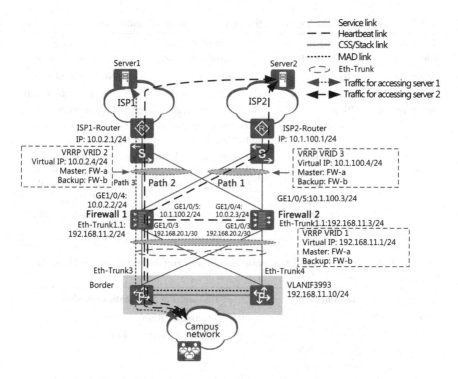

FIGURE 10.13   Intelligent traffic steering.

campus networks, each of which utilizes different authentication principles and applies to different scenarios. For actual applications, we can use one or more authentication modes depending on the situation.

Table 10.14 compares the three authentication modes.

Based on the above comparisons, Table 10.15 lists the recommended authentication modes for common campus network scenarios. If an interface has different requirements, we can use other authentication modes together with the recommended one.

2. Network policy planning

Network policies are deployed at different layers based on the user granularity.

First layer: based on the VNs where users are located (as shown in Figure 10.14). VNs are route-isolated and by default cannot communicate with one another. If no special policy requirement for communication exists, we can configure VN communication directly on

TABLE 10.14 Comparison between the Three User Authentication Modes

Item	802.1X Authentication	MAC Address Authentication	Portal Authentication
Client	Required	Not required	Not required
Advantage	High security	No client required	Flexible deployment
Disadvantage	Rigid deployment	MAC address registration required, complicating management	Low security
Application scenario	Network authentication for office users with high-security requirements	Network authentication for dumb terminals such as printers and fax machines	Network authentication for guests with high mobility and complex terminal types

TABLE 10.15 Recommended Authentication Modes for Common Campus Network Scenarios

Scenario	Terminal Type	Recommended Authentication Mode
Wired	PC	802.1X authentication
	Dumb terminals such as IP phones, printers, and fax machines	MAC address authentication
Wireless	Mobile phone and PC	Depending on the user role, 802.1X authentication for students and faculty, and MAC address-prioritized Portal authentication for guests
	Video terminal	802.1X authentication, or MAC address authentication if 802.1X authentication is not supported

the core switch. VN communication can also be controlled using the firewall. In this case, each VN is connected to a separated security zone on the firewall through an exclusive egress, and interzone policies are configured on the firewall to control VN communication. If security groups belong to different subnets, the border node is also required to implement communication between security groups in a centralized gateway scenario.

Second layer: based on the security groups to which users belong (as shown in Figure 10.15). Switches operate with the SDN controller's free mobility solution in order to divide users with different network access permissions in a VN into security groups. The network access rights of users are determined by the access permissions of

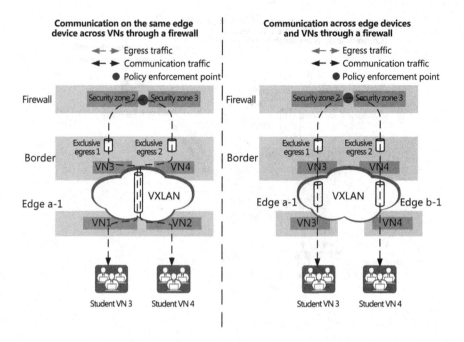

FIGURE 10.14 Traffic model for VN-based security policies.

their homing security groups and intergroup policies. In this solution, security groups and intergroup policies are deployed on the SDN controller, which then delivers the intergroup policies to the corresponding user authentication control points.

10.2.8 SDN Controller Installation and Deployment Planning

Huawei's SDN controller is used in this example and is deployed in on-premises mode. Two network planes are planned: one specifically for internal communication, and the other shared by services and both northbound and southbound interfaces. Figure 10.16 shows the networking mode, and the functions of each plane are described as follows:

- Internal communication plane: used for communication between service nodes of the controller, including communication with the Huawei big data platform and Huawei database system.

- Service plane: used by the controller to provision southbound and northbound services; for example, distributing services to multiple nodes through load balancing.

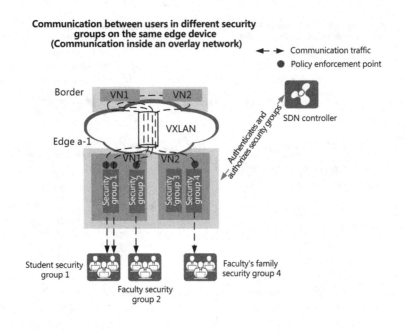

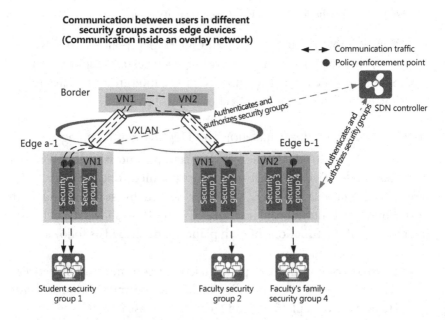

FIGURE 10.15 Traffic model for security policies based on security groups.

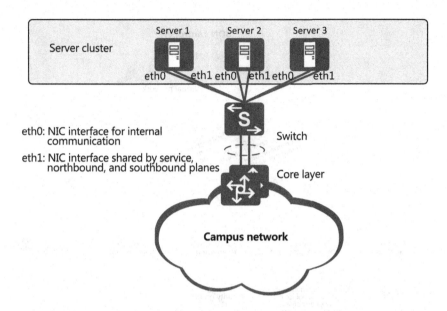

FIGURE 10.16 Huawei SDN controller deployment networking.

- Northbound plane: used by the controller to receive northbound services; for example, accessing the management plane of the controller through the web system.

- Southbound plane: used by the controller to receive southbound services; for example, communicating with devices through Network Configuration Protocol (NETCONF).

10.2.9 SDN Controller Configuration and Provisioning Process

Based on the overall network plan, the SDN controller is responsible for deployment and routine O&M management. For example, University A uses Huawei's SDN controller to automate network lifecycle management, as shown in Figure 10.17.

1. Network device management

    Before the SDN controller can provision services to campus network devices, a management channel must first be established between the SDN controller and network devices. To enable the SDN controller to manage devices:

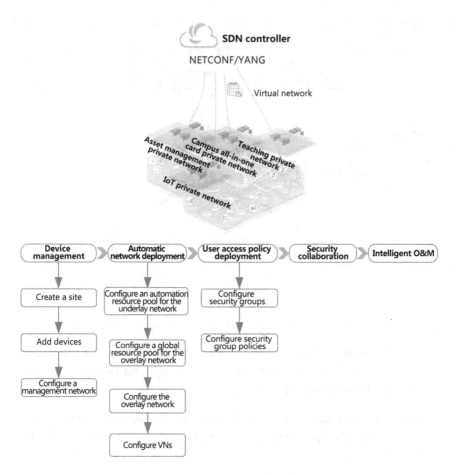

FIGURE 10.17   Automated network lifecycle management.

First, create a network site representing a user network. Using the map system, the administrator can intuitively learn the location of the user network. For this example, we will create a campus network site for University A on the SDN controller and name it Campus Network X of University A.

Next, add devices to the site so that the SDN controller knows which devices need to be managed. In this step, you can import device information (such as equipment serial numbers or ESNs) individually or in batches using a template.

Finally, after network devices are installed and connected, the administrator must establish a management channel between the

SDN controller and network devices. The Huawei SDN controller uses DHCP to establish such a channel.

2. Automatic network deployment

Once the management channel is established, the next step is to build a virtualized campus network:

a. Configure an underlay network automation resource pool, including VLAN resources for Layer 2 interconnection and IP address resources for Layer 3 interconnection. The SDN controller orchestrates OSPF routes on the underlay network based on the topology information obtained by protocols and delivers configurations to network devices to complete automatic deployment. Once this has been completed, an underlay network that carries overlay network packets is established.

b. Configure a global resource pool for the overlay network, including subnets, VLANs, BDs, and VXLAN network identifiers (VNI). When you create a VN, the SDN controller automatically allocates resources from these global resource pools.

c. Configure the overlay network, including VXLAN control plane configuration (such as border and edge node selection, and BGP peer configuration) and wired and wireless network access configuration (including user access authentication control point configuration).

d. Configure VNs. This final step involves the creation of planned service VNs. Each VN represents a service network. For University A, the SDN controller must create a separate VN for the teaching private network, campus all-in-one card private network, asset management private network, and IoT private network. In addition, if different service private networks need to communicate with one another, VN communication must be configured.

3. User access policy deployment

After a virtualized campus network is constructed, network services must be automatically provisioned. This section uses the deployment of typical user access policies as an example to describe

Source Group	Source-to-Destination Group Access			
	Faculty Group	Student Group	Faculty's Family Group	Faculty Group
Faculty Group	Permit	Permit	Permit	Permit
Student Group	Permit	Permit	Deny	Deny
Faculty's Family Group	Permit	Deny	Permit	Deny
Server Group	Permit	Permit	Deny	Permit

FIGURE 10.18 Intersecurity group access permission planning.

how the SDN controller implements automatic policy provisioning based on the innovative free mobility solution.

First, security groups must be configured. Security groups can be classified into two types: dynamic user group and static resource group. A dynamic user group consists of access users, and only the user accounts to be authorized need to be added during configuration (IP addresses are not required). During terminal access authentication, a dynamic mapping table between IP addresses and security groups is generated based on the IP addresses in the protocol packets of the terminals. A static resource group contains resources accessed by users, such as internal servers and the Internet. As the IP addresses of these resources are fixed, you need to add IP addresses when configuring such a group, upon which a static mapping table between IP addresses and security groups is generated.

Next, configure intergroup policies for the security groups. After the policy planning is complete, you need to configure data on the controller, as shown in Figure 10.18.

After the preceding configurations are complete, the SDN controller delivers security groups and inter-group policies to policy enforcement points.

CHAPTER 11

Huawei IT Best Practices

B Y REMAINING CUSTOMER-CENTRIC SINCE its founding, Huawei has grown into a super-large multinational company that has expanded its business all around the world. In addition, throughout the past 30 years, Huawei IT has remained committed to building a leading campus network at Huawei. With this vision, Huawei IT has constantly incorporated cutting-edge technical solutions into the campus network to support the company's rapid business growth.

In the future, Huawei IT looks to actively transform itself from a business supporter to a business enabler in order to play an increasingly important role in Huawei's own digital transformation. This chapter introduces Huawei IT's best practices throughout its own IT transformation journey over the past several decades. Huawei's own campus network covers various application scenarios. The practices explained in this chapter will provide a suitable reference for the ongoing digital transformation of more industries.

11.1 DEVELOPMENT HISTORY OF HUAWEI IT

By the end of 2019, Huawei has operated in 178 countries and regions and established partnerships with more than 60000 companies, with a total workforce of approximately 200000 employees. In 2018 alone, Huawei shipped more than one million base stations and 200 million mobile phones around the world. Huawei has also established 14 R&D centers, 36 joint innovation centers, more than 1000 branches, 1200 spare parts centers, 220000 stores, 8000 conference rooms, and 2000 labs around the world. Huawei employees handle more than 2.8 million emails and hold more than 80000 meetings every day.

The massive amounts of services mentioned above are enabled by more than 600 IT applications, 180000 network devices, and a high-performance backbone network with a total bandwidth of more than 480 Gbit/s. These IT applications and underlying network devices effectively support the smooth, orderly, and efficient operations of Huawei's overall business around the world. As shown in Figure 11.1, Huawei IT has undergone four distinct development phases: localization, internationalization, globalization, and digitalization.

1. Phase 1: Localization

    Before 1997, Huawei's IT network was mainly based on the local campus network in the company's Shenzhen headquarters. Meanwhile, Huawei began building WANs and private lines to meet the needs of branch offices in China, as well as leasing carriers' network resources to facilitate interconnection between the company's branch offices and headquarters in Shenzhen.

2. Phase 2: Internationalization

    During the early 2000s, Huawei introduced large-scale IT office systems into the company's different departments, including R&D, marketing, finance, and supply chain, to support its rapid growth. Huawei IT also started to build data centers and Multiprotocol Label

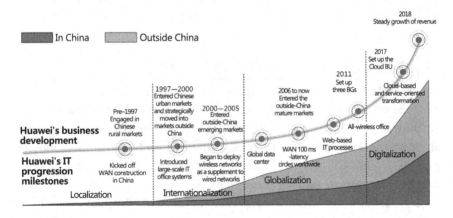

FIGURE 11.1   Four development phases of Huawei IT.

Switching (MPLS) private networks throughout China to realize service data interworking between different regions.

However, due to superior business growth outside China, Huawei's MPLS private network was considered inadequate and therefore unable to support global service data forwarding and communication between different countries and regions. For example, when companies in Latin America communicated with the headquarters in Shenzhen on the MPLS private network, the latency could reach as high as 300 ms, far from meeting the requirements for real-time communications. In addition, latency was even higher in countries without existing network coverage, because these countries had to rely on mainly satellite communication.

Facing this challenge, Huawei started to deploy regional data centers in 2002. Service data were then able to be centrally stored at the regional data centers to support the rapid development of regional services. It also ensured that these regional data centers met applicable business compliance requirements. For example, the General Data Protection Regulation in Europe stipulates that enterprises' data must not be transferred out of the local country; therefore, all data were stored in the regional data center to ensure that Huawei's business operations complied with the region's laws and regulations.

In 2005, due to the rise of Wi-Fi technologies, Huawei IT started to deploy small-scale wireless networks after fully considering the mobility trends and convenience brought by mobile offices to global business expansion. This early Wi-Fi deployment laid the foundation for all-wireless offices in the future.

3. Phase 3: Globalization

In 2006, Huawei began to comprehensively expand into overseas markets. During this period, Huawei IT transformed itself on a large scale to keep pace with the company's globalization.

a. Huawei successively built global data centers in Guizhou, Hong Kong, Johannesburg, London, and Sao Paulo, with each global data center covering business across multiple regions.

b. The global MPLS networks required unified planning and construction, and WANs were needed to cover Huawei's global business premises. Huawei subsequently initiated a network reconstruction project to build eight ultra-low-latency (<100 ms) circles around the world. The MPLS private lines were extended to the nearest data center to achieve equidistant services to ensure employees received consistent experience from anywhere in the world.

c. The steady growth of business outside China led to severe information security risks. In response, Huawei started to plan and design a security policy system to ensure network security, with terminals used as the starting point.

d. In 2009, application distribution technology was considered mature in the IT field. Huawei started to use this technology outside China to upgrade the traditional client/server (C/S) email system to a browser/server (B/S) email system. For traditional office systems in C/S mode, packet requests often cause high latency because they are passed back and forth multiple times between the client and server. In contrast, the B/S system quickly completes interactions on the cloud and results are sent directly to users to significantly reduce the number of interactions. This reduces latency, improves user experience, and lowers bandwidth requirements. In addition, the C/S structure also faces challenges due to the growing number of applications and its poor application scalability. To address this challenge, the web-based system was introduced, allowing servers and storage resources in the data center to be quickly assigned, thereby ensuring smooth application upgrade without affecting user experience.

e. With the pervasive use of portable computers, smartphones, and tablets, the mobile office era represented by Wi-Fi networks came. In 2011, Huawei deployed all-wireless coverage in conference rooms around the world. After receiving positive feedback from employees, Huawei proceeded to extend all-wireless coverage into other areas, such as office spaces, customer reception areas, and canteens. In 2012, Huawei implemented Bring Your Own Device (BYOD) and completed global BYOD coverage in 2014. This enabled Huawei employees to use their own portable devices to process emails, approve common workflow requests, and use enterprise office applications anytime and anywhere.

Between 2006 and 2014, Huawei IT underwent radical changes during the course of the company's globalization. Specifically, internal IT personnel's skills were greatly improved, and IT infrastructure platforms were established. More importantly, Huawei's technical expertise and process standardization accumulated during IT deployment paved the way for the company to more confidently cope with and lead the transformation amid the global digital transformation wave.

4. Phase 4: Digitalization

The entire world is currently transitioning from the information era to the digital era. Like most other companies, Huawei is also undergoing comprehensive digital transformation. To better support the company's digital transformation, Huawei IT has undertaken more responsibilities and is transforming from a supporter to an enabler for digital transformation. Huawei IT is now in the process of considering how to best build standard IT facilities and services in various scenarios to better enable the company's digital transformation; how to best help business teams achieve quick and efficient operations in all scenarios without increasing manpower costs.

Before achieving digitalization, cloudification and servitization first need to be attained. Huawei IT has built an end-to-end (E2E) Huawei IT Service (HIS) platform tailored to all service scenarios. Through this platform, Huawei IT has defined more than 10 standard service modes that can be flexibly combined to serve various typical service scenarios, such as stores, factories, supply centers, exhibitions, conference rooms, offices, and operation centers. Business departments can quickly subscribe to various services on the HIS platform as required. Then, the platform automatically provisions and configures resources, and evaluates and perceives service quality. As a result, service deployment efficiency is significantly improved.

11.2 CHALLENGES TO CAMPUS NETWORKS PRESENTED BY DIGITAL TRANSFORMATION

Throughout Huawei's digital transformation, the IT services powered by cloud computing, big data, AI, and IoT technologies have transformed their role from supporters to enablers. As the underlying infrastructure, the campus network also needs to transform itself to adapt to the changes brought by digital transformation. Key challenges include the following:

- Quickly delivering network services for a large number of sites: Huawei's rapid business growth, especially the rapid growth of its consumer business, has led to an increase in the number of small- and medium-sized campuses. The number of Huawei stores has soared from less than 10000–220000 within five years. In this context, network orchestration services must be suitable for all scenarios, and a large number of new sites must be quickly delivered.

- Simplifying network policy management: Networks for different services are separated using MPLS VPNs; however, static Virtual Private Network (VPN) Routing and Forwarding (VRF) routes should be used if different service VPNs need to communicate with each other. As such, if there are N service VPNs that should communicate with each other, $N \times N$ VRF routes are required, resulting in more complex route cross-connect design. In addition, because firewall policy management systems need to manage tens of thousands of policies, simplified systems are required to promote overall efficiency.

- Building E2E network visibility: In-depth service experience visibility is a requirement that can be directly invoked as a service by upper-layer applications. Take the annual HUAWEI CONNECT event as an example. If there were network problems during the live keynote speech on the morning of the first day of this event, and these problems could not be solved within 1 minute, the entire event would be severely impacted. In similar scenarios, IT network O&M needs to be able to detect network service experience in real time and locate faults efficiently.

- Effectively managing a large number of IoT devices: Huawei's IT network currently has more than one million Wi-Fi access terminals, 400000 wired access terminals, and 100000 IoT devices. In the future, millions of IoT devices will also join the network. This huge number of terminals requires terminal identification and simplified terminal access authentication; however, identifying terminals using traditional approaches results in complex networks that are hard to maintain. In addition, given today's diversified terminal access authentication modes, many terminals need to be authenticated twice before they are granted access to the network.

- Ensuring network security and reliability: In the future, Huawei's IT network will carry more services and support a growing number of terminals, terminal types, and access modes. If a reliability fault or security problem occurs on the network, the impact will be exponentially amplified; therefore, network security and reliability are crucial to ensuring service continuity.

To effectively address the above challenges, the network needed for digital transformation must meet the following four fundamental principles:

- Cloud-based and service-oriented delivery: Proper network services must be provided in cloud-based mode to best suit service scenarios. For example, "one scenario, one policy" implements network as a service.

- Elastic capacity expansion: Different logical service networks must be able to be quickly and efficiently deployed based on intents on a converged network to build an intent-driven, fully connected network.

- Visualized O&M: The network platform must be capable of powerful service orchestration and E2E experience visibility. It must automatically detect applications and terminals, flexibly provision policies for consistent experience, and quickly locate and self-heal from faults.

- Simplified, high-speed, and secure: Devices and links must be deployed in high-availability mode to meet high bandwidth requirements. The basic physical network must also be secure and reliable, with a simplified architecture and simplified protocols.

In accordance with these above-mentioned principles, the overall architecture of Huawei's campus network is divided into the following four layers:

- The first layer (also known as the top layer) is the service layer oriented to each scenario segment and is supported by Huawei's HIS platform. Typical scenario segments include the operation scenario, shared service scenario, and campus office scenario. It supports different applications based on the different requirements of each scenario.

- The second layer is the platform layer. This layer provides various professional application platform services. It covers the big data platform, operation support platform, and security platform. It uses open interfaces to provide basic support for upper-layer applications.

- The third layer is the network layer. This layer implements full connectivity through cloudification. For example, the MPLS private network is constructed through the cloud backbone network and carriers' private line networks, and wireless and wired networks are built on the campuses of each research center and branch office. These deployments all realize full connectivity.

- The bottom layer includes a large number of terminals. Various terminals are on-boarded, authenticated, and managed in a unified manner to meet different service requirements.

11.3 HUAWEI SCENARIO-SPECIFIC CAMPUS NETWORK SOLUTIONS

11.3.1 All-Scenario Wi-Fi: Covering Huawei's Global Offices with Three SSIDs

Huawei provides full Wi-Fi coverage without blind spots for all its office campuses around the world. After first-time Wi-Fi access, Huawei employees can automatically log in to the company's Wi-Fi network and work anywhere and anytime throughout the company's multiple campuses. In addition, employees can use apps on personal mobile devices to efficiently handle routine office tasks, such as processing emails, replying to instant messages, and initiating conference calls, at any time and any place. Huawei has defined three corporate-wide service set identifiers (SSIDs) to ensure consistent network access experience for both employees and guests, as shown in Figure 11.2.

- **Employee 2.0-SSID** applies when employees use company-issued security terminals to access intranet resources. These security terminals have security check software built-in; therefore, employees can access the Wi-Fi network after passing security checks. This approach ensures terminal access security. Employees are granted access to intranet resources after passing 802.1X authentication.

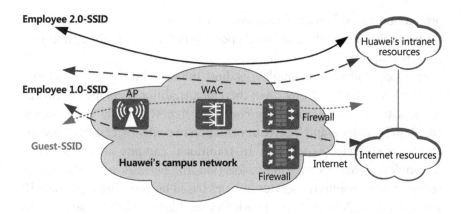

FIGURE 11.2   Three SSIDs for Huawei's all-scenario wireless campus network.

- **Employee 1.0-SSID** applies when employees use their own terminals to access some intranet and Internet resources. Employees are
  required to install the mobile office software WeLink on their own
  terminals; then, after passing 802.1X authentication and travers-
  ing firewalls, employees are granted access to the Internet as well as
  selected intranet resources.

- **Guest-SSID** applies when guests on campus use their own terminals
  to access Internet resources. Guests are allowed to access Internet
  resources after they pass web authentication and traverse firewalls.

In the future, an increasing number of IoT terminals will access the campus
network; therefore, an independent SSID will be planned for IoT access on
the existing Wi-Fi network. In addition, IoT access will go through Media
Access Control (MAC) address authentication as well as equipment serial
number (ESN) verification.

11.3.2 Free Mobility: Providing Consistent Service Experience for Huawei Employees

All-scenario Wi-Fi coverage greatly improves Huawei employees' work
efficiency by allowing employees to work more flexibly and without the
need to deploy network cables.

The campus network is becoming borderless. Employees can easily
and flexibly access the company's intranet anywhere in the world, irrespective of whether they are at research centers, representative offices, or

production bases. However, as wireless offices become more widely used, the borderless network access mode poses great challenges to campus network management and security.

In mobile offices, the employees' host information (IP addresses and VLANs) can change at any time; therefore, traditional campus networks are unsatisfactory because they configure Access Control Lists (ACLs) based on IP address and VLAN information to control access permissions and ensure service experience. In traditional campus networks, when an employee changes their office location, the network administrator is required to re-configure service policies based on the employee's new IP address and VLAN, resulting in a heavy workload for administrators. In addition, when employees' access permissions are incorrectly set due to incorrect policy configurations, resources that employees could access before are no longer accessible. For these reasons, traditional campus networks often lead to unsatisfactory mobile office experience.

Huawei IT tried various methods to prevent frequent service policy changes and ensure that users only use IP addresses on specific network segments to access the network. In this way, administrators could configure IP address-based ACLs on network devices in advance to control the network permissions of different users. This approach solved the policy maintenance headache; however, it restricted employees to only accessing the network in specific office locations.

After analysis, Huawei IT found the main difficulty in managing service policies involved service policies closely coupled with information such as IP addresses, VLANs, network topologies, and office workspaces. Such information, however, keeps changing when employees move between different locations; therefore, Huawei IT concluded that service policies should be frequently changed.

To ensure consistent service policies, Huawei IT had to find a unique piece of information that could be used to define service policies. Huawei IT decided to use an employee's unchangeable identity information (for example, an employee's ID). This approach decouples service policies from ever-changing information such as IP addresses, VLANs, network topologies, and office workspaces. Huawei named this approach the free mobility solution, as shown in Figure 11.3.

Huawei's free mobility solution introduces the centralized control concept of Software-Defined Networking (SDN) to campus networks. It associates network resources with user identities to ensure that network

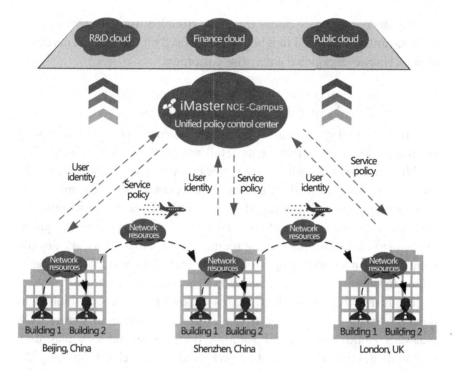

FIGURE 11.3 Architecture of Huawei's free mobility solution for campus networks.

resources can automatically move with users, thereby ensuring user experience and security.

As shown in Figure 11.3, Huawei's free mobility solution deploys a unified policy control center on the campus network, which allows the network administrator to formulate service policies based on employees' identities in advance, without the need to consider employees' IP addresses or network access locations. In this solution, network access devices report employees' identity information to the unified policy control center regardless of whether employees choose to access the network from Beijing, Shenzhen, or London, or from any building in an office space. Then, the policy control center notifies network access devices of the service policies to be executed. In this way, employees' service access rights and user experience remain consistent, and network resources remain accessible to employees. Huawei appropriately named the solution "free mobility" because the solution makes it seem like network resources are moving around freely with employees. Free mobility addresses the poor mobile office experience issues that have troubled users for many years. Such advantages have led to it being widely used on the campus networks of both Huawei and global enterprise customers.

11.3.3 Cloud Management Deployment: Supporting the Rapid Expansion of Huawei Device Stores

Due to the rapid growth of its consumer business, Huawei has rapidly expanded the construction of its retail stores. Since 2014, Huawei has established more than 300 device stores every year on average. To set up the device store's network in traditional network deployment mode, the entire set-up process, including planning, deployment, network optimization, and acceptance, takes at least one week. Engineers are also required to perform onsite commissioning multiple times during network deployment. This conventional approach is both inefficient and labor-intensive, failing to support the rapid expansion of Huawei's device stores.

To address this issue, Huawei uses a cloud management solution to quickly deploy networks for device stores. As shown in Figure 11.4, Huawei's cloud management solution implements network planning, deployment, optimization, and inspection all on the cloud. By using the cloud management solution, the time required to deploy a device store's network is reduced from one week to one day. In addition, all devices used in the solution are plug-and-play, requiring no onsite commissioning by engineers. This approach ensures that network services are available on demand during rapid business expansion.

As shown in Figure 11.5, during cloud management deployment, the network administrator can directly scan the barcodes of network devices using a mobile app (CloudCampus APP). This approach greatly facilitates network deployment and effectively supports the rapid expansion of Huawei's device stores. The cloud management deployment procedure is as follows:

1. The tenant administrator imports the ESNs of network devices (APs in this scenario) for stores in batches and then plans offline configurations on the cloud management platform.

2. Installation engineers connect and power on APs in device stores. They then log in to the CloudCampus APP and use the barcode scanning function to establish links between APs and the cloud

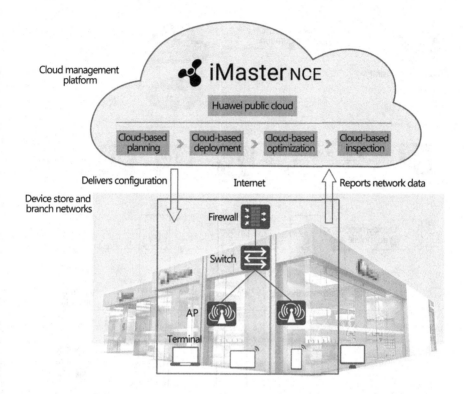

FIGURE 11.4   Cloud management used by Huawei to quickly deploy networks for device stores.

management platform, and deliver public network configurations to APs through the local management SSID. In this way, APs can be successfully discovered and managed by the cloud management platform.

3. APs periodically report performance data to the cloud management platform. Based on this data, the administrator can perform routine maintenance, periodic inspection, and troubleshooting on devices in device stores using the cloud management platform.

11.3.4 Wi-Fi and IoT Convergence: Helping Huawei Campus Network Go Digital

Huawei is a typical multinational corporation that possesses a large number of high-value assets such as lab instruments, network devices, and servers, as well as common assets, such as office computers, desks, and chairs.

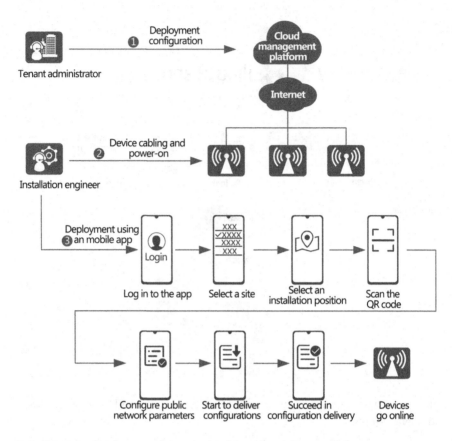

FIGURE 11.5 Deployment by scanning barcodes using a mobile app.

Rough statistics have shown that Huawei possesses more than 400000 assets in total, including over 13000 instruments and meters as well as 120000 electronic devices within the R&D department alone, as shown in Figure 11.6. Effectively managing and utilizing these fixed assets from around the world is a major challenge for Huawei to address.

Since 2014, Huawei has used radio frequency identification (RFID) tags to identify and manage various assets; however, the widespread use of RFID has posed many challenges. For example, it requires the deployment of a large number of RFID tag readers to manage assets. These card readers needed separate power supplies and network cables, leading to a relatively low management efficiency. A management system was also absent for these readers; therefore, it was impossible to detect reader faults quickly. In this case, the relevant administrators did not know whether an RFID reader was faulty or had been relocated until an asset report was

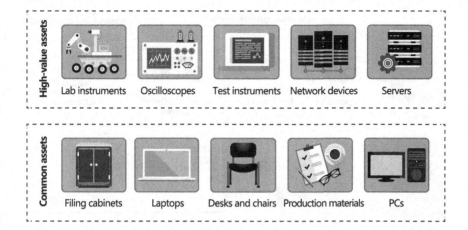

FIGURE 11.6   Examples of Huawei's typical assets.

generated every month. In general, if an RFID reader was faulty or relocated, its data were not found in the monthly asset report.

Huawei IT addressed this problem by using its global Wi-Fi network to manage assets. In addition, Huawei IT also developed an asset management solution that converges Wi-Fi and IoT networks, as shown in Figure 11.7.

In this solution, an AP integrates an RFID card to provide asset identification and management functions. It sends asset information back to the asset management platform through the Wi-Fi network. The asset management system then clearly displays the list and status of each person's key assets in real time, greatly improving asset management efficiency. The biggest advantage of this solution is that it eliminates the need for separate IoT networks; instead, all IoT terminals are managed through the Wi-Fi network, greatly reducing network construction and O&M costs.

The asset management solution proved highly efficient on Huawei's campus network, being quickly applied to various other scenarios. For example, it was used to manage the IoT sensors embedded in conference rooms, vehicles, campus security devices, meters, and even garbage cans and manhole covers. In this case, the solution connects the IoT sensors to an intelligent operation center through the outdoor Wi-Fi network to achieve ubiquitous connections throughout the campus and ultimately improve the digital level of Huawei's campus network. Using this solution, Huawei's campus network becomes a true cornerstone with all-scenario awareness and always-on service continuity.

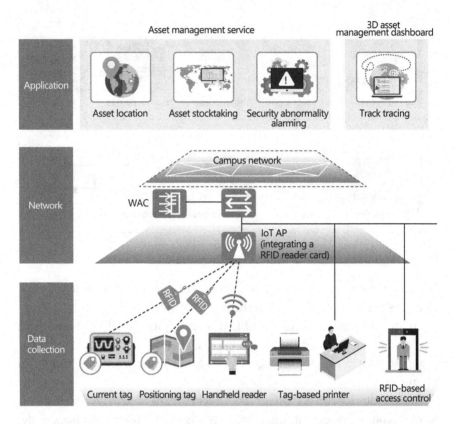

FIGURE 11.7 Networking architecture of the asset management solution with Wi-Fi and IoT convergence.

11.3.5 Intelligent O&M: Driving Huawei IT to Build a Proactive O&M System

Huawei has designed and developed an intelligent O&M solution that has been used and proven by Huawei IT.

Large-scale Wi-Fi adoption brings unprecedented convenience to daily office work. At the same time, it also leads to more O&M problems such as a large number of faults and difficult fault locating. In the past, Huawei IT received approximately 400 help-seeking calls every month on average, 40% of which were related to wireless networks. To address this, Huawei IT collaborated with the SDN controller product team to find innovative ways to improve Wi-Fi network O&M and fault locating. The two sides optimized the Wi-Fi network O&M algorithm model based on real service data for the field-proven intelligent analysis engine of the SDN controller. Specifically, they summarized four typical Wi-Fi network issue types (connection, air interface performance, roaming, and device issues) from fault data, defined the data criteria for each issue type, and specified how to design intelligent O&M algorithms based on these data criteria.

The comprehensive O&M scenarios on Huawei's campus network promote improvement on the O&M algorithms of the SDN controller. This, in turn, helps Huawei build a proactive O&M system for campus networks. Specifically, a visualized quality evaluation system is used to constantly drive Huawei IT to proactively improve network quality. For example, the controller dashboard can display severe latency and packet loss rate occurring on an employee's terminal, and the terminal-associated AP and its wireless signal coverage status can both be viewed based on the AP deployment diagram. If the problem is caused by poor signal coverage, administrators can then make corresponding adjustments in real time. In addition, Huawei's O&M solution helps achieve automatic fault identification, root cause location, and potential exception prediction, radically changing the passive response of traditional O&M, as shown in Figure 11.8. In the first year after the introduction of the intelligent

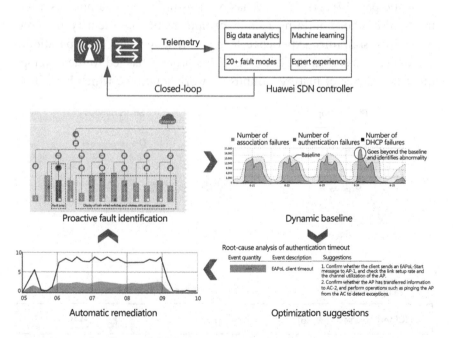

FIGURE 11.8   Intelligent O&M scenarios on Huawei's campus network.

O&M solution, the number of network fault tickets received by Huawei IT decreased by 20%. This figure was then reduced year by year.

11.4 BRINGING DIGITAL TO EVERY CAMPUS

As a large multinational enterprise, Huawei faces a complex operations environment typified by a large business volume, multiple customer groups, diversified service scenarios, global resource configurations, and localized operations. Despite this challenge, Huawei has successfully implemented digital transformation in various fields, such as R&D, sales, manufacturing, delivery, logistics, and campus operations. From 2015 to 2019, digital transformation drove Huawei's rapid growth in business revenue without significantly overwhelming the company's existing workforce.

Huawei operates a wide range of business around the world and boasts a diverse portfolio of campus use cases and scenarios, including office, R&D, manufacturing, hotel, education, and logistics. Therefore, Huawei's own campus network is the mold for an intent-driven campus network.

In the past 10 years, Huawei has been using its own campus as a test field and has been building a smart campus featuring "security and controllability, simplified experience, lean cost management, and operational excellence", as shown in Figure 11.9. This process has led to the development of a series of mature campus network solutions, which have been

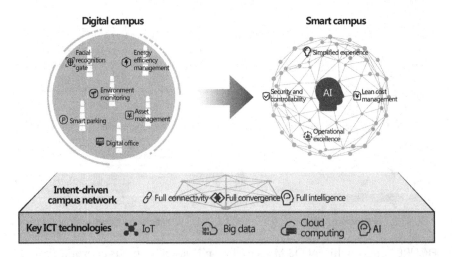

FIGURE 11.9 Campus evolution from the digital era to the intelligence era.

rewarding in Huawei's complex campus service environments. Since their debut, these solutions have been quickly snapped up by customers who want to accelerate their digital transformation and drive business profits.

For the foreseeable future, Huawei will continue to leverage its product portfolio and capitalize on technologies such as big data, IoT, cloud computing, and AI to drive the evolution from digital to intelligent campuses. Throughout this process, Huawei will still adhere to the principle that "we should always test new campus network innovations before releasing them". It is this basic principle that pushes Huawei to work closely with industry partners to build a win-win campus network ecosystem and provide customers with more mature scenario-specific campus network solutions. As such, Huawei's ultimate goal is to bring digital to every campus.

CHAPTER 12

Intent-Driven Campus Network Products

HUAWEI'S INTENT-DRIVEN CLOUD CAMPUS network solution builds on a portfolio of products, including CloudEngine S series campus switches, AirEngine series WLAN products, NetEngine AR series branch routers, HiSecEngine USG series enterprise security products, iMaster NCE-Campus (an all-in-one campus network management, control, and analysis system), and Cybersecurity Intelligence System (CIS) (a security situation awareness system). This chapter details the application scenarios and key functions of each of these products.

12.1 INTENT-DRIVEN CAMPUS NETWORK PRODUCT OVERVIEW

Huawei delivers a full product series, covering the access layer, aggregation layer, and management layer, as shown in Figure 12.1. These products are ideal for building an end-to-end intent-driven campus network.

12.2 CLOUDENGINE S SERIES CAMPUS SWITCHES

Huawei CloudEngine S series campus switches are next-generation Ethernet switches purpose-built for enterprises, governments, education, finance, manufacturing, and other sectors. They are ideal for building a future-proof campus network that features simplified management, high

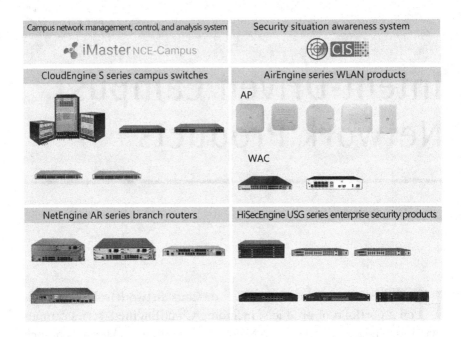

FIGURE 12.1 Huawei's complete range of intent-driven campus network products.

stability, strong reliability, and service intelligence, accelerating digital transformation across industries.

Built on Huawei's high-performance hardware platform and Versatile Routing Platform software platform, CloudEngine S series campus switches stand out with their highly reliable hardware architecture, unmatched data switching performance, high-density GE/10GE/40GE/100GE ports, and abundant Layer 2 and Layer 3 features. CloudEngine S series switches are designed using innovative chips to offer abundant value-added service features, such as wired and wireless convergence, free mobility, Virtual Extensible Local Area Network (VXLAN), telemetry, threat deception, and Encrypted Communications Analytics (ECA). These features pave the way for the evolution of data switching-centric networks to service experience-centric networks.

1. Wired and wireless convergence

CloudEngine S series switches integrate the wireless access controller (WAC) function, so users can manage wireless access points (APs) without purchasing additional WAC hardware. In addition to this, CloudEngine S series switches guarantee Tbit/s-level wireless forwarding, breaking the forwarding performance bottleneck of standalone WACs and embracing the high-speed wireless era.

Unified user management is another highlight of CloudEngine S series switches. Specifically, they authenticate both wired and wireless users, ensuring a consistent user experience regardless of whether they are connected to the network through wired or wireless access devices. These switches support various authentication methods, including 802.1X, MAC address, and Portal authentication, and are capable of managing users based on user groups, domains, and time ranges. These functions visualize user and service management and boost the transformation from device-centric management to user-centric management.

2. Automated deployment

Automated physical network deployment: By using iMaster NCE-Campus, CloudEngine S series switches support graphical user interfaces and Network Configuration Protocol (NETCONF)/Yet Another Next Generation (YANG) for device management and deployment. Switches are plug-and-play, automating physical network deployment.

Automated virtual network (VN) deployment: By using iMaster NCE-Campus, CloudEngine S series switches can quickly create VXLANs through NETCONF/YANG. In this way, multiple service networks or tenant networks are quickly deployed on the same physical network and are securely isolated from each other. This efficiently carries different services and customer data, avoids repeated network construction, and improves network resource utilization.

Automated policy deployment: CloudEngine S series switches support user- and application-based policy deployment, covering permission, bandwidth, and QoS. By using iMaster NCE-Campus, CloudEngine S series switches support automatic translation and delivery of policies and implement refined policy control. As a result, users can enjoy consistent service experience even while moving along the entire network.

3. Intelligent O&M

CloudEngine S series switches with telemetry collect device data in real time and send it to iMaster NCE-Campus. iMaster NCE-Campus then analyzes network data using an intelligent fault identification algorithm, accurately displays the real-time network status, effectively locates and demarcates faults, and identifies then rectifies network problems to guarantee optimal user experience.

CloudEngine S series switches also support enhanced Media Delivery Index (eMDI) to implement intelligent O&M for audio and video services. With the eMDI function, switches function as monitored nodes to periodically collect statistics and report audio and video service indicators to iMaster NCE-Campus. In this way, iMaster NCE-Campus can quickly diagnose audio and video service quality faults based on the data reported by multiple monitored nodes.

4. Big data security collaboration

CloudEngine S series switches use NetStream to sample campus network data and then report the data to the cybersecurity intelligence system (CIS). After receiving this data, the CIS then detects network security threat events, displays network-wide security status, and further processes security threat events. The CIS delivers security collaboration policies to iMaster NCE-Campus, which then forwards these policies to switches for handling security threat events, ultimately ensuring campus network security.

CloudEngine S series switches are also noted for the ECA function. Specifically, switches extract the characteristics of encrypted traffic, generate metadata, and report the metadata to the CIS. The CIS then uses an Artificial Intelligence (AI) algorithm to train traffic models and compare the extracted characteristics of encrypted traffic to identify malicious traffic. Following this analysis, the CIS visually displays the results, provides threat-handling suggestions, and collaborates with iMaster NCE-Campus to automatically isolate threats, ensuring campus network security.

Threat deception is another differentiator of CloudEngine S series switches. Switches detect network-wide threat behavior such as IP address scanning and port scanning, diverting threat traffic to the threat deception system. The threat deception system then performs in-depth interaction with the attacker, records various application-layer attack methods of the attacker, and reports security logs to the CIS. Subsequently, the CIS analyzes security logs, identifies suspicious traffic as an attack, and consequently generates alarms that process the threat accordingly.

12.3 AIRENGINE SERIES WLAN PRODUCTS

Huawei AirEngine series includes a wide range of WLAN products, all of which comply with IEEE 802.11a/b/g/n/ac/ax and are suitable for various scenarios, such as enterprise offices, campuses, hospitals, large shopping malls, exhibition centers, and stadiums. These WLAN products offer a complete WLAN solution and provide customers access to fast, secure, and reliable WLAN connection services.

12.3.1 AirEngine Series WACs

Huawei AirEngine series WACs provide cutting-edge AP management and forwarding capabilities. Combined with Huawei AirEngine series wireless APs, AirEngine WACs are ideal for medium- and large-sized enterprise campuses, enterprise branches, school campuses, and wireless metropolitan area networks.

1. Multi-functional

    AirEngine WACs provide Portal or 802.1X authentication through a built-in server, reducing costs for customers.

2. Built-in application identification server

    With a built-in application identification server, AirEngine WACs provide the following features:

    a. Identify over 6000 Layer 4 to Layer 7 applications, from common office applications to point-to-point download applications, such as Microsoft Lync, FaceTime, YouTube, and Facebook.

    b. Support application-based policy control technologies, including traffic blocking, traffic limiting, and priority-based scheduling policies.

    c. Support online application update in the application signature database, without the need for a software upgrade.

3. Comprehensive high-reliability design

    a. Supports dual AC power supplies for backup, allowing one power supply to work when the other is hot swapped.

    b. Supports 1+1 hot standby and N+1 backup for WACs, ensuring service continuity.

c. Supports port redundancy using Link Aggregation Control Protocol (LACP) and Multiple Spanning Tree Protocol (MSTP).

d. Supports wide area network (WAN) authentication escape. In local forwarding mode, this feature ensures online stations (STAs) stay online and allows access of new STAs when APs are disconnected from WACs, ensuring service continuity.

4. Built-in visualized web system

AirEngine WACs have a built-in web system that facilitates configuration and provides comprehensive monitoring in addition to intelligent diagnosis.

5. Health-centric one-page monitoring, intuitively presenting KPIs

One page includes the summary and real-time statistics, with KPIs displayed in graphs, including user, radio, and AP performance. This way, users can easily obtain useful information from a large amount of data, while being aware of the device and network status.

12.3.2 AirEngine Series Wireless APs

Huawei provides a wide range of wireless APs, including high-density APs, cost-effective APs, and agile distributed APs, to meet different wireless service requirements. This section introduces AirEngine series AP products launched by Huawei for the Wi-Fi 6 era.

AirEngine series APs comply with IEEE 802.11ax (Wi-Fi 6) and have 10GE interfaces. These APs perfectly meet the requirements of bandwidth-hungry services, such as Augmented Reality (AR) or Virtual Reality (VR) interactive teaching, high definition video streaming, multimedia, and desktop cloud applications, enabling users to enjoy high-quality wireless services.

1. Wi-Fi 6 standards compliance

AirEngine APs support 1024-quadrature amplitude modulation (1024-QAM) and 8×8 multiple-input multiple-output (MIMO), reaching an air interface rate of 4.8 Gbit/s. Orthogonal Frequency Division Multiple Access (OFDMA) enables multiple stations (STAs) to receive and send data simultaneously, reducing latency and improving efficiency in the network.

2. SmartRadio for air interface optimization

Load balancing during smart roaming: The load balancing algorithm detects loads on APs after STAs roam and adjusts the STA load on each AP accordingly to improve stability on the network.

• Intelligent Dynamic Frequency Assignment (DFA) technology: The DFA algorithm automatically detects adjacent-channel and co-channel interference and identifies any redundant 2.4 GHz radio. Through automatic inter-AP negotiation, a redundant radio is automatically switched to the 5 GHz band (supported only by dual-5G APs) or is disabled to reduce 2.4 GHz co-channel interference and increase capacity in the system.

Intelligent conflict optimization technology: Dynamic enhanced distributed channel access and airtime scheduling algorithms schedule the channel occupation time and service priority of each STA. This ensures that each STA is assigned as equal amount of time as possible for using channel resources and user services are scheduled in an orderly manner, improving service processing efficiency and user experience.

3. Air interface performance optimization

In high-density scenarios where a large number of STAs access the network, an increased number of low-rate STAs consume more resources on the air interface, reducing AP throughput and worsening user experience. To address this, AirEngine APs check the rate of access STAs and reject low-rate or weak-signal STAs. In addition, the APs monitor the rate and signal strength of online STAs in real time and forcibly disconnect low-rate or weak-signal STAs to enable the STAs to re-associate with APs that have stronger signals. This STA access control technology increases air interface efficiency and enables access of more STAs.

4. 5 GHz-prior access

AirEngine APs can work on both 2.4 and 5 GHz frequency bands. The 5 GHz-prior access function enables APs to steer STAs to the preferred 5 GHz frequency band, reducing loads as well as interference on the 2.4 GHz frequency band and subsequently improving user experience.

5. Analysis on non-Wi-Fi interference sources

AirEngine APs analyze the spectrum of and identify non-Wi-Fi interference sources such as Bluetooth devices, wireless audio transmitters, game controllers, and microwave ovens. With network management software, AirEngine APs accurately locate the source of interference and display their spectrums, facilitating the prompt elimination of interference.

6. Rogue device monitoring

AirEngine APs support the Wireless Intrusion Detection System (WIDS) and Wireless Intrusion Prevention System (WIPS) to monitor, identify, defend, and contain rogue devices, ensuring the security of the air interface environment and wireless transmission.

7. Automatic radio calibration

Automatic radio calibration allows AirEngine APs to collect information such as signal strength, channel, and other parameters of surrounding APs and generate an AP topology according to the collected data. Based on interference from authorized APs, rogue APs, non-Wi-Fi interference sources, and their loads, an AirEngine AP automatically adjusts its transmit power and working channel to optimize the performance of the network, improving network reliability and user experience.

12.4 NETENGINE AR SERIES BRANCH ROUTERS

NetEngine AR series branch routers are next-generation routers developed using innovative chips. They use Solar AX architecture, adopt CPU + NP heterogeneous forwarding, and come with built-in acceleration engines to deliver three times the industry's average performance. They also integrate critical functions, including SD-WAN, routing, switching, Virtual Private Network (VPN), security, and Multiprotocol Label Switching (MPLS). These functions enable the deployment of high-performance NetEngine AR series routers on headquarter or branch networks to provide network egress functions, fully supporting diversified and cloud-based enterprises services.

1. High performance

NetEngine AR series adopts all-new Solar AX architecture and applies innovative CPU + NP heterogeneous forwarding to SD-WAN customer premise equipment (CPE).

NetEngine AR series also comes with a rich set of built-in hardware-based intelligent acceleration engines, including hierarchical quality of service (HQoS), Internet Protocol Security (IPsec), Access Control List (ACL), and application identification. All these capabilities deliver high forwarding performance, efficiently process SD-WAN services, and fully support complex services such as VPN, identification, monitoring, HQoS, traffic steering, optimization, and security.

Solar AX architecture features the following:

a. CPU + NP heterogeneous forwarding delivers high forwarding performance.

b. The built-in hardware acceleration engines (HQoS, IPsec, ACL, and application identification) eliminate service processing bottlenecks.

c. The built-in ultra-fast forwarding algorithm is used to quickly match ACLs and routing rules.

2. High reliability

NetEngine AR series boasts highly reliable features, including the following:

a. Complies with carrier-grade design standards and provides reliable and high-quality services for enterprise users.

b. Offers hot swappable cards and comes with key hardware (such as Main Processing Units (MPUs), power modules, and fans) in redundancy mode, ensuring service security and stability.

c. Provides link backup for enterprise services, improving service access reliability.

d. Detects and determines faults in milliseconds, minimizing service downtime

3. Easy O&M

NetEngine AR series is differentiated for O&M design, such as:

a. Supports multiple management modes, such as SD-WAN management, SNMP-based network management, and web-based network management, simplifying network deployment.

b. Allows for efficient maintenance that is free from onsite commissioning and supports remote centralized management of CPEs, which greatly reduces O&M costs while improving O&M efficiency.

4. Service convergence

NetEngine AR series integrates routing, switching, VPN, security, and Wi-Fi functions to meet diversified enterprise service requirements, save space, and reduce enterprise Total Cost of Ownership (TCO).

5. Security

NetEngine AR series provides comprehensive security protection capabilities with built-in firewall, intrusion prevention system (IPS), URL filtering, and multiple VPN technologies.

6. SD-WAN support

a. NetEngine AR series works with the SD-WAN solution to build cost-effective and business-friendly Internet connections.

b. NetEngine AR series adopts Zero Touch Provisioning (ZTP) to implement one-click deployment via email, USB flash drive, and Dynamic Host Configuration Protocol (DHCP), minimizing skill requirements and provisioning devices in minutes.

c. NetEngine AR series supports first packet identification for Software as a Service (SaaS) applications and service awareness for complex applications.

d. Traffic steering based on bandwidth and link quality ensures optimal experience of key applications and improves bandwidth utilization to 90%.

12.5 HISECENGINE USG SERIES ENTERPRISE SECURITY PRODUCTS

The HiSecEngine USG series products provide next-generation firewall (NGFW) capabilities and collaborate with other security devices to proactively defend against network threats, enhance border detection capabilities, effectively protect networks against advanced threats, and resolve performance deterioration issues.

1. Intelligent protection

Provides built-in Next-Generation Engine (NGE) and Content-based Detection Engine (CDE). As the detection engine of the NGFW, the NGE provides such content security functions as IPS, antivirus, and URL filtering, protecting intranet servers and users against threats. Meanwhile, the brand-new CDE virus detection engine redefines malicious file detection by leveraging AI. In-depth data analysis is available to quickly detect malicious files and improve the threat detection rate, while inclusive AI has been designed to help customers perform more comprehensive network risk assessment, effectively cope with network threats on the attack chain, and implement truly intelligent defense.

2. Simplified O&M

Integrates the cloud-based deployment solution, implementing plug-and-play for more simplified, rapid network deployment. The security controller is deployed as a component and interworks with iMaster NCE-Campus to enable unified management and policy delivery, effectively improving firewall O&M efficiency. The innovative web UI 2.0 provides a new visualized security interface, greatly improving usability and simplifying O&M.

3. Extensive IPv6 capabilities

Provides various IPv6 capabilities, including network switching, policy control, security defense, and service visualization. These advanced techniques enable governments, media and entertainment agencies, carriers, Internet ISPs, and financial services organizations implement IPv6 reconstruction.

4. Intelligent traffic steering

Provides both static and dynamic intelligent traffic steering based on multi-egress links. This function dynamically selects outbound interfaces based on the link bandwidth, weight, priority, or automatically detected link quality set by the administrator, forwards traffic to each link in different link selection modes, and dynamically tunes the link selection result in real time to maximize the efficiency of link resources and improve user experience.

12.6 IMASTER NCE-CAMPUS

iMaster NCE-Campus is an all-in-one management, control, and analysis system developed by Huawei for campus and branch networks. Powered by cloud computing, SDN, and big data analytics technologies, iMaster NCE-Campus automatically and centrally manages underlay and overlay networks to provide greater data collection and analysis capabilities than traditional solutions. In addition, iMaster NCE-Campus centrally controls access permissions, QoS, bandwidth, applications, and security policies of campus users. Driven by services, iMaster NCE-Campus provides simple, fast, and intelligent campus virtualization service provisioning, enabling the network to be more agile for services.

1. Network automation

   a. Automated network deployment: Template-based design and device plug-and-play significantly reduce the operating expense of network deployment.

   b. Automated VN provisioning: Models are abstracted, one multifunctional network is achieved, and VXLAN services are provisioned in minutes.

   c. Automated policy deployment: User policies, such as network access, QoS, bandwidth, application, and security, are centrally configured and adjusted in real time to ensure user access experience.

2. Full-scenario and full-lifecycle management

   a. Comprehensive cloud management: Cloud management is supported on large- and medium-sized campus networks and multibranch networks. This opens up more flexible business models (namely, public cloud and on-premises) to users.

   b. Full-lifecycle management: Full-process management is achieved in one place, integrating network design, service deployment, maintenance, and optimization.

    c. LAN-WAN convergence: End-to-end unified deployment and policy management of enterprise LAN and WAN reduces O&M costs.

3. Openness and cooperation

    Open architecture provides standards-compliant northbound RESTful Application Programming Interfaces (APIs).

4. Full-journey experience visibility of each user, in each application, at each moment

    a. Each moment: Network KPI data are dynamically collected in minutes using telemetry and other data collection technologies. Faults are quickly tracked.

    b. Each user: Data are collected from multiple dimensions, the network profile of each user is presented in real time, and network experience is visualized throughout the full journey.

    c. Each application: Voice and video application experience is detected in real time, faulty devices are quickly and intelligently located, and the root cause of poor Quality of Experience (QoE) is comprehensively analyzed.

5. Automatic identification and proactive prediction of network issues

    a. Big data and AI technologies are used to automatically identify connectivity, air interface performance, roaming, and device issues, improving the identification success rate of potential issues.

    b. Baselines are dynamically generated based on historical data through machine learning. Possible faults are predicted by comparing the baselines with real-time data.

6. Intelligent locating and root-cause analysis of network issues.

    a. The fault modes and impact scopes are intelligently identified using a big data analytics platform and multiple AI algorithms. As such, administrators are able to quickly locate faults.

b. A fault knowledge base is built based on network O&M expertise, which helps accurately infer fault scenarios, analyze root causes, and provide rectification suggestions.

12.7 CIS SECURITY SITUATION AWARENESS SYSTEM

Huawei CIS is based on FusionInsight, the mature commercial big data platform. The system performs multidimensional correlation analysis of huge amounts of data based on the AI detection algorithm, proactively detects various security threat events in real time, and restores the attack behavior of the entire Advanced Persistent Threat (APT) attack chain. In addition, CIS can collect and store multiple types of network information, helping users detect threats, perform forensics, and take the required disciplinary actions. CIS is designed to detect and block threats, collect evidence, trace sources, and implement closed-loop threat management, helping users to effectively deal with threats throughout the entire process.

1. Comprehensive detection: APT attack detection, proactive defense based on integrated deception, and comprehensive asset security awareness

   a. Utilizing information provided by the big data platform, CIS uses machine learning to analyze traffic in each phase of an APT attack chain (including conducting reconnaissance, establishing a foothold, escalating privileges, and causing damage or exfiltrating data), to detect file, mail, Command and Control (C&C), traffic, log, web, and covert tunnel anomalies based on detection models. CIS then correlates these anomalies to detect advanced threats, while also integrating the deception solution for proactive defense.

   b. The asset management module based on proactive scanning technology can comprehensively control asset vulnerabilities.

2. Network-wide collaboration: automatic security response and orchestration, collaboration with security devices to handle detected threats, and reputation sharing on the cloud.

   a. The automatic response orchestration capability enables automatic investigation, evidence collection, and collaborative response to threat events across multiple service scenarios, enabling users to handle events within minutes.

b. CIS can rapidly synchronize detected threat information to Huawei Next-Generation Firewalls (NGFWs) and third-party terminals, which can then detect and eliminate threats.

c. The global threat intelligence center provides a reputation query service based on the threat information detected and uploaded by CIS. In addition, CIS can automatically (or allow users to manually) access the cloud reputation center to query IP reputation, domain reputation, and file reputation based on customer requirements, and then perform advanced analysis based on this information. CIS also provides a web page for users to query intelligence on the cloud for further investigation and analysis based on detected threats.

3. Network-wide visualization: real-time awareness of security situation, enabling search and source tracing of PB-level data within seconds

a. A network-wide security threat map intuitively visualizes enterprise-specific threats, as well as notable threat events in the world. This enables O&M personnel to effectively identify threats and predict security trends for the entire network.

b. Stage mode situation display offers customized development of attack situation displays focused on key areas of customer interest.

c. Rapid smart searches of events and traffic metadata are conducted using keywords, condition expressions, and time ranges to quickly locate detailed threat statistics and context data of interest to security O&M personnel, with over a billion records processed in under 5 seconds.

d. Events are investigated based on the attack chain, and traffic metadata is correlated with different attack stages. Metadata-related Process Characterization Analysis Package (PCAP) files are available for download under the traffic metadata search result list, with security O&M personnel able to efficiently collect and analyze information under a single intuitive interface.

Future Prospects of an Intent-Driven Campus Network

TODAY, IT IS WIDELY recognized in the industry that the future development of campus networks will be largely guided by autonomous driving technologies. As described in Section 2.3, Level 3 autonomous driving networks start to become self-aware and can be context-driven to dynamically adjust and optimize networks based on external environments, implementing intent-based closed-loop management. Level 4 autonomous driving networks will go a step further, unleashing the full potential of digital twins built on campus networks. By that time, networks themselves will be predictable. Upon detecting network parameter changes, networks will take preventive measures before customers can detect problems, avoiding unexpected network incidents.

In the age of connectivity, big data and Artificial Intelligence (AI) technologies will become increasingly mature. As such, networks will further transform towards Level 5 autonomous driving. The biggest difference with Level 5 autonomous driving networks will be that digital twins will not only cover the campus network, but also all things in the entire campus and even in the world at large. Level 5 autonomous driving networks will also run completely independently. This chapter mainly explores the future prospects of Level 4 and Level 5 autonomous driving networks.

13.1 INTELLIGENCE VISION FOR FUTURE CAMPUSES

In the age of connectivity, highly promising technologies, such as Internet of Things (IoT), cloud computing, ultra-broadband, big data, and AI, will become mature. Powered by these technologies, campuses will become fully digitalized, and all things in campuses will generate data. The data will represent the things, statuses of these things, and even various manually-defined concepts.

With ultra-broadband technology, no network bandwidth bottlenecks will exist, and all data will be migrated to the cloud and participate in computing in real time. In addition, with assistance from the super computing power of cloud computing and various mathematical analytics models, AI systems will continuously mine the complex relationships between things represented by big data to achieve a revolutionary leap in cognitive recognition of our world.

In future campuses, the information silos that arise from separately constructed networks on traditional campuses will disappear. All service data will be fully shared, achieving full convergence between the physical and digital worlds. Ultimately, a digital model for the entire campus will be built and updated in real time. By that time, the campus will realize a closed-loop process, starting from sensing the touch points of the physical world, to making decisions in the digital world, and then back to intelligently taking action in the physical world. This process will be fully autonomous, and human participation will not be necessary.

As shown in Figure 13.1, a digitalized campus will be similar to an organic life form. That is, the brain of the organic life form will be an AI

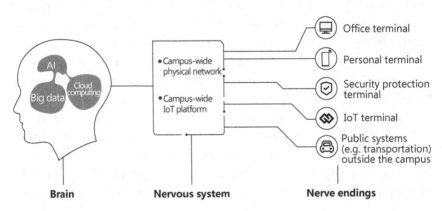

FIGURE 13.1 Intelligent vision for future campuses.

controller, its nervous system will be the physical network of the campus, and its nerve endings will be various service terminals and the digital systems inside and outside the campus.

13.2 NETWORK TECHNOLOGY PROSPECTS FOR FUTURE CAMPUSES

A network's most important aim is to ensure the secure, fast, and accurate transmission of information. An autonomous driving network can effectively ensure the attainment of this goal due to its predictive capabilities. This is particularly true for Level 5 autonomous driving networks that will act upon the big data-based prediction results across the entire campus. On Level 5 autonomous driving networks, all things will be centrally controlled and managed; traffic will be adjusted before network congestion occurs; preventative measures will be taken before network faults occur; and network capacity will be automatically expanded due to insufficient network performance. These measures all ensure the secure, fast, and accurate transmission of information, while, more importantly, also improving network Operations and Maintenance (O&M) efficiency and reducing network deployment and O&M costs.

A Level 5 autonomous driving network, however, can only be achieved through meticulous design. Firstly, we should create an edge intelligence layer that can sense environments in real time on the physical network and greatly simplify network architecture and protocols. Secondly, we need to build a digital twin network through unified modeling to ensure the traceability and predictability of the global situation, and also implement predictive O&M and proactive closed-loop optimization using AI technologies. Lastly, we must establish an open cloud platform to achieve AI algorithm training and optimization, enable the agile development of various applications (covering planning, design, service provisioning, O&M assurance, and network optimization), and facilitate automatic closed-loop operations throughout the whole lifecycle.

1. Intelligent and simplified campus network architecture

   A campus network typically involves three parts: the control system, control mode, and physical network.

   Control system: On a Level 4 autonomous driving network, the previously independent Software-Defined Networking (SDN) controller, Wide Area Network (WAN) controller, and Data Communication

Network controller on Level 3 autonomous driving networks become fully collaborative. That is, the entire campus network will use only one unified AI controller.

As AI technologies become mature, we will most certainly embrace Level 5 autonomous driving networks. On a Level 5 autonomous driving network, the AI controller on the campus network will fully collaborate with all other AI controllers on the campus (for example, those controllers for environment monitoring, logistics, warehousing, and production management). By then, only one unified AI controller will be used throughout the entire campus. This AI controller will even detect data outside the campus upon authorization to better steer campus operations and improve user experience.

Control mode: On a Level 4 autonomous driving network, administrators will no longer need to manage the specific configuration logics and commands of network functions. Instead, they will only need to express their intent to the network, and the network will intelligently adjust itself based on the administrator's intent. When it comes to a Level 5 autonomous driving network, the AI controller will automatically identify administrators' intent, and administrators will be only required to confirm key decisions.

For example, let us say ten guests are planning to come to your city tomorrow by plane, visit your campus, and attend a conference. In this example, Level 4 autonomous driving network administrators will only need to inform the network of the guests' plans on the campus. The AI controller will then be able to automatically determine the network service levels and security levels required for these 10 guests, and deliver related configurations to each network device. The entire process will not require network administrators to know about the different service levels, security levels, and network feature configuration logics implemented.

However, the control mode on Level 4 autonomous driving networks will not be completely without manual intervention; rather, manual intervention will be required if unexpected changes affect the network. For example, in the above example, if the number of guests changes on the day when these guests are expected to visit, network administrators will have to manually submit these changes on the network for adjustment. This is because on a Level 4 autonomous driving network, only the campus network will have achieved control collaboration.

However, the preceding case will be totally different on a Level 5 autonomous driving network. Specifically, the campus network control system will automatically identify and track information changes and adjust network resources accordingly. Guests will only be required to authorize their travel information to the campus AI controller. The campus AI controller will automatically arrange pick-up vehicles based on the updated flight information, weather information, and road traffic status, and adjust campus network resources such as Wi-Fi according to the updated guest arrival time. The entire process will not require manual follow-up and intervention. Instead, network administrators and guest reception personnel will only act according to the guest itinerary information changes provided by the campus AI controller.

Physical network: Currently, we construct service networks separately to ensure convenient O&M and clear responsibilities. For example, we build separate security networks (e.g., video surveillance and fire protection networks), office networks, and production networks. However, with the rise of virtual network technologies such as Virtual Extensible Local Area Network (VXLAN), we can build one single physical network on an intent-driven campus network.

On a Level 5 autonomous driving network, the campus AI controller will take over all manual O&M tasks and automatically detect information such as campus event plans as well as logistics and warehousing data. By that time, terminals on the physical network will not be limited to traditional network terminals such as computers, printers, cameras, and mobile phones, but will also include IoT terminals and various sensors such as curtain sensors, light sensors, wind sensors, robots, and drones. If the physical network cannot support these upcoming services, the campus AI controller will automatically adjust the physical network devices and topology, including purchasing and leasing new devices. In a similar way, if the campus AI controller detects that spare network devices are overstocked in the warehouse, the campus AI controller will sell or lease out devices to effectively control the physical network's construction and O&M costs. In addition, future network devices will most likely become modularized. That is, network devices will become similar to desktop computers, with each component being flexibly removed and replaced. In this case, the campus AI controller will manage each component in a refined manner.

2. Intelligent and simplified campus network design, deployment, and O&M

In the future, the campus network architecture will become more intelligent and simplified. This, in turn, will greatly reduce the required investment and costs in the design, deployment, and O&M phases of the campus network.

A Level 4 autonomous driving network will independently design the campus network based on the requirements and related information provided by technical personnel. However, manual adjustment will still be required before network deployment, and manual configuration and commissioning will also be required during network deployment. However, on a Level 5 autonomous driving network, enterprises will only need to set their budget range. Then, the AI controller will automatically determine the most appropriate network architecture and select the optimal devices for the campus. The AI controller will also arrange robots to complete site surveying as well as device configuration and commissioning.

The following takes an enterprise that intends to create a branch in city A as an example. After the enterprise sets its budget range, the AI controller automatically collects and analyzes all information about the branch. The collected information would include the enterprise's internal information (for example, business and staffing planning for the branch's next 10 years), the climate information of city A, topographical information of the branch, public security situation of city A, logistics and transportation information of city A, inventory material information of the enterprise, and recently purchased material information. Based on this comprehensive information, the AI controller will automatically produce multiple campus network design solutions for the enterprise, with detailed descriptions of the differences between the different solutions. The designed solutions will not only involve office and production network services, but also all possible digital services in the campus, covering living and entertainment, security defense, and environment ecosystem construction. What's more, the AI controller will automatically generate physical network plans (for example, bandwidth planning and Wi-Fi network management) based on future service planning, mobile office requirements, and IoT requirements.

Once an enterprise finalizes its design solution, the AI controller will start to deploy the campus network based on the finalized design solution. Specifically, the AI controller automatically purchases goods, tracks goods delivery and logistics, and arranges robots for site survey, civil engineering, and cabling, as well as device installation and commissioning. Network administrators can view the detailed data and reports of the entire process, and they only need to adjust the solution upon any force majeure events. After deployment, the campus network is continuously optimized and expanded based on the network's actual service running status. Service adjustment is linked with the IT system to ensure that when the IT system detects new service requirements, it automatically triggers the AI controller to provision new services.

The following uses today's Wi-Fi network optimization and management pain points as an example to envision the future design and deployment process. Strictly speaking, a Wi-Fi network is a self-interference system. To ensure best results, Access Point (AP) interference in the campus should be constantly detected and then continuous optimizations should be carried out accordingly. The current solution involves importing the building drawings of each area in the campus to the software and then using the software to simulate the signal strength of APs. Then, the AP installation position is determined using the software. After that, personnel are assigned to perform a site survey to evaluate the impact of software installation, environment, and electromagnetic interference on Wi-Fi signals, and then determine the final locations of APs one by one based on personal experience. After APs are installed and go online, we need to manually check, perform acceptance tests on, and optimize AP signals. In subsequent network O&M, Wi-Fi network optimization or reconstruction affects many aspects. For this reason, network administrators do not optimize or reconstruct Wi-Fi networks in an area until users submit complaints about network quality. This will no longer happen on Level 5 autonomous driving networks.

On a Level 5 autonomous driving network, the Wi-Fi network design will not be based on the personal experience of an engineer or the model recommended by an expert. Instead, an optimal design solution will be automatically obtained through big data analytics by the AI controller that features powerful computing and learning capabilities. A wireless interference avoidance mechanism will be designed for all things that may cause interference on the entire campus, rather than only focusing on network devices. In addition, the location of wireless terminals will depend on the simulation data model obtained after the entire campus is constructed, rather than on campus construction drawings.

In the deployment phase, robots will perform a site survey and adjust the digital campus model in real time. Then, based on the optimal construction sequence provided by the AI controller, robots will directly perform campus network deployment and related engineering tasks. During the construction phase, the digital campus model will also be continuously updated to ensure that the network's construction is aligned with the design. Then, once deployment is complete, the Level 5 autonomous driving network will use probe data to simulate the running of real service data on the Wi-Fi network. Meanwhile, the AI controller will collect, analyze, and evaluate the probe data.

In the daytime, the network will constantly self-diagnose and self-analyze itself; meanwhile, in the evening, the network will self-optimize and self-adjust as well as deploy and debug new policies. For example, if network bandwidth is insufficient, the AI controller will automatically adjust network bandwidth, policy, and architecture based on the company's budget, actual number of users and devices, and real campus environment. If capacity expansion is required, the AI controller will check the inventory and logistics capabilities, automatically determine the cost-effectiveness of each capacity expansion solution, and provide the optimal capacity expansion solution for enterprise managers. Then, with the manager's approval, network capacity will be automatically expanded.

In the O&M phase, administrators on a Level 3 autonomous driving network need to manually operate the SDN controller's Graphical User Interfaces (GUIs) to detect and locate faults. However, on a Level 5 autonomous driving network, the network will analyze user service experience in real time. When service experience deteriorates or potential faults are predicted, the network will record the anomalies, analyze their root causes, and optimize the network in real time. Once optimization is complete, the network will continuously follow up the optimization results until user experience returns to normal. What is more? The entire process will not require any manual intervention.

If a network fault occurs without any precursor, the network will automatically analyze information such as logs and alarms generated along with the fault, and rectify the fault in real time. All related spare parts will be sent to the fault point immediately. Then, once the consumption of spare parts is confirmed, the warehouse will automatically purchase and supplement spare parts by itself.

Let us use frame freezing in a video conference as an example. Generally speaking, frame freezing is caused by packet loss. With regards to autonomous driving networks at lower levels, after receiving a frame freezing fault report from a user, the administrator then has to find the packet loss point based on the source and destination. However, this fault locating method is inefficient, because it is difficult to reproduce the frame freezing fault.

This situation is entirely different on autonomous driving networks at higher levels. Specifically, the AI controller collects all traffic on the network in real time, analyzes the traffic types, and determines the flow quality in real time for each traffic type. If the quality of a flow is poor, the diagnosis mechanism is automatically triggered to locate the packet loss point. Then, once the packet loss point is found, the AI controller automatically analyzes the running status and logs of the packet loss point and performs optimization accordingly. Following optimization, the AI controller continues to verify whether the corresponding service experience has been improved and then decides whether to further optimize. In this way, fault diagnosis is changed from passive to proactive, and fault rectification is shifted from post-event to real-time, effectively enhancing the satisfaction of campus network users.

The information that can be sensed and referenced by the network will not be limited to the information of the network itself. Therefore, the success rate of network fault prediction and defense will be greatly improved, and services will run uninterrupted due to reliability assurance that has been planned during the network design phase.

3. Intelligent and simplified campus network security

After a campus goes digital, security-related services still remain the top priority. Before the arrival of Level 4 autonomous driving networks, network security is primarily implemented by deploying security components such as firewalls on security borders to prevent network attacks. Such security components fend off network attacks by relying on a sound antivirus database. However, these security components are ineffective towards new security threats.

On a Level 4 autonomous driving network, security will be implemented by focusing on security protection. That is, the network will be capable of real-time threat awareness. In addition to building a security protection system with the global virus database and attack modes, the network will also monitor and analyze traffic behaviors in real time and determine whether to take preventative security actions based on attack characteristics. This approach will not only detect known security threats, but also effectively identify unknown security threats.

In terms of security, a Level 5 autonomous driving network will demonstrate even further progress. The network will predict possible attacks based on the data of all people, incidents, things, and environments in the campus in real time, and even simulate to automatically perform attack-defense drills on the campus data model.

For example, after predicting that a typhoon is about to land, the AI controller will use a campus digital model to simulate the possible impacts of the typhoon on the campus and then update data in real time based on typhoon tracking and wind speed changes. If it determines that the typhoon will cause damage to houses or articles, the AI controller will assign robots to perform reinforcement and apply for a budget to purchase spare parts to be used once the typhoon passes. When someone passes through a dangerous area, the AI controller will remind them to minimize the impact of force majeure factors such as typhoons on the campus.

4. Intelligent and simplified campus network team

A digital campus requires enterprise Information and Communication Technology (ICT) teams to shift their focus from hardware features and performance to the overall solution. This is truly a radical change that will pose higher requirements on products and services from ICT vendors, as well as lead to deeper collaboration between ICT teams while most likely combining the network, application, and security teams.

When purchasing products for a digital campus, enterprises will not simply evaluate vendors based on function fulfillment or performance indicators. Instead, they will evaluate the end-to-end problem-solving capabilities of vendors' solutions according to their service scenarios. As such, vendors will need to have a deeper understanding of enterprise services. Only then can they perform scenario-specific abstraction, which can be converted into a simple, easy-to-use man-machine interface. And in doing so, they can truly help enterprises improve production efficiency internally and enhance customer experience externally.

In the future, Huawei will lead industry development in five directions: redefining the technical architecture, reshaping the product architecture, setting the industry pace, resetting the industry direction, and opening up new industry space. Huawei will also strive to break the limits in four aspects to create a better future:

a. Redefine the Moore's Law and challenge the Shannon limit to build the best connections in the world

b. Reshape the computing architecture to make computing power greater and more economical

c. Create the best hybrid cloud to enable industry digitalization

d. Enable full-stack, all-scenario AI for ubiquitous intelligence

Partners and developers are the key to AI technology development. Considering this, Huawei released the Developer Enablement Plan and Shining Star Plan in 2018, which provide solid support in terms of resources, platforms, training courses, and joint solutions, thereby building a foundation for partner applications. The road to autonomous driving networks will be a long one for the ICT industry. Given this, all parties in the industry should work together to move forward in a common direction. Indeed, Huawei is committed to providing leading ICT solutions through continuous innovation, taking on complexity while creating simplicity for customers, and embracing a fully connected, intelligent world together with enterprises and partners.

Acronyms and Abbreviations

1024-QAM	1024-Quadrature Amplitude Modulation
256-QAM	256-Quadrature Amplitude Modulation
3G	Third Generation
4G	Fourth Generation
5G	Fifth Generation
AAA	Authentication, Authorization, and Accounting
ABR	Area Border Router
ACK	Acknowledgment
ACL	Access Control List
AD	Active Directory
AES	Advanced Encryption Standard
AGV	Automated Guided Vehicle
AI	Artificial Intelligence
AN	Autonomous Network
AP	Access Point
API	Application Programming Interface
APT	Advanced Persistent Threat
AR	Augmented Reality
ARP	Address Resolution Protocol
AS	Autonomous System
ASIC	Application Specific Integrated Circuit
ASN.1	Abstract Syntax Notation One
ASP	Authorized Service Partner
ATM	Asynchronous Transfer Mode
B/S	Browser/Server
BBS	Bulletin Board System

BD	Bridge Domain
BFD	Bidirectional Forwarding Detection
B-frame	Bidirectional Predicted Frame
BGP	Border Gateway Protocol
BI	Business Intelligence
BLE	Bluetooth Low Energy
BMS	Bare Metal Server
BP	Backpropagation
BSC	Base Station Controller
BYOD	Bring Your Own Device
C&C	Command and Control
C/S	Client/Server
CA	Certificate Authority
CAPEX	Capital Expenditure
CAPWAP	Control and Provisioning of Wireless Access Points
CCA	Clear Channel Assessment
CDE	Content-Based Detection Engine
CDMA	Code Division Multiple access
CIO	Chief Information Officer
CIS	Cybersecurity Intelligence System
CLI	Command Line Interface
CMF	Configuration Manager Frame
CNN	Convolutional Neural Network
CPE	Customer Premise Equipment
CRC	Cyclic Redundancy Check
CSMA/CA	Carrier Sense Multiple Access with Collision Avoidance
CSMA/CD	Carrier Sense Multiple Access with Collision Detection
CSP	Certified Service Partner
CSS	Cluster Switch System
DBS	Dynamic Bandwidth Selection
DC	Data Center
DCA	Dynamic Channel Assignment
DCN	Data Center Network
DF	Delay Factor
DFA	Dynamic Frequency Assignment
DFBS	Dynamic Frequency Band Selection
DGA	Domain Generation Algorithm
DHCP	Dynamic Host Configuration Protocol

DL MU-MIMO	Downlink MU-MIMO
DMZ	Demilitarized Zone
DNS	Domain Name System
DPI	Deep Packet Inspection
DSS	Data Security Standard
E2E	End-to-End
EAP	Extensible Authentication Protocol
ECA	Encrypted Communications Analytics
EDCA	Enhanced Distributed Channel Access
eMDI	Enhanced MDI
ERP	Enterprise Resource Planning
ERT	Export VPN Target
ESDP	Electronic Software Delivery Platform
ESL	Electronic Shelf Label
ESN	Equipment Serial Number
ETC	Electronic Toll Collection
EVPN	Ethernet Virtual Private Network
FE	Fast Ethernet
FEC	Forward Error Correction
FFT	Fast Fourier Transformation
FHSS	Frequency Hopping Spread Spectrum
FPI	First Packet Identification
GCI	Global Connectivity Index
GDP	Gross Domestic Product
GDPR	General Data Protection Regulation
GE	Gigabit Ethernet
GI	Guard Interval
GIS	Geographic Information System
GPB	Google Protocol Buffers
GPT	General Purpose Technology
GPU	Graphics Processing Unit
GRE	Generic Routing Encapsulation
GUI	Graphical User Interface
HD	High Definition
HDG	Huawei Developers Gathering
HEW SG	High Efficiency WLAN Study Group
HF	High Frequency
HIPERLAN	High Performance Radio LAN

HIS	Huawei IT Service
HSB	Hot Standby
HTTP	Hypertext Transfer Protocol
HTTPS	Hypertext Transfer Protocol Secure
IAB	Internet Architecture Board
ICMP	Internet Control Message Protocol
ICT	Information and Communication Technology
IDE	Integrated Development Environment
IDN	Intent-Driven Network
IEEE	Institute of Electrical and Electronics Engineers
IETF	Internet Engineering Task Force
I-frame	Intra-Coded Frame
IGP	Interior Gateway Protocol
IMAP	Interactive Mail Access Protocol
IoT	Internet of Things
IPC	Inter-Process Communication
IPS	Intrusion Prevention System
IPsec	Internet Protocol Security
IRB	Integrated Routing and Bridging
IRT	Import VPN Target
IS-IS	Intermediate System to Intermediate System
ISM	Industrial, Scientific and Medical
ISP	Internet Service Provider
ITU-T	International Telecommunication Union-Telecommunication Standardization Sector
JSON	JavaScript Object Notation
KPI	Key Performance Indicator
L2TP	Layer 2 Tunneling Protocol
LACP	Link Aggregation Control Protocol
LAN	Local Area Network
LBS	Location-Based Service
LDAP	Lightweight Directory Access Protocol
LF	Low Frequency
LLDP	Link Layer Discovery Protocol
LoS	Line-of-Sight
LSA	Link State Advertisement
LSDB	Link State Database
LTE	Long-Term Evolution

MAC	Media Access Control
MAN	Metropolitan Area Network
MD5	Message-Digest Algorithm 5
MDI	Media Delivery Index
MIB	Management Information Base
MIMO	Multiple-Input Multiple-Output
MITM	Man-in-the-Middle
MLR	Media Loss Rate
MOS	Mean Opinion Score
MPDU	MAC Protocol Data Unit
MPEG	Moving Picture Experts Group
MPLS	Multiprotocol Label Switching
MPU	Main Processing Unit
MSP	Managed Service Provider
MSTP	Multiple Spanning Tree Protocol
MTU	Maximum Transmission Unit
MU-MIMO	Multiuser MIMO
NA	Neighbor Advertisement
NAC	Network Access Control
NAS	Network Attached Storage
NAT	Network Address Translation
ND	Neighbor Discovery
NE	Network Element
NETCONF	Network Configuration Protocol
NFC	Near Field Communication
NFV	Network Functions Virtualization
NGE	Next-Generation Engine
NGFW	Next-Generation Firewall
NHTSA	National Highway Traffic Safety Administration
NMS	Network Management System
NP	Network Processor
NS	Neighbor Solicitation
NUD	Neighbor Unreachability Detection
NVE	Network Virtualization Edge
NVGRE	Network Virtualization Using Generic Routing Encapsulation
NVO3	Network Virtualization over Layer 3
O&M	Operations and Maintenance

OA	Office Automation
OFDM	Orthogonal Frequency Division Multiplexing
OLT	Optical Line Terminal
ONU	Optical Network Unit
OPEX	Operating Expense
OSI	Open System Interconnection
OSPF	Open Shortest Path First
P2P	Point-to-Point
PAM4	Four-Level Pulse Amplitude Modulation
PCAP	Process Characterization Analysis Package
PCI	Peripheral Component Interconnect
PCIe	Peripheral Component Interconnect Express
PD	Powered Device
PES	Packetized Elementary Stream
P-frame	Predicted Frame
PHY	Physical Layer
PKI	Public Key Infrastructure
PMSI	P-Multicast Service Interface
PON	Passive Optical Network
POP3	Post Office Protocol Version 3
PPTP	Point-to-Point Tunneling Protocol
PSE	Power Sourcing Equipment
PTZ	Pan-Tilt-Zoom
QoE	Quality of Experience
QR	Quick Response
QSFP	Quad Small Form-Factor Pluggable
RADIUS	Remote Authentication Dial-In User Service
RAT	Remote Access Trojan
REST	Representational State Transfer
RET	Retransmission
RF	Radio Frequency
RFC	Requirement for Comments
RFID	Radio Frequency Identification
RPC	Remote Procedure Call
RR	Route Reflector
RSA	Rivest-Shamir-Adleman
RSSI	Received Signal Strength Indicator
RTP	Real-time Transport Protocol

RU	Resource Unit
SAE	Society of Automotive Engineers
SD	Standard Definition
SDK	Software Development Kit
SDN	Software-Defined Networking
SFP	Small Form-Factor Pluggable
SHF	Super-High Frequency
SIEM	Security Information and Event Management
SIG	Special Interest Group
SIP	Session Initiation Protocol
SISO	Single-Input Single-Output
SLA	Service Level Agreement
SME	Small- and Medium-Sized Enterprise
SMI	Structure of Management Information
SMS	Short Message Service
SMTP	Simple Mail Transfer Protocol
SNMP	Simple Network Management Protocol
SNR	Signal-to-Noise Ratio
SOHO	Small Office Home Office
SPF	Shortest Path First
SPT	Shortest Path Tree
SSH	Secure Shell
SSID	Service Set Identifier
SSL	Secure Sockets Layer
STA	Station
STP	Shielded Twisted Pair
STT	Stateless Transport Tunneling
SVF	Super Virtual Fabric
TCP/IP	Transmission Control Protocol/Internet Protocol
TDMA	Time Division Multiple Access
TLS	Transport Layer Security
TS	Transport Stream
TTM	Time to Market
TWT	Target Wakeup Time
UDP	User Datagram Protocol
UHF	Ultra High Frequency
UI	User Interface
UL MU-MIMO	Uplink MU-MIMO

UML	Unified Modeling Language
URI	Uniform Resource Identifier
UTP	Unshielded Twisted Pair
UWB	Ultra-Wideband
VAS	Value-Added Service
VHT	Very High Throughput
VLAN	Virtual Local Area Network
VM	Virtual Machine
VMOS	Video Mean Opinion Score
VN	Virtual Network
VNI	VXLAN Network Identifier
VNIC	Virtual Network Interface Card
VPN	Virtual Private Network
VR	Virtual Reality
VRF	VPN Routing and Forwarding
VRP	Versatile Routing Platform
VRRP	Virtual Router Redundancy Protocol
VTEP	VXLAN Tunnel End Point
VXLAN	Virtual Extensible LAN
WAC	Wireless Access Controller
WAN	Wide Area Network
WIDS	Wireless Intrusion Detection System
WIPS	Wireless Intrusion Prevention System
WLAN	Wireless Local Area Network
WWW	World Wide Web
XJTU	Xi'an Jiaotong University
XML	Extensible Markup Language
XSLT	Extensible Stylesheet Language Transformations
YANG	Yet Another Next Generation
YIN	YANG Independent Notation
ZC	ZigBee Coordinator
ZED	ZigBee End Device
ZR	ZigBee Router
ZTP	Zero Touch Provisioning

Printed in the United States
by Baker & Taylor Publisher Services